CALCULUS PROBLEMS FOR A NEW CENTURY

Robert Fraga, Editor

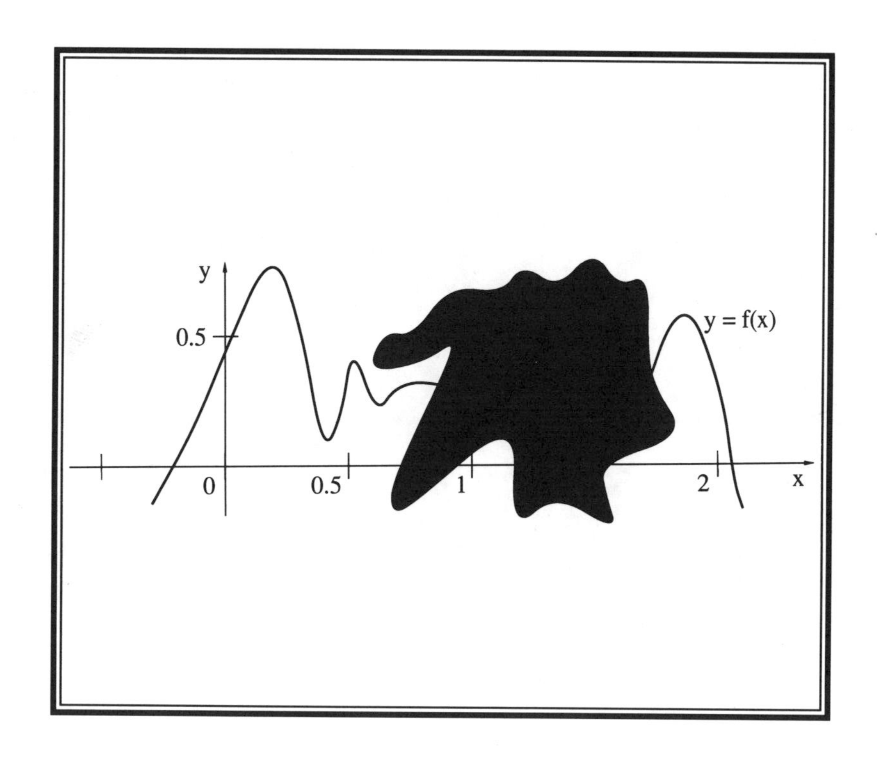

Resources for Calculus Collection

Volume 2

CALCULUS PROBLEMS FOR A NEW CENTURY

Robert Fraga, Editor

A Project of
The Associated Colleges of the Midwest and
The Great Lakes Colleges Association

Contributors to this Volume

Robert Fraga
Ripon College

Deborah Hart
Knox College

Eugene Herman
Grinnell College

James Stein
California State University, Long Beach

Lyle Welch
Monmouth College

Supported by the National Science Foundation
A. Wayne Roberts, Project Director

MAA Notes Volume 28

Published and Distributed by
The Mathematical Association of America

MAA Notes and Reports Series

The MAA Notes and Reports Series, started in 1982, addresses a broad range of topics and themes of interest to all who are involved with undergraduate mathematics. The volumes in this series are readable, informative, and useful, and help the mathematical community keep up with developments of importance to mathematics.

MAA Notes

1. Problem Solving in the Mathematics Curriculum, *Committee on the Teaching of Undergraduate Mathematics,* a subcommittee of the Committee on the Undergraduate Program in Mathematics, *Alan H. Schoenfeld,* Editor
2. Recommendations on the Mathematical Preparation of Teachers, *Committee on the Undergraduate Program in Mathematics, Panel on Teacher Training.*
3. Undergraduate Mathematics Education in the People's Republic of China, *Lynn A. Steen,* Editor.
5. American Perspectives on the Fifth International Congress on Mathematical Education, *Warren Page,* Editor.
6. Toward a Lean and Lively Calculus, *Ronald G. Douglas,* Editor.
8. Calculus for a New Century, *Lynn A. Steen,* Editor.
9. Computers and Mathematics: The Use of Computers in Undergraduate Instruction, *Committee on Computers in Mathematics Education, D. A. Smith, G. J. Porter, L. C. Leinbach, and R. H. Wenger,* Editors.
10. Guidelines for the Continuing Mathematical Education of Teachers, *Committee on the Mathematical Education of Teachers.*
11. Keys to Improved Instruction by Teaching Assistants and Part-Time Instructors, *Committee on Teaching Assistants and Part-Time Instructors, Bettye Anne Case,* Editor.
13. Reshaping College Mathematics, *Committee on the Undergraduate Program in Mathematics, Lynn A. Steen,* Editor.
14. Mathematical Writing, by *Donald E. Knuth, Tracy Larrabee, and Paul M. Roberts.*
15. Discrete Mathematics in the First Two Years, *Anthony Ralston,* Editor.
16. Using Writing to Teach Mathematics, *Andrew Sterrett,* Editor.
17. Priming the Calculus Pump: Innovations and Resources, *Committee on Calculus Reform and the First Two Years,* a subcomittee of the Committee on the Undergraduate Program in Mathematics, *Thomas W. Tucker,* Editor.
18. Models for Undergraduate Research in Mathematics, *Lester Senechal,* Editor.
19. Visualization in Teaching and Learning Mathematics, *Committee on Computers in Mathematics Education, Steve Cunningham and Walter S. Zimmermann,* Editors.
20. The Laboratory Approach to Teaching Calculus, *L. Carl Leinbach et al.,* Editors.
21. Perspectives on Contemporary Statistics, *David C. Hoaglin and David S. Moore,* Editors.
22. Heeding the Call for Change: Suggestions for Curricular Action, *Lynn A. Steen,* Editor.
23. Statistical Abstract of Undergraduate Programs in the Mathematical Sciences and Computer Science in the United States: 1990–91 CBMS Survey, *Donald J. Albers, Don O. Loftsgaarden, Donald C. Rung, and Ann E. Watkins.*
24. Symbolic Computation in Undergraduate Mathematics Education, *Zaven A. Karian,* Editor.
25. The Concept of Function: Aspects of Epistemology and Pedagogy, *Guershon Harel and Ed Dubinsky,* Editors.

26. Statistics for the Twenty-First Century, *Florence and Sheldon Gordon,* Editors.
27. Resources for Calculus Collection, Volume 1: Learning by Discovery: A Lab Manual for Calculus, *Anita E. Solow,* Editor.
28. Resources for Calculus Collection, Volume 2: Calculus Problems for a New Century, *Robert Fraga,* Editor.
29. Resources for Calculus Collection, Volume 3: Applications of Calculus, *Philip Straffin,* Editor.
30. Resources for Calculus Collection, Volume 4: Problems for Student Investigation, *Michael B. Jackson and John R. Ramsay,* Editors.
31. Resources for Calculus Collection, Volume 5: Readings for Calculus, *Underwood Dudley,* Editor.

MAA Reports

1. A Curriculum in Flux: Mathematics at Two-Year Colleges, *Subcommittee on Mathematics Curriculum at Two-Year Colleges,* a joint committee of the MAA and the American Mathematical Association of Two-Year Colleges, *Ronald M. Davis,* Editor.
2. A Source Book for College Mathematics Teaching, *Committee on the Teaching of Undergraduate Mathematics, Alan H. Schoenfeld,* Editor.
3. A Call for Change: Recommendations for the Mathematical Preparation of Teachers of Mathematics, *Committee on the Mathematical Education of Teachers, James R. C. Leitzel,* Editor.
4. Library Recommendations for Undergraduate Mathematics, *CUPM ad hoc Subcommittee, Lynn A. Steen,* Editor.
5. Two-Year College Mathematics Library Recommendations, *CUPM ad hoc Subcommittee, Lynn A. Steen,* Editor.

These volumes may be ordered from the Mathematical Association of America,
1529 Eighteenth Street, NW, Washington, DC 20036.
202-387-5200 FAX 202-265-2384

Third Printing

ISBN 0-88385-087-7
Library of Congress Catalog Number 92-62283
Printed in the United States of America
Current Printing
10 9 8 7 6 5 4 3

INTRODUCTION
RESOURCES FOR CALCULUS COLLECTION

Beginning with a conference at Tulane University in January, 1986, there developed in the mathematics community a sense that calculus was not being taught in a way befitting a subject that was at once the culmination of the secondary mathematics curriculum and the gateway to collegiate science and mathematics. Far too many of the students who started the course were failing to complete it with a grade of C or better, and perhaps worse, an embarrassing number who did complete it professed either not to understand it or not to like it, or both. For most students it was not a satisfying culmination of their secondary preparation, and it was not a gateway to future work. It was an exit.

Much of the difficulty had to do with the delivery system: classes that were too large, senior faculty who had largely deserted the course, and teaching assistants whose time and interest were focused on their own graduate work. Other difficulties came from well intentioned efforts to pack into the course all the topics demanded by the increasing number of disciplines requiring calculus of their students. It was acknowledged, however, that if the course had indeed become a blur for students, it just might be because those choosing the topics to be presented and the methods for presenting them had not kept their goals in focus.

It was to these latter concerns that we responded in designing our project. We agreed that there ought to be an opportunity for students to discover instead of always being told. We agreed that the availability of calculators and computers not only called for exercises that would not be rendered trivial by such technology, but would in fact direct attention more to ideas than to techniques. It seemed to us that there should be explanations of applications of calculus that were self-contained, and both accessible and relevant to students. We were persuaded that calculus students should, like students in any other college course, have some assignments that called for library work, some pondering, some imagination, and above all, a clearly reasoned and written conclusion. Finally, we came to believe that there should be available to students some collateral readings that would set calculus in an intellectual context.

We reasoned that the achievement of these goals called for the availability of new materials, and that the uncertainty of just what might work, coupled with the number of people trying to address the difficulties, called for a large collection of materials from which individuals could select. Our goal was to develop such materials, and to encourage people to use them in any way they saw fit. In this spirit, and with the help of the Notes editor and committee of the Mathematical Association of America, we have produced five volumes of materials that are, with the exception of volume V where we do not hold original copyrights, meant to be in the public domain.

We expect that some of these materials may be copied directly and handed to an entire class, while others may be given to a single student or group of students. Some will provide a basis from which local adaptations can be developed. We will be pleased if authors ask for permission, which we expect to be generous in granting, to incorporate our materials into texts or laboratory manuals. We hope that in all of these ways, indeed in any way short of reproducing substantial segments to sell for profit, our material will be used to greatly expand ideas about how the calculus might be taught.

Though I as Project Director never entertained the idea that we could write a single text that would be acceptable to all 26 schools in the project, it was clear that some common notion of topics essential to any calculus course would be necessary to give us direction. The task of forging a common syllabus was managed by Andy Sterrett with a tact and efficiency that was instructive to us all, and the product of this work, an annotated core syllabus, appears as an appendix in Volume 1. Some of the other volumes refer to this syllabus to indicate where, in a course, certain materials might be used.

This project was situated in two consortia of liberal arts colleges, not because we intended to develop materials for this specific audience, but because our schools provide a large reservoir of classroom teachers who lavish on calculus the same attention a graduate faculty might give to its introductory analysis course. Our schools, in their totality, were equipped with most varieties of computer labs, and we included in our consortia many people who had become national leaders in the use of computer algebra systems.

We also felt that our campuses gave us the capability to test materials in the classroom. The size of our schools enables us to implement a new idea without cutting through the red tape of a larger institution, and we can just as quickly reverse ourselves when it is apparent that what we are doing is not working. We are practiced in going in both directions. Continual testing of the materials we were developing was seen as an integral part of our project, an activity that George Andrews, with the title of Project Evaluator, kept before us throughout the project.

The value of our contributions will now be judged by the larger mathematical community, but I was right in thinking that I could find in our consortia the great abundance of talent necessary for an undertaking of this magnitude. Anita Solow brought to the project a background of editorial work and quickly became not only one of the editors of our publications, but also a person to whom I turned for advice regarding the project as a whole. Phil Straffin, drawing on his association with UMAP, was an ideal person to edit a collection of applications, and was another person who brought editorial experience to our project. Woody Dudley came to the project as a writer well known for his witty and incisive commentary on mathematical literature, and was an ideal choice to assemble a collection of readings.

Our two editors least experienced in mathematical exposition, Bob Fraga and Mic Jackson, both justified the confidence we placed in them. They brought to the project an enthusiasm and freshness from which we all benefited, and they were able at all points in the project to draw upon an excellent corps of gifted and experienced writers. When, in the last months of the project, Mic Jackson took an overseas assignment on an Earlham program, it was possible to move John Ramsay into Mic's position precisely because of the excellent working relationship that had existed on these writing teams.

The entire team of five editors, project evaluator and syllabus coordinator worked together as a harmonious team over the five year duration of this project. Each member, in turn, developed a group of writers, readers, and classroom users as necessary to complete the task. I believe my chief contribution was to identify and bring these talented people together, and to see that they were supported both financially and by the human resources available in the schools that make up two remarkable consortia.

A. Wayne Roberts
Macalester College
1993

THE FIVE VOLUMES OF THE RESOURCES FOR CALCULUS COLLECTION

1. Learning by Discovery: A Lab Manual for Calculus
Anita E. Solow, editor

The availability of electronic aids for calculating makes it possible for students, led by good questions and suggested experiments, to discover for themselves numerous ideas once accessible only on the basis of theoretical considerations. This collection provides questions and suggestions on 26 different topics. Developed to be independent of any particular hardware or software, these materials can be the basis of formal computer labs or homework assignments. Although designed to be done with the help of a computer algebra system, most of the labs can be successfully done with a graphing calculator.

2. Calculus Problems for a New Century
Robert Fraga, editor

Students still need drill problems to help them master ideas and to give them a sense of progress in their studies. A calculator can be used in many cases, however, to render trivial a list of traditional exercises. This collection, organized by topics commonly grouped in sections of a traditional text, seeks to provide exercises that will accomplish the purposes mentioned above, even for the student making intelligent use of technology.

3. Applications of Calculus
Philip Straffin, editor

Everyone agrees that there should be available some self-contained examples of applications of the calculus that are tractable, relevant, and interesting to students. Here they are, 18 in number, in a form to be consulted by a teacher wanting to enrich a course, to be handed out to a class if it is deemed appropriate to take a day or two of class time for a good application, or to be handed to an individual student with interests not being covered in class.

4. Problems for Student Investigation
Michael B. Jackson and John R. Ramsay, editors

Calculus students should be expected to work on problems that require imagination, outside reading and consultation, cooperation, and coherent writing. They should work on open-ended problems that admit several different approaches and call upon students to defend both their methodology and their conclusion. Here is a source of 30 such projects.

5. Readings for Calculus
Underwood Dudley, editor

Faculty members in most disciplines provide students in beginning courses with some history of their subject, some sense not only of what was done by whom, but also of how the discipline has contributed to intellectual history. These essays, appropriate for duplicating and handing out as collateral reading aim to provide such background, and also to develop an understanding of how mathematicians view their discipline.

ACKNOWLEDGEMENTS

Besides serving as editors of the collections with which their names are associated, Underwood Dudley, Bob Fraga, Mic Jackson, John Ramsay, Anita Solow, and Phil Straffin joined George Andrews (Project Evaluator), Andy Sterrett (Syllabus Coordinator) and Wayne Roberts (Project Director) to form a steering committee. The activities of this group, together with the writers' groups assembled by the editors, were supported by two grants from the National Science Foundation.

The NSF grants also funded two conferences at Lake Forest College that were essential to getting wide participation in the consortia colleges, and enabled member colleges to integrate our materials into their courses.

The projects benefited greatly from the counsel of an Advisory Committee that consisted of Morton Brown, Creighton Buck, Jean Callaway, John Rigden, Truman Schwartz, George Sell, and Lynn Steen.

Macalester College served as the grant institution and fiscal agent for this project on behalf of the schools of the Associated Colleges of the Midwest (ACM) and Great Lakes Colleges Association (GLCA) listed below.

ACM	GLCA
Beloit College	Albion College
Carleton College	Antioch College
Coe College	Denison University
Colorado College	DePauw University
Cornell College	Earlham College
Grinnell College	Hope College
Knox College	Kalamazoo College
Lake Forest College	Kenyon College
Lawrence University	Oberlin College
Macalester College	Ohio Wesleyan University
Monmouth College	Wabash College
Ripon College	College of Wooster
St. Olaf College	
University of Chicago	

I would also like to thank Stan Wagon of Macalester College for providing the cover image for each volume in the collection.

Table of Contents

V. The Definite Integral

VI. The Definite Integral Revisited

VII. Sequences and Series of Numbers

VIII. Sequences and Series of Functions

IX. The Integral in $\mathbb{R}^2$ and $\mathbb{R}^3$

X. Vectors and Vector Geometry

XI. The Derivative in Two and Three Variables

XII. Line Integrals

I. Functions and Graphs (commentary)

II. The Derivative (commentary)

III. Extreme Values (commentary)

IV. Antiderivatives and Differential Equations (commentary)

V. The Definite Integral (commentary)

VI. The Definite Integral Revisited (commentary)

VII. Sequences and Series of Numbers (commentary)

VIII. Sequences and Series of Functions (commentary)

IX. The Integral in $\mathbb{R}^2$ and $\mathbb{R}^3$ (commentary)

X. Vectors and Vector Geometry (commentary)

XI. The Derivative in Two and Three Variables (commentary)

XII. Line Integrals (commentary)

Foreword

This volume of problems has been produced under the auspices of a calculus reform project funded by the National Science Foundation. Its aim is to address the need, expressed by many mathematicians, for a fresh approach to the study of elementary calculus at the college level. In particular, we seek to provide a collection of textbook problems which stimulates students to understand the power and the beauty of calculus rather than to regard the subject as a hodgepodge of unrelated techniques.

The current calculus reform movement has developed a number of themes and ideas which this collection attempts to incorporate. It is now apparent that the existence of computer algebra systems (CAS) transforms the study of calculus by providing students with a powerful resource for computations, graphical representation, and symbolic manipulation. In this environment, many of the problems calculus students traditionally see become trivialities, and we have, therefore, avoided such problems. Our exercises stress conceptual understanding over rote drill. Ocassionally this is done by asking students to use the computer's output to make a conjecture. Sometimes we ask students to formulate the answer to a problem and to leave the calculation to a computer package or a calculator. In this way, interesting problems that were once thought too hard for students because of their computational difficulties are now accessible. More often, however, our exercises neither require the use of a CAS nor are they trivialized by one. An effort has been made to convey functions by graphs and tables, rather than by rules in the belief that "real world" data generally come that way.

Our exercises are designed to help students achieve a better understanding of this rich and useful subject. We have tried to frame questions that require students to grapple with ideas rather than techniques, and we have shunned problems involving tricky computations. We have written, in fact, some questions that may seem, at first, too easy, but not all students will find them trivial. These questions may be especially good for in-class discussions.

The writers of this volume are: Deborah Hart of Knox College, Eugene Herman of Grinnell College, James Stein of California State University Long Beach, Lyle Welch of Monmouth College, and its editor, Robert Fraga of Ripon College. Besides creating our own problems, we have drawn on a wealth of existing material, a code for which is given on the page following this foreword. In some instances, problems have been adapted or revised to accord with the authors' goals.

A word about the source of material is in order. Indication of source in this volume means that the writers used a form of the problem available in the source cited in square brackets at the end of the problem commentary. It has been our goal to assemble a set of problems, both old and new, which we feel will be useful to teachers of calculus. Sometimes this has meant rescuing some problems which have been jettisoned during the transition from one edition to another of a standard textbook. Such classics are designated by [CP].

In other cases, the problems are new insofar as they reflect the resources now available to calculus students. Problems which require computer or calculator assistance are marked by the symbol [C] at the beginning of the statement of the problem.

The problem sets are structured around syllabi for a two-semester course in elementary calculus worked out by colleagues in the project. The commentaries which accompany the problems frequenty contain more than answers to the questions posed. Often there is an indication about a way in which a question can be extended, or how it can be construed in a different (frequently physical) context, or something about its history as well as the reason why the problem was selected. This volume gives teachers a variety of exercises suitable for calculus students in the 1990's.

Acknowledgements for Volume II

The code on the following page identifies the sources from which some of the problems in this collection were drawn. Modifications in the problems have sometimes been made to suit the writers' purposes. In addition to the code citations, we wish to acknowledge using problems developed at the mathematics departments of St. Olaf College, Grinnell College, and Miami University. We also gratefully acknowledge the assistance of Aparna W. Higgins of the University of Dayton in drawing to our attention problems that have been lost between editions of calculus books and Barry J. Arnow of Kean College, whose presentation of the Buffon Needle Problem inspired the version which appears in this book. Kent Merryfield of California State University Long Beach read the entire book, made useful comments about it, and wrote a significant number of the problems. Formatting the book in TEX was done by Robert Messer of Albion College. Many editorial changes were keyed in by David Weitermann, and many of the figures were drawn electronically by Helen Pelster, both students of Ripon College.

The map of Lake Ontario in Chapter V and the map of Iowa in Chapter XI are from the Rand McNally Road Atlas (C) 1986, greatly reduced from original size. R.L. 92-S-172.

Advanced Placement Examinations [AP]
Reproduced by permission of
Educational Testing Service,
Copyright owner

University of Arizona [AZ]
Calculus Projects

Charlene Beckman & Ted Sundstrom [CB]
Grand Valley State University

Caspar Curjel [CC]
Multivariable and Vector Calculus
Copyright © 1990 by McGraw Hill
Reprinted by permission by McGraw Hill,Inc.

Calculus for a New Century [CNC]
MAA Notes, No. 8

David Appleyard [DA]
Carleton College

Goldstein, Lay & Schneider [GL& S]
Calculus and Its Applications
Copyright © 1984 by Prentice Hall
Reprinted by permission by Prentice-Hall,Inc.

Gilbert Strang [GS]
Calculus
Wellesley Cambridge Press

Joseph Mandlebaum [JM]
Frankfurt High School

James Stewart [JS]
Calculus 2nd Edition
Copyright © 1991,1987 by Wadsworth,Inc.
Reprinted by permission of Brooks Cole Publ. Co.

Morris Kline [MK]
Calculus
Brooks/Cole

Harvey B. Keynes [MN]
Sourcebook of Calculus Problems
NSF Teacher Renewal Project

Maple *Calculus Workbook* [MP]
Calculus Projects

Paul Foerster [PF]
Alamo Heights High School

Brian J. Winkel [RH]
Rose-Hulman Institute of Technology

Tom Apostol [TA]
Calculus: Volume I
Copyright © 1967 John Wiley & Sons
Reprinted by permission of John Wiley & Sons

Thomas & Finney [TF]
Calculus and Analytic Geometry
Copyright © 1988
Addison Wesley Publ. Co.

Thomas Tucker [TT]
Colgate University

Zaven Karian [ZK]
Denison University

Suggestions to the Student

The problems in this book are a bit different from the usual calculus textbook problems. They are not intended to be harder although some may well be. They are intended, instead, to help you better understand the concepts of calculus and how to apply them. None of these problems asks simply for a computation, and some ask for no computation at all. Instead, they may ask you to do one of the following: Apply a concept or technique you have just learned in a mildly novel context; combine concepts or techniques that you have seen only in isolation before; give a graphical interpretation of the behavior of a function; make an inference, from a graph or a table of data, about a function or a physical relationship.

When you begin working on these problems, you may feel that you do not know how to get started on a problem or where you should end up. That's only natural. In fact, some of the problems can be approached in a variety of ways and have no single answer. Since the purpose of all the problems in this volume is to help you develop a better understanding of calculus, a good way to get started is to see if you understand the question. Talk it over with a classmate and see if the two of you have the same interpretation. If you don't, check in the textbook to see if you have the right meanings for the crucial words in the problem. Draw a picture, if possible, to illustrate the problem. If you encounter a function that is hard to graph, use a computer or a graphing calculator to draw the graph. In fact, *all* uses of computers and calculators are legitimate in working on these problems. If you are still stuck, talk it over some more with a classmate or ask for a discussion in class, but be prepared to offer the thoughts you have developed about the problem.

The keys to getting the most out of these problems are thinking, discussing, and writing. When you recognize a concept or technique that is likely to be involved in a problem, ask yourself what you know about it and how it might be applied, and be prepared to reread your textbook or lecture notes to refresh your understanding. Then test your ideas by discussing them with a classmate or in class. Finally, write up your conclusions in complete English sentences that convey your understanding as clearly as you know how. With practice, you will discover that discussing and writing promote clear thinking and thus help you develop a better understanding of the material that you are studying.

Suggestions to the Instructor

We hope you find, as we have, that assigning problems from this volume has a beneficial effect on your calculus course. By a judicious use of these problems, you will help your students to develop a deeper understanding of the main concepts, a more flexible and imaginative approach to problem-solving, and a greater appreciation of the ideas and techniques of calculus.

An injudicious use of these exercises, however, could have an unhappy outcome. The greatest danger, we have found, is that if you assign too many of these problems or too high a proportion of the more challenging ones, many of your students will get discouraged and quit trying. You may be surprised to discover, for example, that some problems that you and we would believe to be easy will cause students to think long and hard.

We recommend, therefore, that you experiment until you find the right balance of problems for your students. A scheme that has worked well for many instructors and that you might try is the following: Assign four or five of our problems each week. Include a range of levels of difficulty and types of problems in each of these weekly assignments. Give students a week or so to work on the problems, urge them to start working well before the due date, and encourage them to ask questions in class about the problems before they are due.

Just four or five problems a week may seem too few to have much effect, but creating the right environment will magnify their impact greatly. First, we suggest that you give students adequate time to treat the problems seriously by significantly reducing the number of regular textbook problems you assign. Then tell them the importance you attach to the problems from our sets, and follow through by showing them you mean it. Devote class time to discussion of some of the problems. Make sure the problems are graded carefully, either by yourself or a well-trained student assistant. Count this grade in their final grade, and give similar problems on all exams, warning students that you will do so. Your students will soon get the idea that working on these weekly problem sets is some of the most important work they are doing for the course.

Even more can be done to ensure that the work done by your students on these weekly problem sets will be beneficial. Often the way we teach mathematics leads students to believe that they learn mathematics only by watching us solve problems and then mimicking what we and the textbook authors do. The benefits of discussing and writing mathematics are rarely apparent to them. Here is an opportunity to change that. Encourage students to work in pairs on the weekly problem sets as a mechanism for stimulating discussion at least within the pairs. Some ways to foster such teamwork are: Allow each pair to hand in a joint assignment rather than two separate assignments. Select pairs yourself, perhaps pairing students who live near one another, and point out to students that two brains will be better than one, especially on the harder problems. To get students to write more, set high standards for their solutions, including requirements that they give explanations for their conclusions and that they do so in complete sentences. Although many of our problems ask for an explanation, few students will comply unless they see that you insist on it. Soon your students will discover that discussing and writing mathematics really does help them understand, especially in the context established by the problems in this volume.

Chapter I: Functions and Graphs

Section 1: Domain and Range. Elementary Functions

1. Use function notation to describe the way the second variable (abbreviated DV) depends on the first variable (abbreviated IV) in each of the following. What are the domain and range of each function?

   a. IV: an acute angle X in a right triangle
   DV: the area A of the triangle if the hypotenuse is a fixed length H

   b. IV: one leg L of a right triangle
   DV: the hypotenuse H of the right triangle if the other leg is 2

   c. IV: the hypotenuse H of a right triangle
   DV: the other leg L of the right triangle if one leg is 5

   d. IV: temperature C in degrees Centigrade
   DV: temperature F in degrees Fahrenheit

   e. IV: any real number x
   DV: y, the larger of x and $1 - x$

   f. IV: the angle X of a sector in a circle of a constant radius R
   DV: the area A of the sector

2. Suppose the following: The function f is periodic with period 3, $f(-2) = 0$, $f(-1.5) = 1$, f is linear between -2 and -1.5, and f is linear between -1.5 and 1. Draw the graph of f. For which values of x is $f(x) = 1$? For which values of x is $f(x) = 0.5$? What is the range of f?

3. Explain why there is no value of k for which the graph of $y = x^2 + kx$ is tangent to the line $y = 5$.

4. The equation $2x + ky = -k$ is the equation of a line. In each of the parts below, determine which values of the constant k, if any, will allow the line to have the property listed in that part.

   a. The line has slope 3.

   b. The line passes through $(0, 1)$.

   c. The line is horizontal.

   d. The line is vertical.

   e. The line passes within 1 unit of the origin.

5. A function f has the values $f(0) = 3$, $f(2) = 1$, is piecewise linear, and has the slope -1 if $x < 0$ and 1 if $x > 2$. Sketch the graph of the function g defined by each of the following rules.

   a. $g(x) = f(x)$

   b. $g(x) = -f(-x)$

   c. $g(x) = f(x + 2)$

   d. $g(x) = f(2x)$

   e. $g(x) = f(3x - 6)$

Section 2: Trigonometric Functions

1. Give a reason why the cosine function is not a polynomial.

2. The red stripe on a barber pole makes one complete revolution around the pole. If the pole is 1 meter high and 30 centimeters in radius, what angle does the stripe make with the pole, and how long is the stripe? Assume that the stripe is a thin line.

3. Let a, b, and c be positive constants. Describe what happens to the graph of $y = a\sin(bx + c)$ if:

   a. a and b are fixed and c is doubled.

   b. a and c are fixed and b is doubled.

   c. b and c are fixed and a is doubled.

4. a. Express $\sin(\frac{\pi}{2} - x)$ in terms of $\cos x$.

   b. Express $\cos(\frac{\pi}{2} - x)$ in terms of $\sin x$.

   c. Express $\cos^2 x$ in terms of $\cos 2x$.

   d. Express $\sin^2 x$ in terms of $\cos 2x$.

   e. Express $\cos x$ in terms of $\tan \frac{x}{2}$.

   f. Express $\sin x$ in terms of $\tan \frac{x}{2}$.

   Hint: In the last two parts, express x as $2(x/2)$ and use double angle formulas.

5. Show why, for angles in the first quadrant, the definition of the sine and cosine as circular functions, using the unit circle as the basis for definition, is equivalent to the definition of sine and cosine in terms of the legs and hypotenuse of a right triangle. What major theorem from geometry is being used? Why is the relationship $\sin^2 x + \cos^2 x = 1$ sometimes called the trigonometric version of the Pythagorean Theorem?

6. Determine the smallest positive numbers x for which the function f defined by $f(x) = -4\sin(4x + \frac{\pi}{6})$ has:

   a. the value zero;

   b. its maximum value;

   c. its minimum value.

7. While leaning out of the window of your house, you notice a) that it is a nice day, b) that the line of sight to the top of a nearby tree makes an angle of $22°$ with the horizontal, and c) that the line of sight to the base of the tree makes an angle of $7°$ with the horizontal. Taking advantage of a), you go outside and discover that it is 180 feet from the house to the base of the tree. How tall is the tree?

8. A lighthouse beam rotates three times per minute. From a point on shore nearest the lighthouse, the beam moves 300 feet down shore in 4 seconds. How far is the lighthouse from the shore? Discuss qualitatively (*without* actually solving the problem) whether the lighthouse will be nearer or further from shore, relative to the answer you have just obtained, if the *only* change to the information above is:

   a. the beam moves 500 feet down shore;

   b. the beam moves 100 feet down shore;

   c. the beam rotates twice per minute;

   d. the beam takes 3 seconds to move 300 feet down shore.

Section 3: Exponential and Logarithmic Functions

1. If $b > 1$, what can be said about the relative magnitudes of x^b, b^x, and $\log_b x$ for large values of x?

2. The figure below shows the graphs of $y = e^x$ and $y = x^{144}$ on the interval $[990, 1000]$. Which graph is which and why?

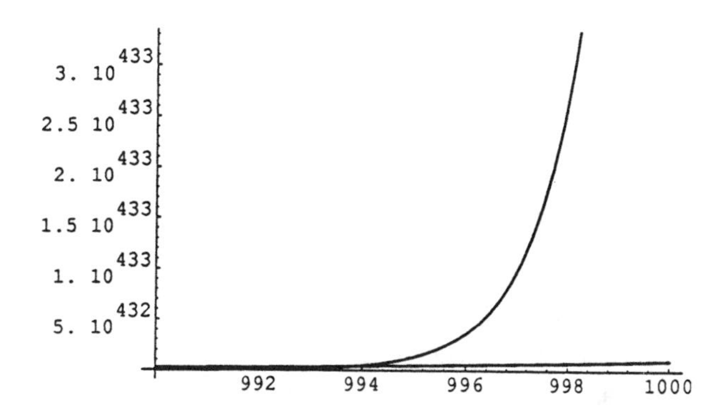

3. Superman has a violent reaction to red kryptonite, which decays into green kryptonite, fortunately with a half-life of 15 hours. It is no longer dangerous to Superman when 90% of the red kryptonite has decayed. If Superman is exposed to pure red kryptonite, for how long is he in danger?

4. What is the domain of $\ln(x^2 - 8x + 15)$?

5. Given the graph of $y = 5x$, describe how the following graphs can be obtained from this graph by means of translations and reflections about various axes.

   a. $y = .2^x$

   b. $y = 3 - 5^x$

   c. $y = 7(5^x)$

   d. $y = 5(7 + 5^{x-4})$

6. Start with the graph of $y = 4^x$. What is the equation of the graph which results from:

   a. reflecting the graph about the x-axis;

   b. reflecting the graph about the y-axis;

   c. reflecting the graph about the line $y = 5$;

   d. reflecting the graph about the line $x = -2$;

   e. reflecting the graph first about the x-axis, then about the y-axis;

   f. reflecting the graph first about the y-axis, then about the x-axis.

7. A colony of bacteria living on a Petri dish under optimal conditions doubles in size every ten minutes. At noon on a certain day, the Petri dish is completely covered with bacteria. At what times (to the nearest hour, minute, and second) was the percentage of the plate covered by bacteria:

   a. 50% b. 25% c. 5% d. 1%

8. a. What is the relation between the half-life H of a radioactive substance, and the third-life T of the same substance? (You should be able to guess the meaning of the term third-life from context).

   b. What is the relation between the doubling period D of a colony of rabbits and the tripling period T of the same colony?

Section 4: Composite Functions

1. Let $f(x) = 3x + 2$, $g(x) = \sin 4x$, and $h(x) = \ln x$. Express the following composite functions as functions of x, and indicate their domains.

   a. $f \circ f$ b. $f \circ g$ c. $f \circ h$

   d. $g \circ f$ e. $g \circ g$ f. $g \circ h$

   g. $h \circ f$ h. $h \circ g$ i. $h \circ h$

2. a. Express the function h defined by $h(x) = \sqrt{\ln x + 4}$ as a composite function $h = g \circ f$. What are the domains of f, g, and h?

   b. Is the composite of two linear functions a linear function?

   c. Is the composite of two quadratic functions a quadratic function?

   d. Is the composite of two polynomials a polynomial?

   Explain carefully your answers to parts b, c, and d.

3. a. Simplify the trigonometric composite function $f(x) = \sin(\arctan x)$.

   b. Simplify the trigonometric composite function $g(x) = \cot(\arccos(2x + 5))$.

   c. What is the domain of $\arcsin(\tan x)$?

   d. What is the domain of $\operatorname{arcsec}(\sin x)$?

4. If $f(g(x)) = \ln(x^2 + 4)$, $f(x) = \ln(x^2)$, and $g(x) > 0$ for all real x, what is $g(x)$?

5. A function h is said to be *even* if $h(x) = h(-x)$ for all real x. A function h is said to be *odd* if $h(x) = -h(-x)$ for all real x. What can be said about the oddness and evenness of the composites of two functions which are themselves either odd or even?

6. Show that g is the inverse of f if and only if the graph of g is the reflection of the graph of f about the line $y = x$.

7. Let $f(x) = ax + b$ and $g(x) = cx + d$.

   a. When does $f \circ g = g \circ f$?

   b. When does $f \circ g = f$?

   c. When does $f \circ g = g$?

8. Does $f \circ (g + h) = f \circ g + f \circ h$?

9. Show that $f(x) = \sin \frac{1}{x}$ defined on the set $\{x \mid x > 0\}$ has no inverse on the interval $(0, c)$ for any $c > 0$.

10. If $f(x)$ has an inverse and f is increasing, show that f^{-1} is increasing.

11. Show that, if $f(x) = x^2 - 4x$, there is a number $x = a$ such that $f(x)$, with domain $\{x \mid x < a\}$, has an inverse $g(x)$, and that $f(x)$, with domain $\{x \mid x > a\}$, has an inverse $h(x)$, but that $g(x)$ and $h(x)$ are different.

Section 5: Functions Described by Tables or Graphs

1. The picture below shows the graph of $y = f(x)$. Answer the following questions.
   a. What is the domain of f?
   b. What is the range of f?
   c. On which interval or intervals is f increasing?
   d. On which interval or intervals is f decreasing?
   e. Does f have an inverse on $\{x \mid x < -2\}$? If so, what are the domain and range of the inverse?
   f. Does f have an inverse on $\{x \mid x < 0\}$? If so, what are the domain and range of the inverse?
   g. Does f have an inverse on $\{x \mid x > -2\}$? If so, what are the domain and range of the inverse?
   h. Does f have an inverse on $\{x \mid x > 0\}$? If so, what are the domain and range of the inverse?

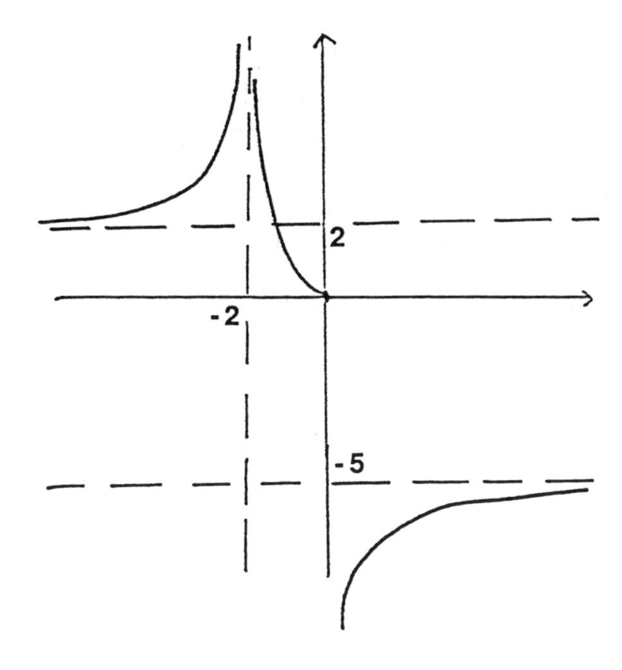

2. A sociological study was made to examine the process by which doctors decide to adopt a new drug. The doctors were divided into two groups. The doctors in group A had little interaction with other doctors and so received most of their information via mass media. The doctors in group B had extensive interaction with other doctors and so received most of their information via word of mouth. For each group, let $f(t)$ be the number of doctors who have learned about a new drug after t months. Match the graph of $f(t)$ for group A and group B to the graphs below. Explain your choice.

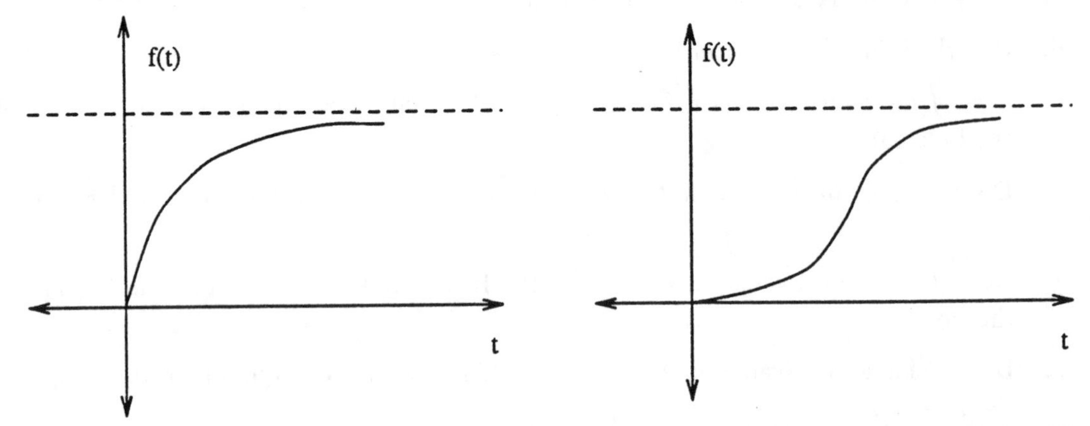

3. The graphs above also describe the learning curves of two different types of jobs. If t is the time on the job, $f(t)$ describes how much of the required job skills the individual possesses. One type of job is skilled or semi-skilled, such as the operator of a word-processor. The other type of job is primarily unskilled, such as the french-fry chef at the Golden Arch Room (McDonald's). Which of the above graphs describes which learning curve, and why?

4. Consider the situation described in the following sentence:

   The child's temperature has been rising for the last two hours, but not as rapidly since we gave her the antibiotic an hour ago.

   Sketch an example of a graph of temperature as a function of time which would be consistent with the situation described in the sentence above.

5. What can you deduce about the shape of the container from the measurements in the table below?

Depth of Water	5	10	15	20	25
Volume of Water	400	740	960	1100	1200

6. Differently shaped bowls are used to hold water. Four possible graphs of the volume V (in cublic inches) of water contained, as a function of the depth D (in inches) of the water, are given below.

A.

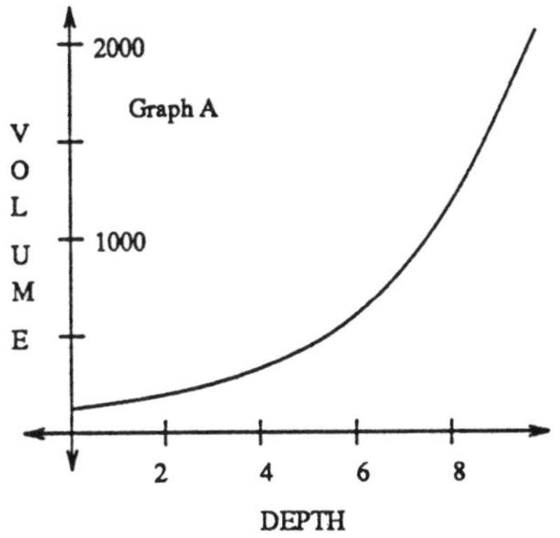

Graph A

VOLUME: 1000, 2000

DEPTH: 2, 4, 6, 8

B.

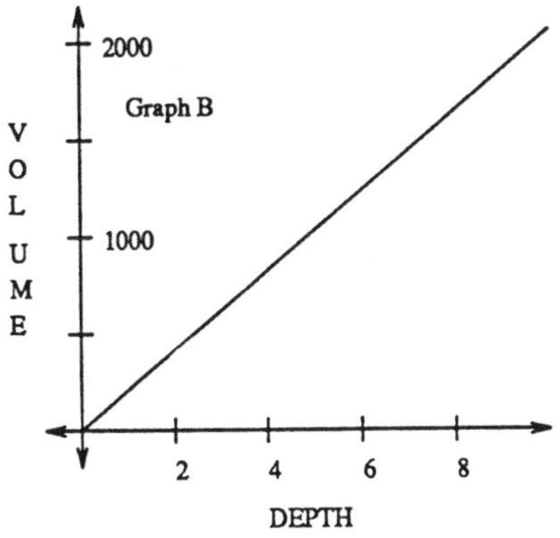

Graph B

VOLUME: 1000, 2000

DEPTH: 2, 4, 6, 8

C.

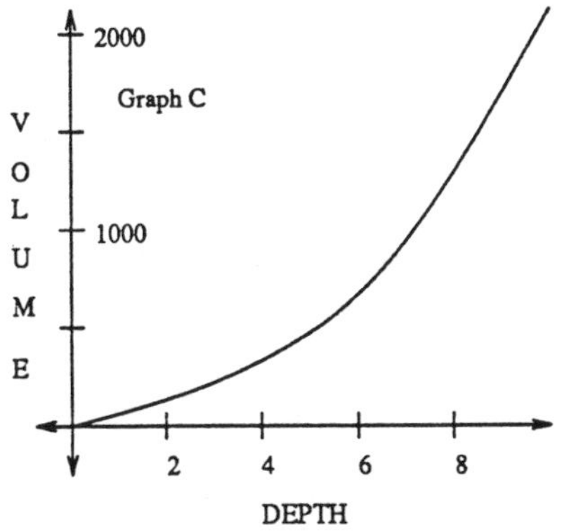

Graph C

VOLUME: 1000, 2000

DEPTH: 2, 4, 6, 8

D.

Graph D

VOLUME: 1000, 2000

DEPTH: 2, 4, 6, 8

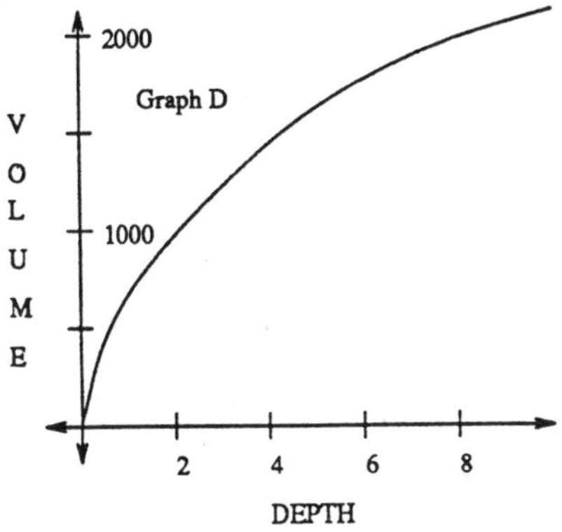

a. Explain what $V(0)$ and $V(5)$ signify.

b. Describe possible bowl shapes for each of the graphs.

7. Complete the following table, which gives the number of unindicted co-conspirators (UCC) in a recent national scandal as a function of the year. Based on this table, what curve would you fit to the data? What general conclusion do you draw?

Year	1983	1984	1985	1986	1987
Number of UCC	27	31	35	39	43
First Differences	—				

8. The number of deer in a forest at time t years after the beginning of a conservation study is shown in the graph below.

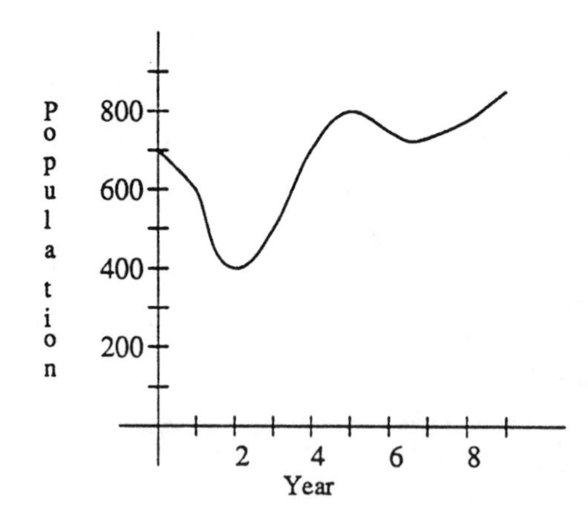

a. Over which of the following periods does the population of deer decline at the rate of 50 deer per year?

(i) $[1,2]$ (ii) $[1,3]$ (iii) $[1,4]$ (iv) $[2,3]$ (v) $[5,6]$

b. When is the population of deer increasing most rapidly?

c. Approximately how fast is the deer population increasing or decreasing at time $t = 1.5$? The answer should include appropriate units.

9. The following table shows the number of lawyers and the total annual income of those lawyers by year in a certain town.

Year	1982	1983	1984	1985	1986
Number of Lawyers	40	45	50	40	50
Total Annual Income	$2,500,000	$2,700,000	$3,400,000	$3,000,000	$3,600,000

Assume that, during the five year period given above, the rate of inflation is 5% per year.

a. What statistic best describes the financial state of a typical lawyer? What formula computes it?

b. Construct a table corresponding to this statistic.

c. When are lawyers most well off?

d. Which year sees the greatest improvement?

e. Which year sees the worst deterioration?

Section 6: Parametric Equations

1. Describe the graphs given parametrically by the following expressions, and, in particular, how the point $P(t) = (x(t), y(t))$ moves as t goes from $-\infty$ to $+\infty$.

   a. $x = t, \quad y = t^2.$

   b. $x = -t, \quad y = t^2.$

   c. $x = t^2, \quad y = t^4.$

   d. $x = \sin t, \quad y = \sin^2 t.$

   e. $x = \sin^2 t, \quad y = \sin^4 t.$

2. Two concentric circles of radii a and b, $a < b$, are centered at the origin. A ray is drawn from the origin outward. This ray makes an angle of t radians with the positive x-axis. Let $P = P(t)$ denote the intersection of the ray with the circumference of the inner circle, and let $Q = Q(t)$ denote the intersection of the ray with the circumference of the outer circle.

   a. A line is drawn through P parallel to the x-axis, and a line is drawn through Q parallel to the y-axis. Let $R = R(t)$ denote the intersection of the two lines. What are the parametric equations of $R(t)$?

   b. A line is drawn through P parallel to the y-axis, and a line is drawn through Q parallel to the x-axis. Let $S = S(t)$ denote the intersection of the two lines. What are the parametric equations of $S(t)$?

   c. What are the paths described by $R(t)$ and $S(t)$?

3. A string is wound around a circle of radius r, and then unwound while being held taut. The curve traced by the point P at the end of the string is called the *involute* of the circle. Find parametric equations for the involute if the initial position of the point P is $(r, 0)$.

4. a. Draw the upper half of a circle of radius a centered at the origin. Let $A = (-a, 0)$ and $B = (a, 0)$. Let Q be a point on the circumference of the upper semi-circle, and let t be the angle $\angle QAB$, so $Q = Q(t)$. Let $h(t) > 0$, and let $P = P(t)$ be a point on the extension of AQ such that the length of $QP = h(t)$. In your sketch, let R be the intersection of a line through P parallel to the y-axis and one through Q parallel to the x-axis. Let X denote the intersection of the x-axis and a line through Q parallel to the y-axis.

   b. What are the parametric equations of the point $P(t)$ for $0 \leq t < \frac{\pi}{2}$? Note that $h(t)$ is, in a sense, a third parameter.

   Hint: Show that $\angle AQO = t$ and that $\angle QOX = 2t$. If $P = (x(t), y(t))$, then $x(t) = OX + QR$ and $y(t) = QX + PR$. Make substitutions in the right-hand sides of the equations for $x(t)$ and $y(t)$.

5. If a projectile is launched from the origin with an initial velocity v at an angle A inclined upward from the horizontal, the parametric equations of the projectile t seconds after launching are given by $x(t) = (v\cos A)t$, $y(t) = (v\sin A)t - \frac{1}{2}gt^2$ where g is the acceleration due to gravity (32 feet per second per second). Air resistance is neglected in these equations.

   a. Eliminate the parameter t to obtain the equation of the projectile in xy-coordinates. What geometric figure is it?

   b. Where will the projectile hit the ground?

   c. What launch angle A will give the maximum distance away from the launch position that the projectile will hit?

   d. If the angle of launch A is such as to give the maximum horizontal distance, how long will the projectile remain in the air?

   e. If a centerfielder 300 feet from home plate throws a baseball at the optimal angle, with what velocity must he throw the ball in order that it just reach home plate? How long will it be in the air?

   f. Suppose that the centerfielder in part e throws the ball with an initial velocity of 90 mph. At what angle should he throw it in order that it just reach home plate? How long will it be in the air?

Section 7: Polar Coordinates

1. Describe all pairs of polar coordinates (r, θ) which represent the point P whose Cartesian coordinates are (x, y).

2. Let f be a function that is periodic with period 2π, i.e. $f(u + 2\pi) = f(u)$ for all real numbers u.

   a. If $y = f(x)$, what is the graphical consequence of periodicity in Cartesian coordinates?

   b. If $r = f(\theta)$, what is the graphical consequence of periodicity in polar coordinates?

3. A function f is said to be *even* if $f(x) = f(-x)$ for all real x.

   a. If $y = f(x)$, what is the graphical consequence of evenness in Cartesian coordinates?

   b. If $r = f(\theta)$, what is the graphical consequence of evenness in polar coordinates?

4. A function f is said to be *odd* if $f(-x) = -f(x)$ for all real x.

   a. If $y = f(x)$, what is the graphical consequence of oddness in Cartesian coordinates?

   b. If $r = f(\theta)$, what is the graphical consequence of oddness in polar coordinates?

5. Without actually graphing the functions, decide whether the graph of $r = \cos\theta + 3$ is a translate of the graph of $r = \cos\theta$. Hint: Change both equations to Cartesian coordinates.

6. What is the graph of $r = 5/(\sin\theta + \cos\theta)$, and how is it traced out as $-\infty < \theta < +\infty$?

7. What restriction of the graph of $r = f(\theta)$ will guarantee that the graph is symmetric about the origin?

8. a. In rectangular coordinates, how do you obtain the graph of $y = -f(x)$ from the graph of $y = f(x)$?

   b. In polar coordinates, how do you obtain the graph of $r = -f(\theta)$ from the graph of $r = f(\theta)$?

   c. Let c be a fixed number. In rectangular coordinates, how would you obtain the graph of $y = f(x - c)$ from the graph of $y = f(x)$?

   d. Let c be a fixed number. In polar coordinates, how would you obtain the graph of $r = f(\theta - c)$ from the graph of $r = f(\theta)$?

   e. Use parts a and c to explain how the graph of $y = \sin x$ can be obtained from the graph of $y = \cos x$.

   f. Use parts b and d to explain how the graph of $r = \sin\theta$ can be obtained from the graph of $r = \cos\theta$.

9. Describe situations in which the use of polar coordinates is more natural that the use of rectangular coordinates.

Chapter II: The Derivative

Section 1: Average Rates of Change

1. Amy takes a trip from Chicago to Milwaukee. Due to road construction, she drives the first 10 miles at a constant speed of 20 MPH. For the next 30 miles she maintains a constant speed of 60 MPH and then stops at McDonald's for 10 minutes for a snack. She drives the next 45 miles at a constant speed of 45 MPH.

   a. Draw a graph which shows her distance along the road from her starting point as a function of time.

   b. Draw a graph which shows her velocity as a function of time.

   c. What is her average speed for the trip (including the stop at McDonald's)?

2. The position $p(t)$ of an object moving along a line is given by the graph below.

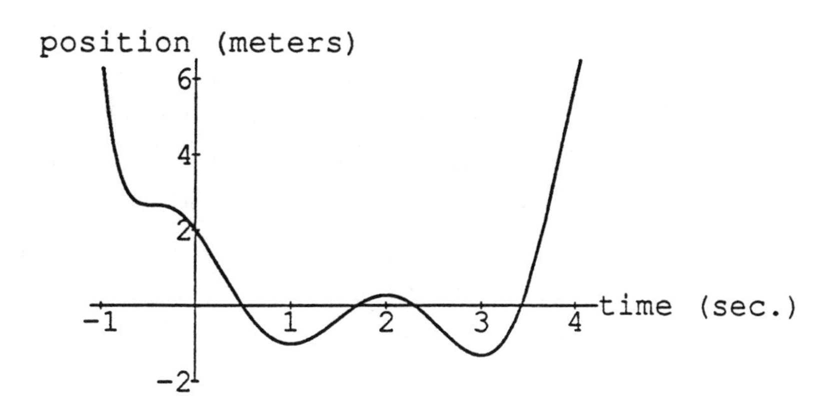

   a. Find the average velocity of the object between times $t = 1$ and $t = 4$.

   b. Find the equation of the secant line of $p(t)$ between times $t = 1$ and $t = 4$, and sketch the graph of the function together with this secant line on the same axes.

   c. For what times t is the object's velocity positive? For what times is it negative?

3. Suppose $f(1) = 2$ and the average rate of change of f between 1 and 5 is 3. Find $f(5)$.

4. The position $p(t)$ (in meters) of an object at time t (in seconds) along a line is given by

$$p(t) = 3t^2 + 1.$$

 a. Find the change in position of the object between time $t = 1$ and $t = 3$.

 b. Find the average velocity of the object between time $t = 1$ and $t = 3$.

 c. Find the average velocity of the object between any time t and another time $t + \Delta t$.

5. Let $y = f(x) = x^2 + x - 2$.

 a. Find the average rate of change of y with respect to x between $x = -1$ and $x = 2$.

 b. Draw the graph of f and the graph of the secant line through $(-1, -2)$ and $(2, 4)$.

 c. Find the slope of the secant line graphed in part b and then find an equation of this secant line.

 d. Find the average rate of change of y with respect to x between any point x and another point $x + \Delta x$.

6. A car travels for 20 minutes with an average velocity of 30 MPH, and then for 30 minutes with an average velocity of 50 MPH. What is the average velocity of the car for the 60-mile trip?

7. a. A car travels for 30 miles with an average velocity of 40 MPH and then for 30 minutes at 60 MPH. What is the average velocity of the car for the 60-mile trip?

 b. Another car travels for 30 minutes at 40 MPH and then for 30 minutes at 60 MPH. Find the average velocity of this car in the 1-hour period.

 c. A car is to travel two miles. It goes the first mile at an average velocity of 30 MPH. The driver wishes to average 60 MPH for the entire two-mile trip. Is this possible? Explain.

8. The weight $w(t)$ (in grams) of a tumor t weeks after it forms is given by $w(t) = t^2/15$. Find the average rate at which the tumor is growing during the fifth week after it was formed.

9. The owners of a flour mill estimate that it costs them $c(x) = 500 + 2x^{2/3} + (x/50)$ dollars to produce x pounds of flour per day. Find the average rate (per pound) at which their cost is changing if they increase their daily production from $x = 27000$ pounds to $x = 64000$ pounds.

10. Answer the following questions true or false and justify your answers:

    a. Velocity is the rate of change of position.

    b. If the average rate of change of a function between two points is zero, then the function must be constant between these two points.

    c. If the average rate of change of a function between any two points is always the same, then the graph of the function is a straight line.

Section 2: Introduction to the Derivative

1. Water is flowing into a large spherical tank at a constant rate. Let $V(t)$ be the volume of water in the tank and $H(t)$ be the height of the water level at time t.

   a. Give a physical interpretation of $\frac{dV}{dt}$ and $\frac{dH}{dt}$.

   b. Is $\frac{dV}{dt}$ positive, negative or zero when the tank is one quarter full? Justify your answer.

   c. Is $\frac{dH}{dt}$ positive, negative or zero when the tank is one quarter full? Justify your answer.

   d. Which of $\frac{dV}{dt}$ and $\frac{dH}{dt}$ is constant? Explain your answer.

2. Interpret what each of the following sentences says about the derivative of the given function. In each case, be sure to indicate what the function and any variables are. Then sketch the graph of a function which satisfies the properties indicated in the sentence.

   a. The price of a product decreases as more of it is produced.

   b. The child's temperature has been rising the last three hours, but not as rapidly since we gave her an antibiotic an hour ago.

   c. The cost of health insurance is rising at an ever increasing rate.

   d. The car is gradually slowing to a stop.

3. Let $f(x) = 2^x$.

   a. Find the average rate of change of f between 0 and 1.

   b. Find the average rate of change of f between $-1/2$ and $1/2$.

   c. Use a calculator to estimate $f'(0)$, the instantaneous rate of change of f at 0.

   d. Sketch the graph of f and use it to explain why the answer to part b is a better estimate of $f'(0)$ than the answer to part a. Can you suggest a generalization?

4. The position $p(t)$ of an object at time t is given by $p(t) = 3t^2 + 1$ (c.f. Problem 4 of Section II.1).

   a. Find the instantaneous velocity of the object at time $t = -1$.

   b. Find the instantaneous velocity of the object at an arbitrary time t.

5. Let $f(x) = x^2 + x - 2$ (c.f. Problem 5 of Section II.1).
   a. Use the definition of the derivative to find $f'(x)$.
   b. Find an equation of the tangent line to the graph of f at the point $(-1, -2)$.
   c. Sketch the graph of f together with the tangent line found in part b on the same axes.

6. Some values of a function f are given in the table below.

x	1	1.9	1.97	2	2.02	2.2	3	3.9	3.99	4	4.01	4.1
$f(x)$	2.5	6.6	6.905	7	7.059	7.5	8	8.82	8.98	9	9.2	11

   a. Use this table to estimate the value of $f'(2)$.
   b. Judging from the values in this table, do you think that $f(x)$ is differentiable at $x = 4$? Justify your answer.

7. Let f be a function which satisfies $f(x + y) = f(x) + f(y) + 2xy$ for all real numbers x and y and suppose $\lim_{h \to 0} \frac{f(h)}{h} = 7$.
   a. Find $f(0)$.
   b. Use the definition of the derivative to find $f'(x)$.

8. Let f be a function which satisfies $f(1+h) - f(1) = 3h + 4h^2 - 5h^3$ for all real numbers h. Show that f is differentiable at 1 and find $f'(1)$.

9. For the following functions f, graph $g(\Delta x) = \dfrac{f(1 + \Delta x) - f(1)}{\Delta x}$ as a function of Δx and use these graphs to make a conjecture about the value of $f'(1)$.
   a. $f(x) = x$
   b. $f(x) = x^2$
   c. $f(x) = x^3$
   d. $f(x) = x^4$
   e. $f(x) = 5x + 3$
   f. $f(x) = 3x^2 + 1$
   g. $f(x) = x + (1/x)$
   h. $f(x) = \sin(\pi x)$
   i. $f(x) = \cos(\pi x)$
   j. $f(x) = \log_{10} x$
   k. $f(x) = 10^x$
   l. $f(x) = |x - 1|$

10. It is claimed that, for any differentiable function f,

$$f'(x) = \lim_{\Delta x \to 0} \frac{f(x+\Delta x) - f(x-\Delta x)}{2\Delta x} \qquad (*)$$

Investigate this claim by graphing $w(\Delta x) = \dfrac{f(x+\Delta x) - f(x-\Delta x)}{2\Delta x}$ as a function of Δx for various function rules $f(x)$ and various numbers x and comparing them with the graphs of $g(\Delta x) = \dfrac{f(x+\Delta x) - f(\Delta x)}{\Delta x}$. If you do not think that formula $(*)$ is correct, indicate the relationship of $\lim_{\Delta x \to 0} w(\Delta x)$ to the derivative of $f(x)$.

11. Repeat the previous problem if instead it is claimed that

$$f'(x) = \lim_{\Delta x \to 0} \frac{f(x-\Delta x) - f(x)}{\Delta x}$$

12. Suppose f is a function for which $\lim_{x \to 2} \dfrac{f(x) - f(2)}{x-2} = 0$. Which of the following questions MUST be true, MIGHT be true or can NEVER be true?

a. $f'(2) = 2$.

b. $f(2) = 0$.

c. $\lim_{x \to 2} f(x) = f(2)$.

d. $f(x)$ is continuous at $x = 0$.

e. $f(x)$ is continuous at $x = 2$.

13. For the following functions f, use the definition of the derivative to determine whether f is differentiable at 0. If it is differentiable, find $f'(0)$.

a. $f(x) = \begin{cases} x^2 & \text{if } x \geq 0, \\ x^3 & \text{if } x < 0. \end{cases}$

b. $f(x) = \begin{cases} x & \text{if } x \geq 0, \\ 2x & \text{if } x < 0. \end{cases}$

c. $f(x) = \begin{cases} x & \text{if } x \geq 0, \\ x+1 & \text{if } x < 0. \end{cases}$

d. $f(x) = \begin{cases} x^2 \sin(1/x) & \text{if } x \neq 0, \\ 0 & \text{if } x = 0. \end{cases}$

14. Suppose $f(1) = 2$ and $f'(1) = 3$. Estimate the values of $f(1.1)$ and $f(.95)$.

Section 3: Graphical Differentiation Problems

1. The graph of a function f is given below. Estimate the values of $f'(x)$ at the following points.

   a. $x = -2$ b. $x = -1$ c. $x = 0$

   d. $x = 1.5$ e. $x = 2$ f. $x = 3$

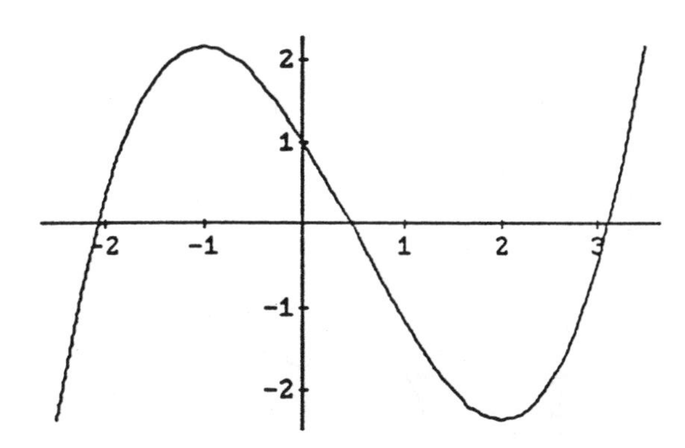

2. The graphs of two functions f and g are given below. What is the derivative of $h(x) = f(x) - g(x)$?

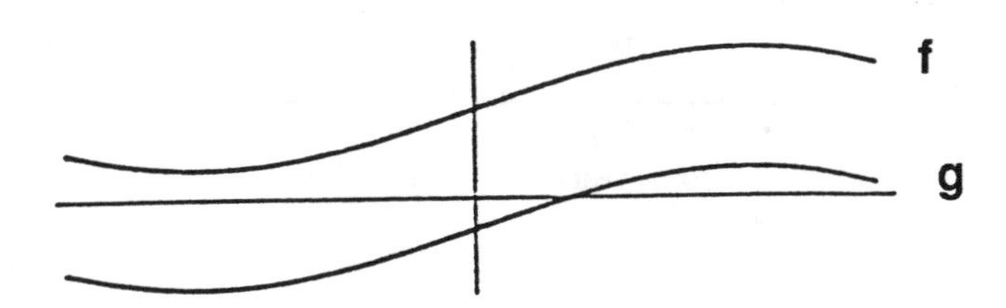

3. A graph of a function f is given below. Sketch the graph of a function g which satisfies both of the following conditions:

   (i) $g'(x) = f'(x)$ for all real numbers x and
   (ii) $g(-1) = 0$.

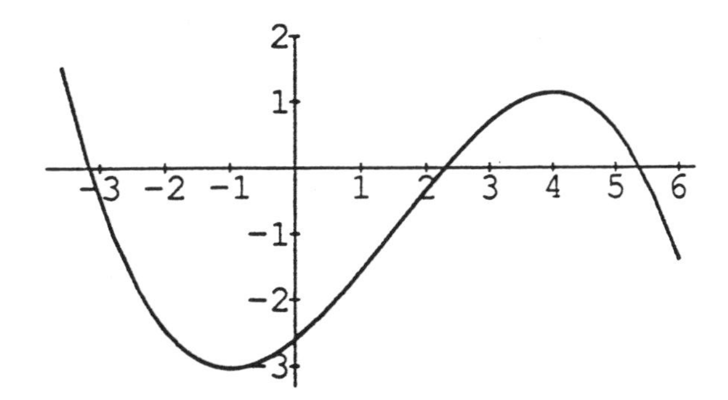

4. For each of the following functions f, sketch the graph of the function and use this graph to sketch the graph of its derivative. Then compare this to the graph of g and make a conjecture about the relationships between f and g.

a. $f(x) = x^2$, $g(x) = 2x$

b. $f(x) = x^3$, $g(x) = 3x^2$

c. $f(x) = \sqrt{x}$, $g(x) = 1/(2\sqrt{x})$

d. $f(x) = \sin x$, $g(x) = \cos x$

e. $f(x) = \cos x$, $g(x) = -\sin x$

f. $f(x) = \ln x$, $g(x) = 1/x$

g. $f(x) = e^x$, $g(x) = e^x$

5. The number of deer in a forest t years after the beginning of a population study is shown by the graph below.

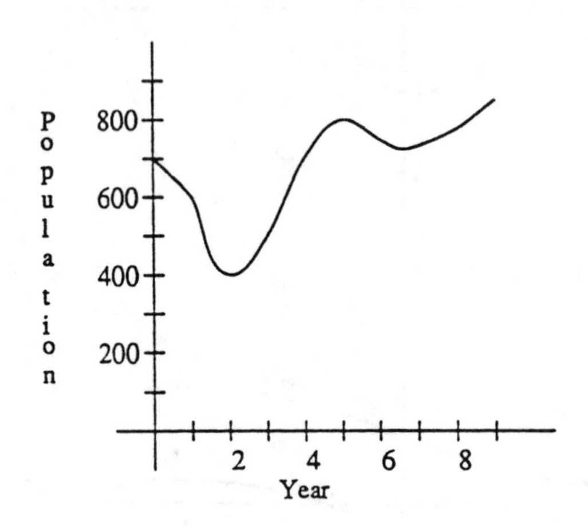

a. Over which of the following time intervals did the population of deer decline at an average rate of 50 deer per year?

(i) $[0,1]$ (ii) $[1,2]$ (iii) $[1,3]$ (iv) $[1,4]$ (v) $[5,6]$

b. When was the population of deer increasing most rapidly?

c. Approximately how fast was the population of deer increasing (or decreasing) $1\frac{1}{2}$ years after the study began?

6. Match the five functions a–e given below with their derivatives (i)–(v). Explain your reasoning.

a.

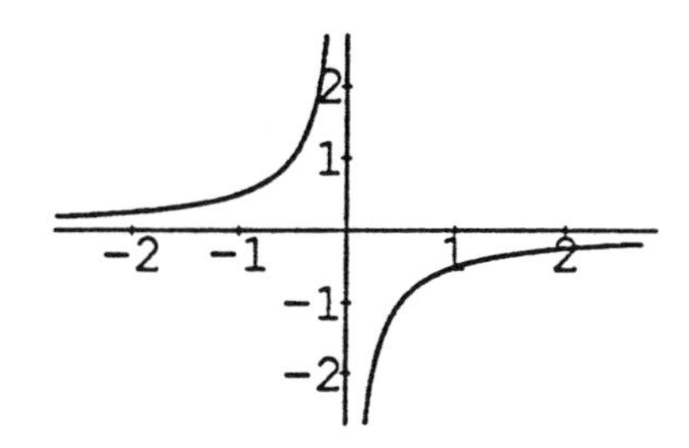

(i)

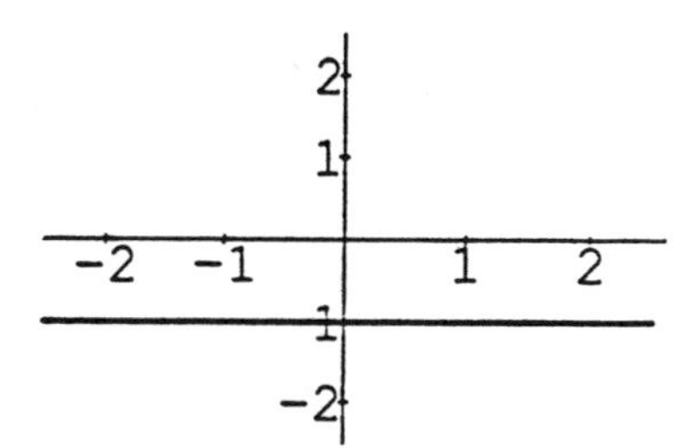

b.

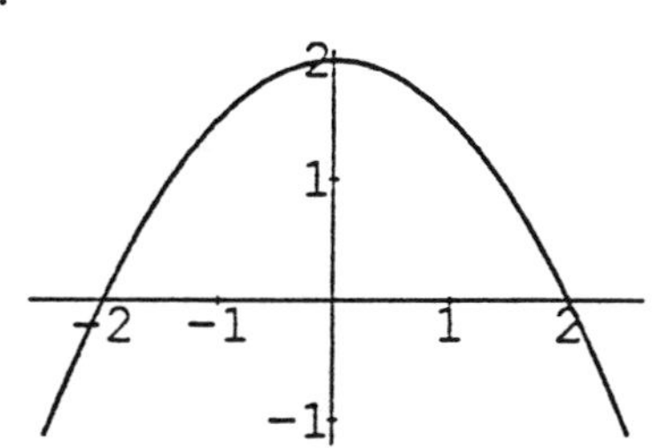

(ii)

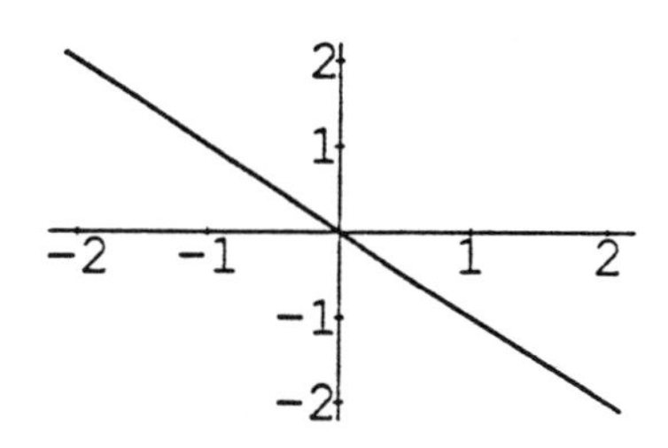

c.

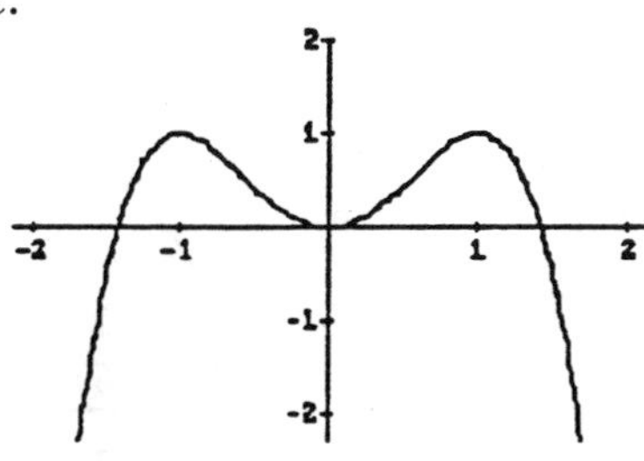

(iii)

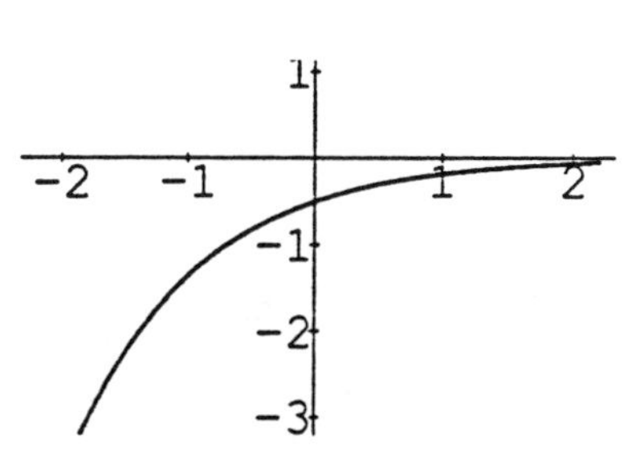

d.

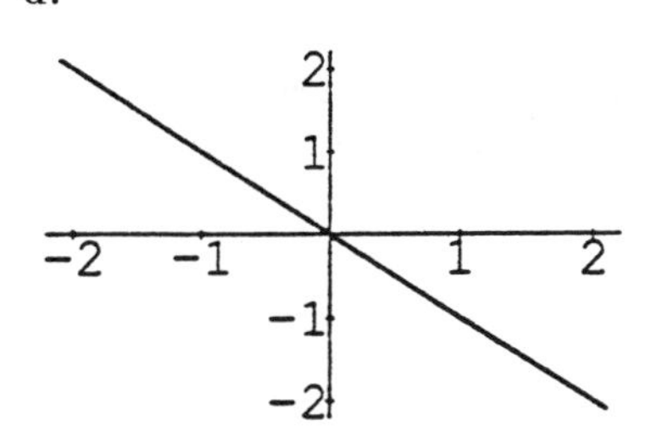

(iv)

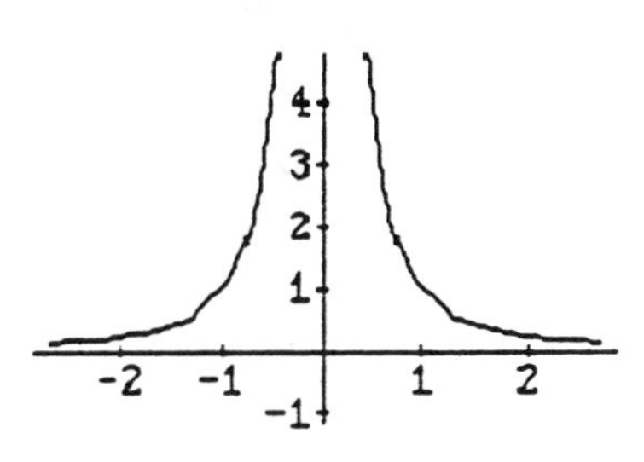

e.

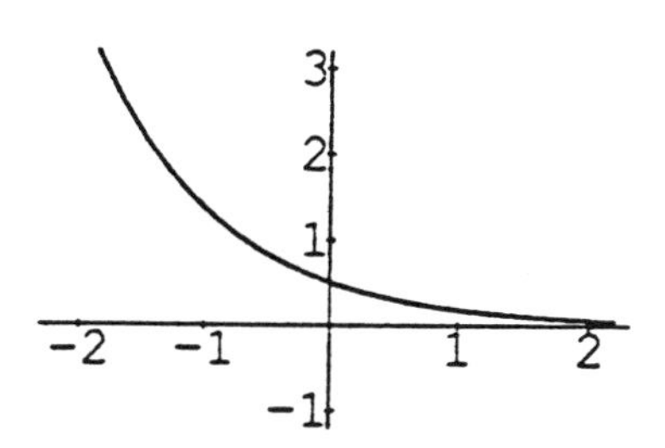

(v)

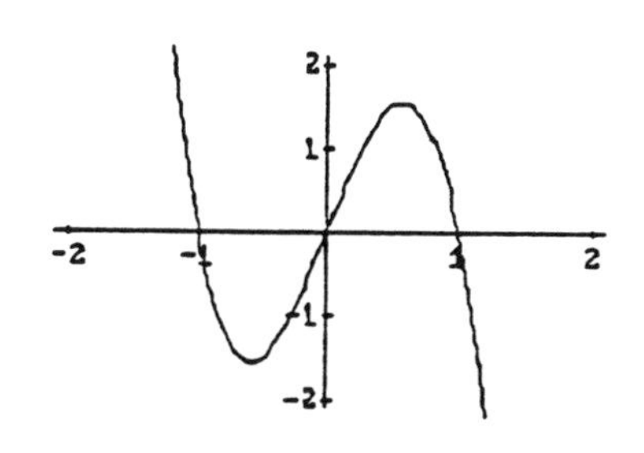

7. The graphs of some functions are given below. Indicate on what intervals the functions are increasing and on what intervals they are decreasing, and then sketch the graphs of their derivatives.

a.

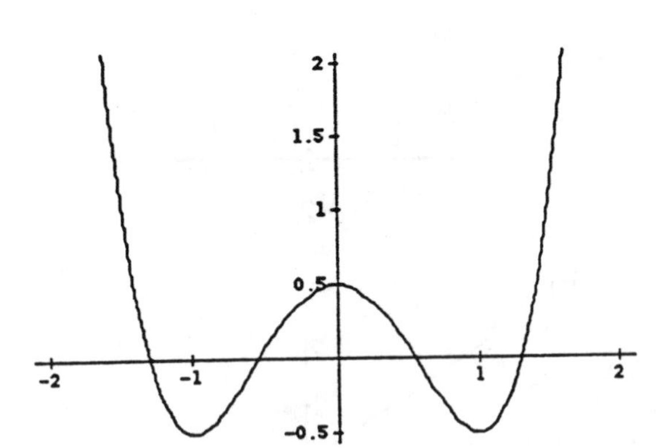

b.

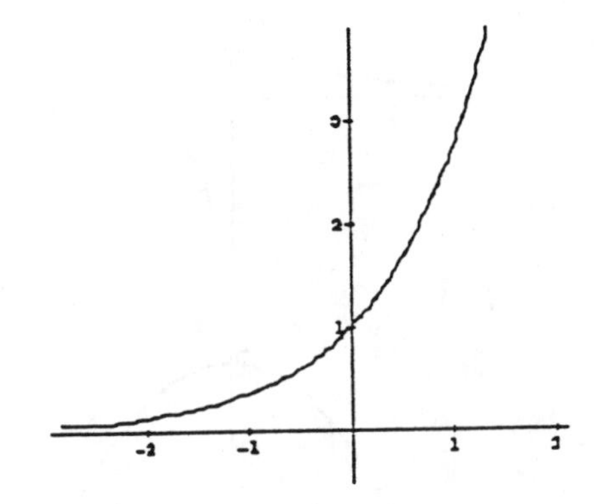

c.

d.

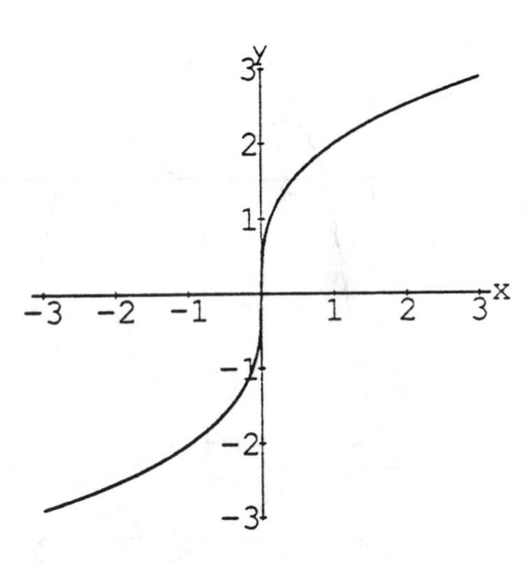

7. (continued) Proceed as in the first four parts of this question.

e.

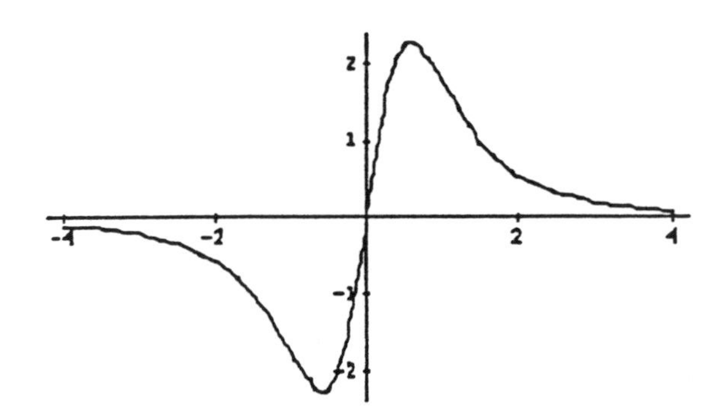

f.

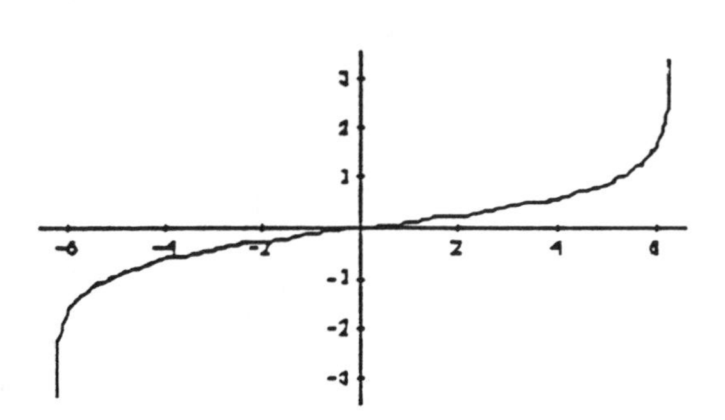

g.

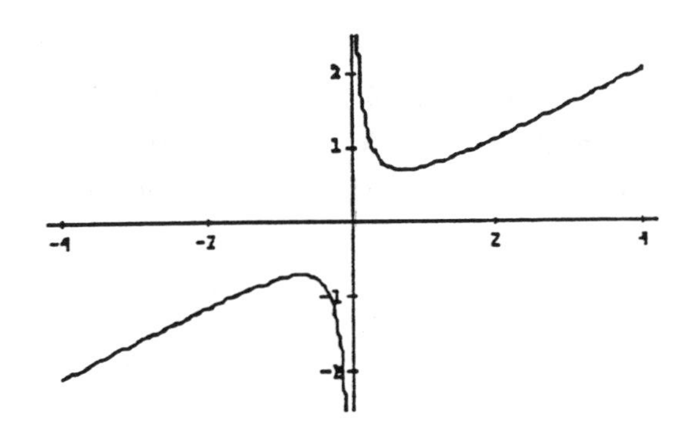

h.

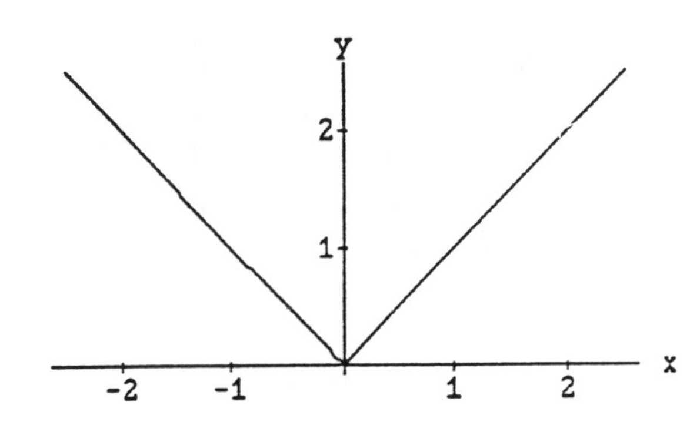

8. The graph of a function f is given below. Find all points x in $[-4, 4]$ at which f is *not* differentiable at x. Explain your answer.

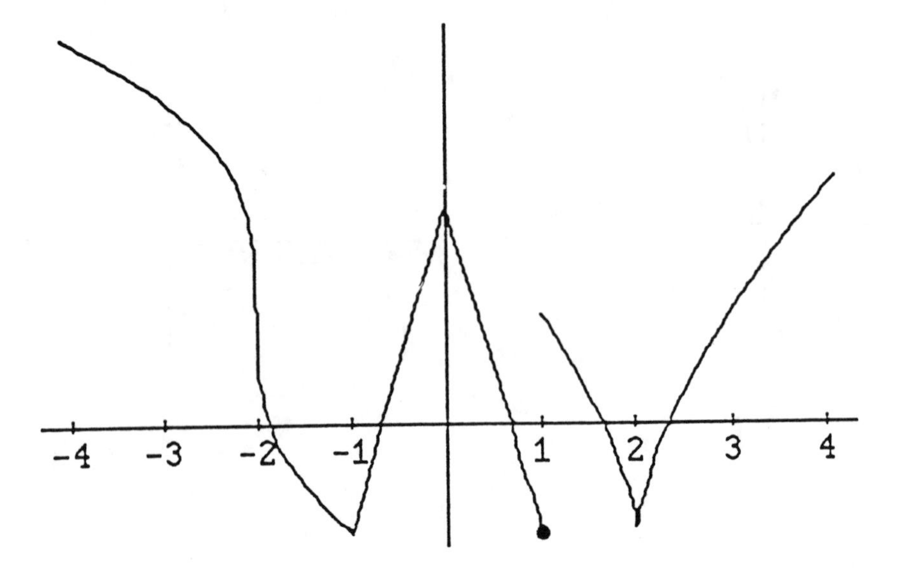

9. The consumer price index for the years 1976–1987 is graphed below. This index measures how much a basket of commodities that cost \$100 in the beginning of 1976 would cost at any given time. Sketch the derivative of this function, more commonly known as the inflation rate. In what years was the inflation rate the greatest and the least?

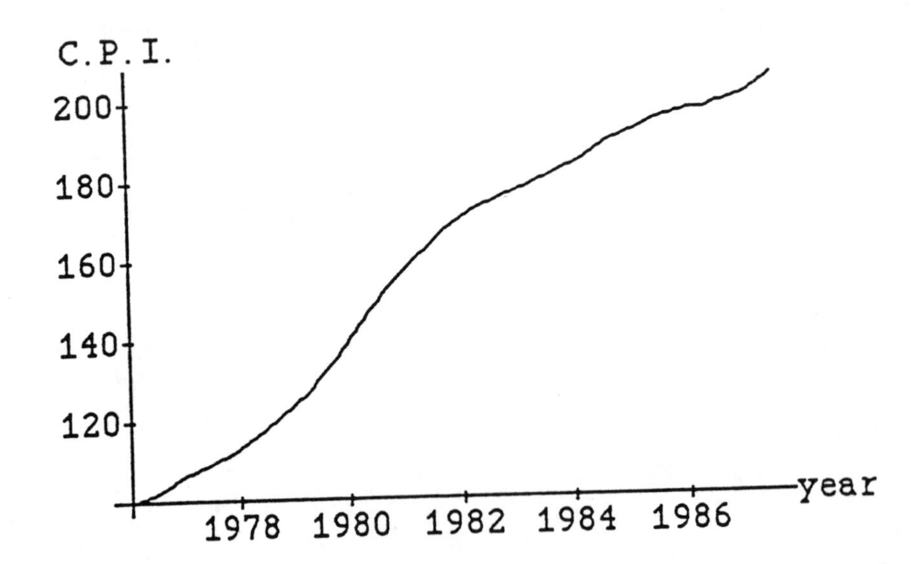

Section 4: Limits

1. The graphs of the functions f and g are given below.

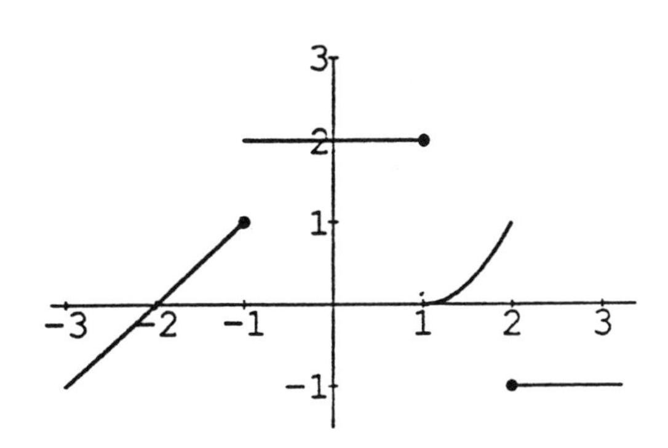

graph of f

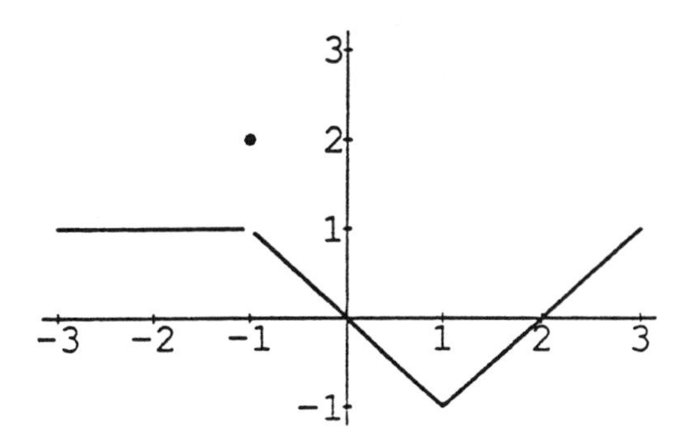

graph of g

Determine whether the following limits exist. If they do, then find the limit.

a. $\lim_{x \to -1} f(x)$

b. $\lim_{x \to 1} f(x)$

c. $\lim_{x \to -1} g(x)$

d. $\lim_{x \to 1} g(x)$

e. $\lim_{x \to -1} f(x) + g(x)$

f. $\lim_{x \to 0} 2f(x) + 3g(x)$

g. $\lim_{x \to -1} f(x)g(x)$

h. $\lim_{x \to 2} f(x)g(x)$

i. $\lim_{x \to 0} \dfrac{f(x)}{g(x)}$

j. $\lim_{x \to 0} \dfrac{g(x)}{f(x)}$

k. $\lim_{x \to -2} g(f(x))$

l. $\lim_{x \to -1} f(g(x))$

2. The graphs of functions f and g are those given in Problem 1 above. Determine whether the following limits exist and find the limit when it exists.

a. $\lim_{x \to -1^-} f(x)$

b. $\lim_{x \to -1^+} f(x)$

c. $\lim_{x \to -1^-} g(x)$

d. $\lim_{x \to -1^+} g(x)$

e. $\lim_{x \to 0^-} f(x+2)$

f. $\lim_{x \to -1^-} f(x^2)$

3. a. Explain what is wrong with the equation $\dfrac{x^2-1}{x-1} = x+1$.

   b. In light of part a, explain why the equality $\lim_{x\to 1} \dfrac{x^2-1}{x-1} = \lim_{x\to 1}(x+1)$ *is* nonetheless correct.

4. Use a calculator to help guess whether the following limits exists, and what their values are if they do exist.

   a. $\lim_{x\to 0} \dfrac{\sin x}{x}$

   b. $\lim_{x\to 0} \dfrac{1-\cos x}{x}$

   c. $\lim_{x\to 0} |x|^x$

   d. $\lim_{x\to 0} (1+x)^{1/x}$

   e. $\lim_{x\to 2} \dfrac{x^2-4}{x-2}$

   f. $\lim_{x\to 1} \dfrac{\sin(x-1)}{(x-1)^2}$

   g. $\lim_{x\to 0} \dfrac{2^x-1}{x}$

   h. $\lim_{x\to 0} x \log_{10} x$

   i. $\lim_{x\to 0} \dfrac{5000x^2}{\sin x + 5000x^2}$

   j. $\lim_{x\to 0} \sin \dfrac{1}{x}$

   k. $\lim_{x\to 0} \sin \dfrac{\pi}{x}$

   l. $\lim_{x\to 0} x \sin \dfrac{1}{x}$

5. Is there a number a such that the limit $\lim_{x\to 3} \dfrac{2x^2-3ax+x-a-1}{x^2-2x-3}$ exists? Explain your answer.

6. Determine whether the following limits exist and find the limits if they do exist.

   a. $\lim_{x\to 0} \dfrac{\sin x}{|x|}$

   b. $\lim_{x\to 0} \dfrac{1-\cos^2(3x)}{x^2}$

   c. $\lim_{x\to 0} \cos\left(\dfrac{1-\cos x}{x}\right)$

7. Let $f(x) = \begin{cases} 0 & \text{if } x \neq 0, \\ \pi & \text{if } x = 0. \end{cases}$

   a. Find $\lim_{x \to 0} f(x)$.

   b. Find $\lim_{x \to 0} \cos(f(x))$.

   c. Find $\lim_{x \to 0} f(\sin(x))$.

   d. Find $\lim_{x \to 0} f(f(x))$.

8. Find the positive integer k for which $\lim_{x \to 0} \dfrac{\sin(\sin(x))}{x^k}$ exists, and then find the value of this limit.

9. Let $f(x) = \dfrac{(x^2 - 2x + 3)(x - 2)}{x - 2}$.

   a. Sketch the graph of f.

   b. Find $\lim_{x \to 2} f(x)$.

   c. Find the smallest interval which contains all values of $f(x)$ where x is in the interval $(1.9, 2.1)$ but $x \neq 2$.

   d. Find an interval I containing 2 which satisfies $2.9 < f(x) < 3.1$ for all x in this interval except $x = 2$.

10. Which of the following statements MUST be true, MIGHT be true or is NEVER true about a function f which is defined for all real numbers? Justify your answers.

    a. $\lim_{x \to a} f(x) = f(a)$.

    b. If $\lim_{x \to 0} \dfrac{f(x)}{x} = 1$, then $f(0) = 0$.

    c. If $\lim_{x \to 0} \dfrac{f(x)}{x} = 1$, then $\lim_{x \to 0} f(x) = 0$.

    d. If $\lim_{x \to 0} \dfrac{f(x) - f(0)}{x} = 3$, then $f'(0) = 3$.

    e. If $\lim_{x \to 1^-} f(x) = 1$ and $\lim_{x \to 1^+} f(x) = 3$, then $\lim_{x \to 1} f(x) = 2$.

Section 5: Continuity

1. The graph of a function f is given below.

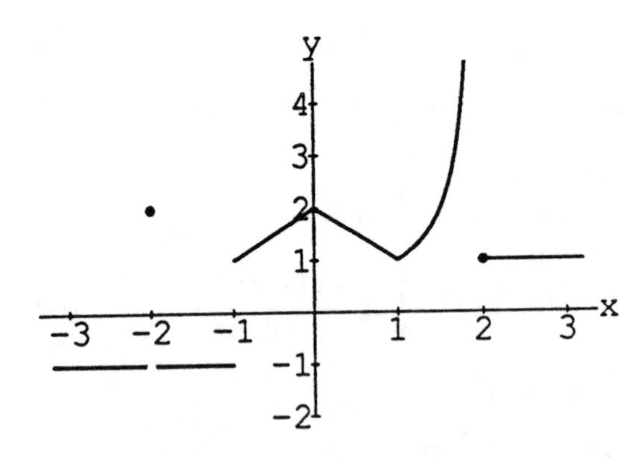

a. For what numbers x in $[-3, 3]$ is f *not* continuous at x? Explain your answer.

b. For what numbers x in $[-3, 3]$ is f *not* differentiable at x? Explain your answer.

2. For some given non-zero number a, define

$$f(x) = \begin{cases} (x^2 - a^2)/(x - a) & \text{if } x \neq a, \\ 0 & \text{if } x = a. \end{cases}$$

a. Is f defined at a?

b. Does $\lim_{x \to a} f(x)$ exist? Justify your answer.

c. Is f continuous at a? Justify your answer.

d. Is f differentiable at a? Justify your answer.

3. For $x \neq 0$ define $f(x) = \dfrac{\sin x}{x}$. Is it possible to define $f(0)$ so that f is continuous at 0? Explain your answer.

4. If $\lim_{x \to a} f(x) = L$, which of the following statements, if any, MUST be true? Justify your answer.

(i) f is defined at a.

(ii) $f(a) = L$.

(iii) f is continuous at a.

(iv) f is differentiable at a.

5. If a function f is continuous at a, which of the following statements, if any, MUST be true? Justify your answer.

   (i) f is defined at a.

   (ii) $\lim_{x \to a} f(x)$ exists.

   (iii) $\lim_{x \to a} f(x) = f(a)$

   (iv) f is differentiable at a.

6. Suppose f is continuous and $\lim_{x \to a} f(x) = L$. Find $f(a)$. Justify your answer.

7. Let $f(x) = \begin{cases} 2 & \text{if } x < 0, \\ 3 - x & \text{if } 0 \leq x \leq 1, \\ x^2 + 1 & \text{if } x > 1. \end{cases}$

   a. Is f continuous at $x = 0$? Justify your answer.

   b. Is f continuous at $x = 1$? Justify your answer.

8. Determine whether the following statements MUST be true or are at least SOMETIMES false. Justify your answers.

   a. If f is continuous at a point x, then it is differentiable at x.

   b. If f is differentiable at a point x, then it is continuous at x.

9. Let $f(x) = \begin{cases} ax & \text{if } x \leq 1, \\ bx^2 + x + 1 & \text{if } x > 1. \end{cases}$

   a. Find all choices of a and b such that f is continuous at $x = 1$.

   b. Draw the graph of f when $a = 1$ and $b = -1$.

   c. Find values of a and b such that f is differentiable at $x = 1$.

   d. Draw the graph of f for the values of a and b found in part c.

10. George takes a trip from St. Louis to Chicago. He leaves at 9 AM on Monday and arrives at 2 PM that day. He returns on Tuesday, leaving at 9 AM and arriving back in St. Louis at 2 PM, retracing exactly the same route. Show that there is a point on the road through which he passes at the same time both days.

11. Determine whether the following statements are always true or are at least sometimes false. Justify your answers.

    a. If $f(1) < 0$ and $f(2) > 0$, then there must be a point z in $(1, 2)$ such that $f(z) = 0$.

    b. If f is continuous on $[1, 2]$, $f(1) < 0$ and $f(2) > 0$, then there must be a point z in $(1, 2)$ such that $f(z) = 0$.

    c. If f is continuous on $[1, 2]$ and there is a point z in $(1, 2)$ such that $f(z) = 0$, then $f(1)$ and $f(2)$ must have different signs.

    d. If f has no zeros and is continuous on $[1, 2]$, then $f(1)$ and $f(2)$ have the same sign.

Section 6: Power, Sum and Product Rules

1. Suppose $f'(2) = 4$, $g'(2) = -3$, $f(2) = -1$ and $g(2) = 1$. Find the derivative at 2 of each of the following functions.

   a. $s(x) = f(x) + g(x)$

   b. $p(x) = f(x)g(x)$

   c. $q(x) = f(x)/g(x)$

2. If $f(x) = x$, find $f'(137)$.

3. A ship is moving directly away from shore at a speed of 30 knots. A person on the ship is moving towards the bow of the ship at a speed of 3 knots.

   a. How fast is this person moving away from shore?

   b. What differentiation rule does this problem demonstrate?

4. The width of a rectangle is increasing at a rate of 2 cm/sec, and its length is increasing at a rate of 3 cm/sec. At what rate is the area of the rectangle increasing when its width is 4 cm. and its length is 5 cm?

5. Find a formula for the nth derivative of the function f defined by each of the following rules. Do this first for $n = 1, 2, 3$ and then try to find an expression for the nth derivative.

   a. $f(x) = 1/x$

   b. $f(x) = 1/(1 + x)$

   c. $f(x) = \sqrt{x}$

6. Suppose 1) that f and g are functions which are differentiable at $x = 1$ and 2) that $f(x)g(x) = x$ for all real numbers x. Use the product rule to show that either $f(1) \neq 0$ or $g(1) \neq 0$.

7. Consider a function f which satisfies the following properties:

   (i) $f(x+y) = f(x)f(y)$,
   (ii) $f(0) \neq 0$,
   (iii) $f'(0) = 1$.

   a. Show that $f(0) = 1$. Hint: Let $x = y = 0$ in formula (i).

   b. Show that $f(x) \neq 0$ for all x. Hint: Let $y = -x$ in formula (i).

   c. Use the definition of the derivative to show that $f'(x) = f(x)$ for all real numbers x.

   d. Let g be another function that satisfies properties (i)–(iii) and let $k(x) = f(x)/g(x)$. Show that k is defined for all x and find $k'(x)$. Use this to discover the relationship between f and g.

   e. Can you think of a function which satisfies (i)–(iii). Can there be more than one such function?

8. Let $h(x) = f(x)g(x)$, where f and g are differentiable functions. Find formulas (which do not refer to h or its derivatives) for:

   a. $h''(x)$

   b. $h^{(3)}(x)$

   c. $h^{(n)}(x)$

   Hint: The formula for $h^{(n)}(x)$ can be expressed nicely in terms of the binomial coefficients $\binom{n}{k}$.

9. Let $h(x) = f(x)g(x)$ where the graphs of the functions f and g are given below.

   a. Evaluate $h(-2)$ and $h(3)$.

   b. Estimate $f'(-2)$, $f'(3)$, $g'(-2)$, and $g'(3)$.

   c. Estimate $h'(-2)$ and $h'(3)$.

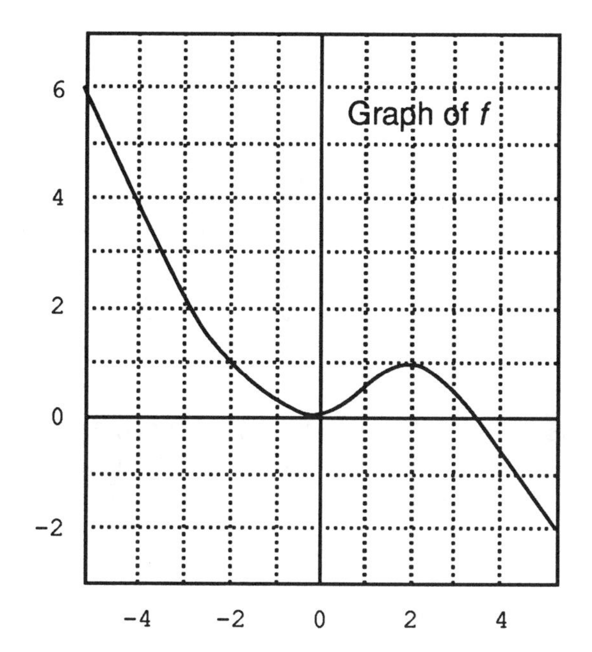

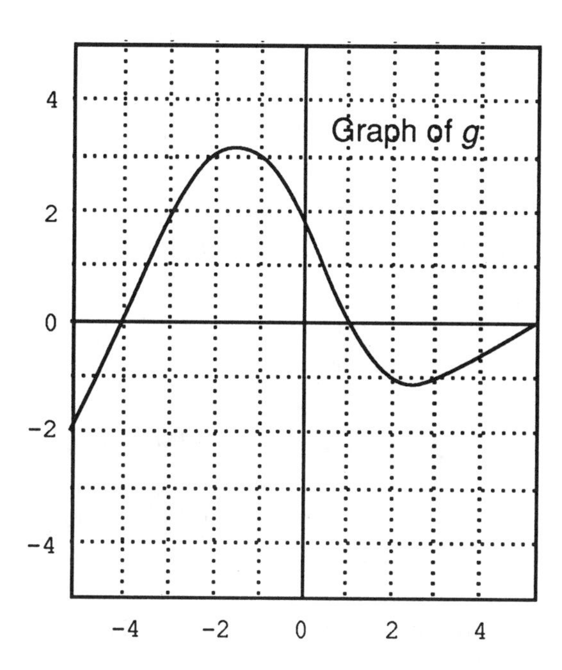

Section 7: The Chain Rule

1. Suppose $f'(2) = 4$, $g'(2) = -3$, $f(2) = -1$, $g(2) = 1$, $f'(1) = 2$ and $g'(-1) = 5$. Find the derivative at $x = 2$ of:

   a. $h(x) = f(g(x))$

   b. $k(x) = g(f(x))$

2. Find $\dfrac{d}{dx}\sin(|x|)$ for all x where this function is differentiable.

3. If y is a positive and increasing function of x, for what value of y is the rate of change of y^3 with respect to x twelve times the rate of change of y with respect to x?

4. If $f'(x) = g(x)$ and $h(x) = x^2$, then find a formula for the derivative of $f(h(x))$ in terms of $g(x)$ and x.

5. Let f be a differentiable function and define $f_2(x) = f(f(x))$, $f_3(x) = f(f(f(x)))$ and so on. Find formulas for the derivatives of $f_2(x)$ and $f_3(x)$. Do you see a pattern here? Try to find an expression for the derivative of $f_n(x)$, where n is any integer greater than or equal to 2.

6. Suppose f is a function which is differentiable for all real numbers. Justify the following statements:

   a. If f is an even function, then f' is an odd function.

   b. If f is an odd function, then f' is an even function.

7. Let $h(x) = f(g(x))$, where f and g have the graphs given below.

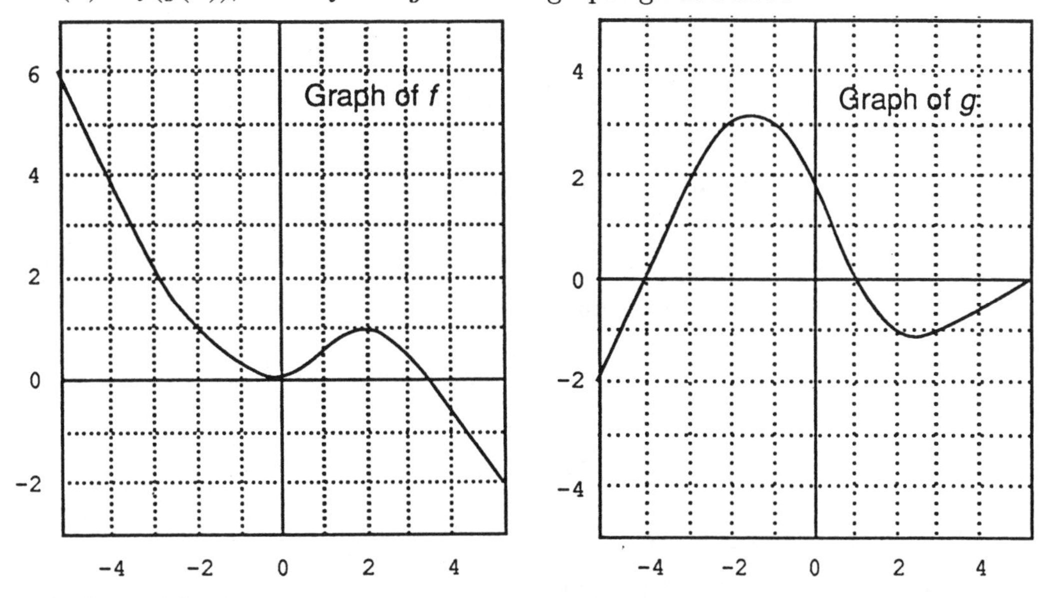

a. Evaluate $h(-2)$ and $h(3)$.

b. Is $h'(-3)$ positive, negative or zero? Explain how you know this.

c. Is $h'(-1)$ positive, negative, or zero? Explain how you know this.

8. Suppose f and g are differentiable functions for which:

(i) $f(0) = 0$ and $g(0) = 1$
(ii) $f'(x) = g(x)$ and $g'(x) = -f(x)$.

a. Let $h(x) = f^2(x) + g^2(x)$. Find $h'(x)$, and use this to show that $f^2(x) + g^2(x) = 1$ for all x.

b. Suppose $F(x)$ and $G(x)$ are another pair of differentiable functions which satisfy properties (i) and (ii) and let

$$k(x) = [F(x) - f(x)]^2 + [G(x) - g(x)]^2$$

Find $k'(x)$ and use this to discover the relationship between $f(x)$ and $F(x)$, and $g(x)$ and $G(x)$.

c. Think of a pair of functions f and g which satisfy properties (i) and (ii). Can there be any others? Justify your answer.

Section 8: Implicit Differentiation and Derivatives of Inverses

1. If $x = t^3 - t$ and $y = \sqrt{3t+1}$, find $\dfrac{dy}{dx}$ at $t = 1$.

2. Consider the curve $x^2 + y^2 = c^2$ where c is some non-zero constant.

   a. Sketch the graph of this curve and use the graph to predict in what quadrants $\dfrac{dy}{dx}$ will be positive and in what quadrants it will be negative.

   b. Find $\dfrac{dy}{dx}$ and confirm the predictions you made in part (a).

3. Consider the curve $x^2 - y^2 = c^2$ where c is some non-zero constant.

   a. Sketch the graph of this curve and use the graph to predict in what quadrants $\dfrac{dy}{dx}$ will be positive and in what quadrants it will be negative.

   b. Find $\dfrac{dy}{dx}$ and confirm the predictions you made in part a.

4. Consider the curve (called a *lemniscate*) defined by

$$(x^2 + y^2)^2 = 12(x^2 - y^2)$$

   a. Find $\dfrac{dy}{dx}$ for this curve.

   b. Graph this equation (a computer graphing package might help) and check at several points that your answer to part a agrees qualitatively with this graph.

5. Consider the curve defined by the equation $y^5 - y - x^2 = -1$.

   a. Find the equations of three different lines which are tangent to the curve at $x = 1$.

   b. Find the coordinates of the points at which the tangent line to the curve is vertical. Express the coordinates numerically to three decimal places.

   c. Use a graphing package to sketch this curve. Where does the curve intersect the y-axis?

6. Let $f(x) = x^3 + x$. If h is the inverse function of f, find $h'(2)$.

Section 9: Derivatives of Trigonometric, Log, and Exponential Functions

1. If $\sin x = e^y$, find $\dfrac{dy}{dx}$ in terms of x.

2. For $-\dfrac{\pi}{2} < x < \dfrac{\pi}{2}$, define $f(x) = \dfrac{x + \sin x}{\cos x}$.

   a. Is f an even function, an odd function or neither? Justify your answer.

   b. Find $f'(x)$.

   c. Find an equation of the line tangent to the graph of f at the point where $x = 0$.

3. Which of the following functions satisfy the equation $f'''(x) = f'(x)$?

   a. $f(x) = 2e^x + 1$ b. $f(x) = e^{-x}$ c. $f(x) = \sin x$

   d. $f(x) = \sqrt{2}$ e. $f(x) = \ln x$ f. $f(x) = e^{2x}$

4. If $f(x) = e^x$, which of the following is equal to $f'(e)$?

   a. $\displaystyle\lim_{\Delta x \to 0} \frac{e^{x+\Delta x}}{\Delta x}$ b. $\displaystyle\lim_{\Delta x \to 0} \frac{e^{x+\Delta x} - e^e}{\Delta x}$ c. $\displaystyle\lim_{\Delta x \to 0} \frac{e^{e+\Delta x} - e}{\Delta x}$

   d. $\displaystyle\lim_{\Delta x \to 0} \frac{e^{x+\Delta x} - 1}{\Delta x}$ e. $\displaystyle\lim_{\Delta x \to 0} e^{e} \frac{e^{\Delta x} - 1}{\Delta x}$

5. Which of the following is part of the graph of the function $f(x) = \sin 2x + 2e^{-x}$? Justify your answer. You may want to check your answer with a graphing calculator or a graphics package.

    a. b.  c.

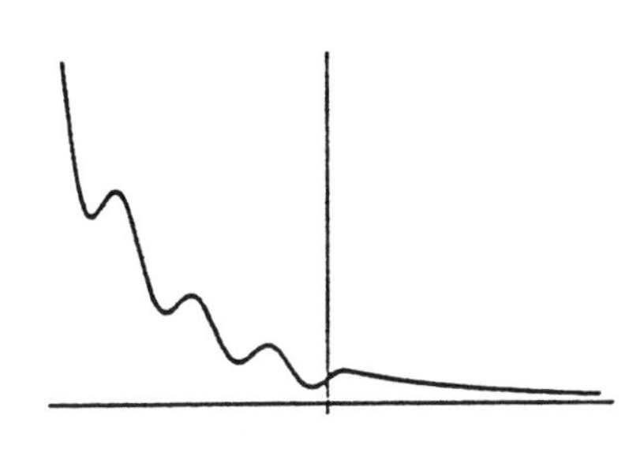

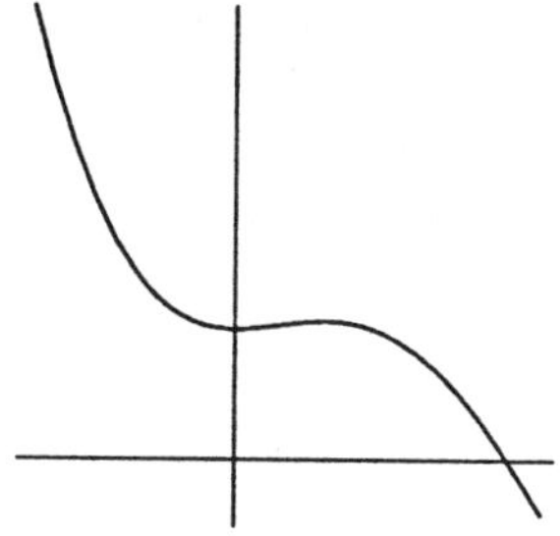

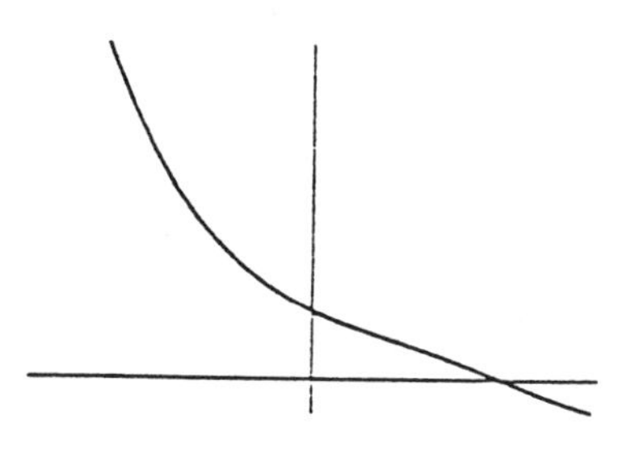

6. Which of the following graphs is the graph of the function f defined by the rule $f(x) = \sin x + 5\ln(x+1)$? Justify your answer. You may want to check your answer with a graphing calculator or a graphics package.

a.

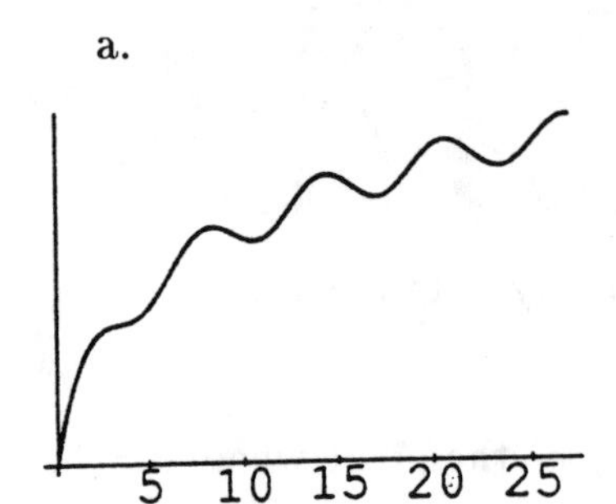

b.

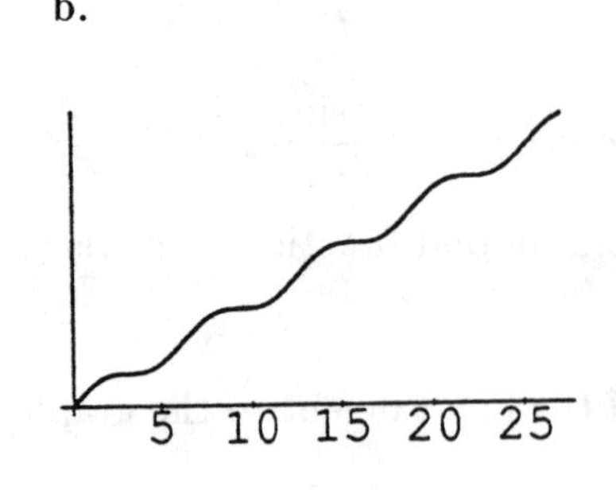

c.

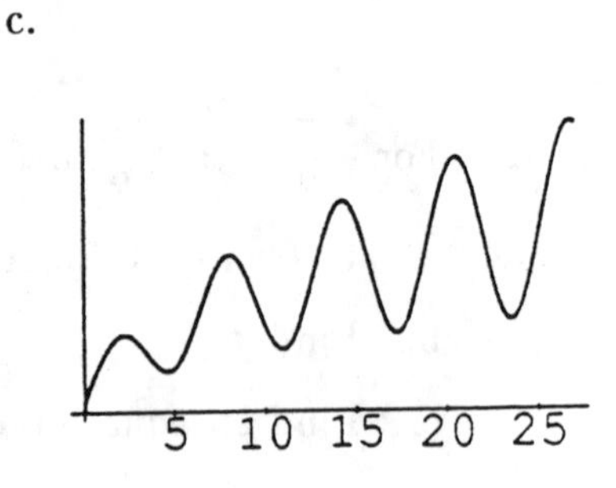

7. Find $\lim_{x \to 0} \dfrac{e^{3+x} - e^3}{x}$.

8. Graph 2^x, e^x and 5^x on the same axis. In what ways are these graphs the same? In what ways are they different? In other words, how does the choice of the base b affect the graph of b^x?

9. Find a number a such that the graphs of the two functions defined by the rules $y = a^x$ and $y = \log_a x$ have a common tangent line. Hint: Since each of these functions is the inverse of the other, their graphs are reflections, one of the other, in the line $y = x$ which is their common tangent line.

10. Graph $\log_2 x$, $\ln x$ and $\log_5 x$ on the same axis. In what ways are these graphs the same? In what ways are they different? In other words, how does the choice of the base b affect the graph of $\log_b x$?

11. Find all positive numbers x that satisfy the inequality $x^e < e^x$.

    Hint: Take the natural logarithm of both sides of the inequality which yields $e \ln x < x$. Confirm this and justify taking the logarithms. This inequality is equivalent to the inequality $\dfrac{\ln x}{x} < \dfrac{1}{e}$. Let $f(x) = \dfrac{\ln x}{x}$, compute $f'(x)$, and determine where f is increasing and where it is decreasing.

12. For $-\dfrac{3\pi}{4} < x < \dfrac{\pi}{4}$, define $f(x) = \sin x + \cos x$.

    a. Show that f is increasing on its domain and therefore has an inverse.

    b. Find the derivative of $f^{-1}(x)$. A useful formula: $(\cos x - \sin x)^2 = 2 - (\cos x + \sin x)^2$.

Section 10: Root Finding Methods

These are miscellaneous problems involving finding solutions to equations of the form $f(x) = 0$. They include problems using Rolle's Theorem, the Intermediate Value Theorem and Newton's method. The sections on continuity and on increasing functions also have some problems on root finding. More problems on Newton's Method can be found in the chapter on sequences and series.

1. Sketch the graph of a function f for which $f(0) = 1$, $f(3) = -1$ and the equation $f(x) = 0$ has:

   a. exactly one solution on $[0, 2]$.

   b. exactly two solutions on $[0, 2]$.

   c. exactly three solutions on $[0, 2]$.

2. a. If f is differentiable and $f'(x) \neq 0$ for all real numbers x, how many solutions can the equation $f(x) = 0$ possibly have? Explain your answer.

   b. If f is differentiable and the equation $f'(x) = 0$ has only one solution, how many solutions can the equation $f(x) = 0$ have? Explain your answer.

   c. If f is differentiable and the equation $f'(x) = 0$ has n solutions, how many solutions can the equation $f(x) = 0$ have? Explain your answer.

3. a. If $p(x)$ is a polynomial of odd degree, must the equation $p(x) = 0$ have a (real) solution? Explain your reasoning.

   b. If $p(x)$ is a polynomial of even degree, must the equation $p(x) = 0$ have a (real) solution? Explain your reasoning.

4. Explain why the equation $x^3 - 3x + b = 0$ can have at most one solution in the interval $[-1, 1]$ for any given real number b.

5. Show that the equation $\cos x + x \sin x - x^2 = 0$ has exactly two (real-valued) solutions.

6. A function f has the graph given below. Let x_1 and x_2 be the estimates of a root of f obtained by applying Newton's method once and twice respectively using an initial estimate of $x_0 = 1$. Estimate the numerical values of x_1 and x_2 and draw the appropriate tangent lines used to obtain x_1 and x_2.

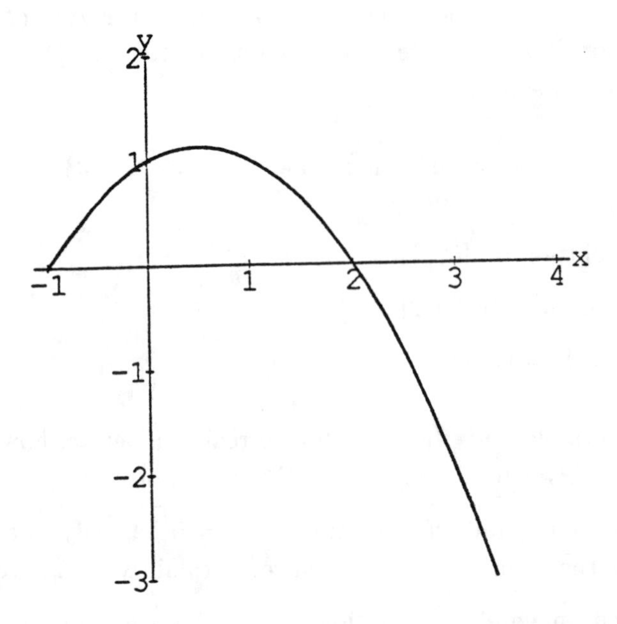

7. Suppose f is always concave up and that there is a number z such that $f(z) = 0$ but $f(x) > 0$ for all $x > z$. If an initial estimate x_0 is chosen which is greater than z, explain why all further estimates x_n obtained by Newton's method will also be greater than z.

8. The hare and the tortoise compete in a 1-kilometer race. Remarkably, the distance of both competitors t minutes from the start of the race can be expressed by simple formulas: the hare runs $f(t) = \frac{500}{3}(2\sqrt{t} + \sqrt[3]{t})$ meters while the tortoise runs $g(t) = 100t + 250\sqrt{t}$ meters after t minutes.

   a. Which racer passes the halfway point of the race first? How long does it take this racer to get to the halfway point?

   b. After the start of the race, the competitors meet exactly once. When and where does that happen?

   c. Who wins the race? How long does it take each competitor to finish the race?

Section 11: Related Rates

Generic Advice for Solving Story Problems

There are many "game plans" for solving story problems. All contain good advice although none can be regarded as an algorithm for solving story problems because there is no such algorithm. That is what makes solving story problems so challenging.

1. Read the problem carefully! In particular, ask yourself what constitutes an answer to the problem.

2. Draw a picture if possible.

3. Label quantities associated with the problem.

4. What relationships exist between the quantities you have labeled? Do enough relationships (equations) exist to supply you with an answer to the problem?

5. Try to identify steps which bring you closer to a solution.

6. If the problem seems too difficult, try solving a simpler problem of the same type.

7. When you have found an answer, check, if possible, to see that it satisfies the conditions of the problem. Try, at least, to decide whether or not the answer is *plausible.* Does it make sense?

1. A baseball diamond is a square 90 feet on a side. A runner starts from home plate towards first base at 20 ft/sec. How fast is the runner's distance from second base changing when the runner is halfway to first base? Is this distance increasing or decreasing? Why? (i)

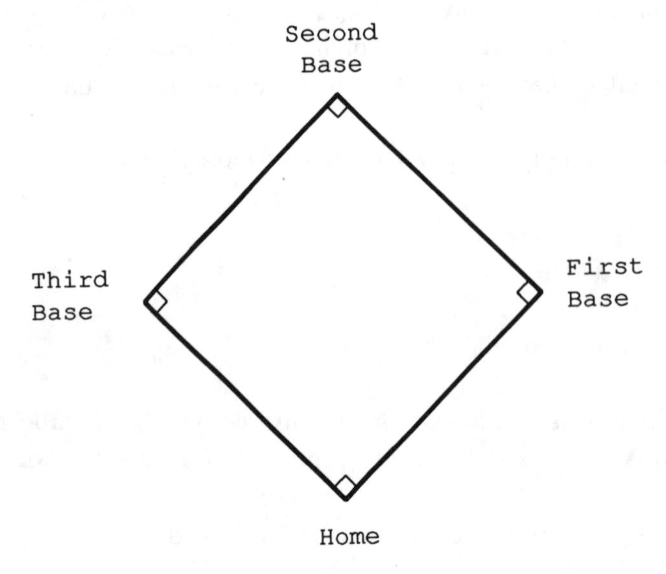

Figure for Problem 1

2. Experimentation shows that when air expands adiabatically (without losing or gaining heat), the volume V in cubic centimeters and the pressure P in kilopascals (kPa) are related by the equation

$$PV^k = C.$$

where k and C are constants. When the volume is $10\,\text{cm}^3$ and increasing by $4\,\text{cm}^3/\text{min}$, the pressure is 25 kPa and is decreasing by 14 kPa/min. What is the value of k?

3. (The Secret of the Ooze) An inverted cone with height 10 cm and radius 2 cm is partially filled with liquid, which is oozing through the sides at a rate proportional to the area of the cone in contact with the liquid. (The area of a cone is given by πrl where r is the radius and l is the slant height of the cone.) Liquid is also being poured in the top of the cone at a rate of $1\,\text{cm}^3/\text{min}$. When the liquid depth is 4 cm, the depth is decreasing at .1 cm/min. At what rate must liquid be poured into the top of the cone to keep the liquid at a constant depth of 4 cm?

4. The bad news: A tanker accident has spilled oil in Pristine Bay. The (moderately) good news: oil-eating bacteria are gobbling up 5 cubic feet per hour. The slick takes the form of a cylinder, whose height is the thickness of the slick. When the radius of the cylinder is 500 feet, the thickness of the slick is .01 ft, and decreasing at .001 ft/hr. At this time, what is the rate at which the area of the slick is increasing?

5. Soup is being poured at a rate of 3 cubic inches per second into a hemispherical bowl with a radius of 5 inches. How fast is the level of soup rising when there is 2 inches of soup in the bowl? Hint: Let D be the depth of the soup, and let R be the radius of the soup surface. By the Chain Rule,

$$\frac{dV}{dt} = \frac{dV}{dD}\frac{dD}{dt}.$$

Note that increasing the soup depth by a differential amount dD increases the soup volume by a differential amount dV, which is approximated by $\pi R^2 dD$.

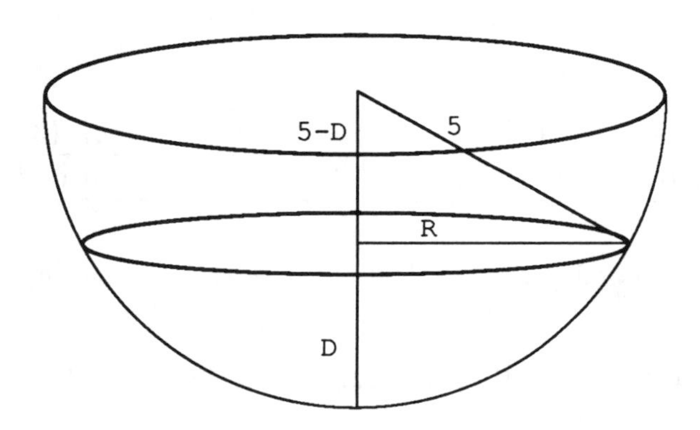

6. An athlete is running around a circular track of radius 100 m at 5 m/sec. A spectator is 300 m from the center of the track. How fast is the distance between the two changing when the runner is approaching the spectator and the distance between them is 250 m?

7. A wheel of radius 10 cm is revolving 4 times per second. Attached to the rim is one end of a rod of length 30 cm. The other end of the rod moves a piston in a straight line extending through the center of the wheel. How fast is the piston moving when the angle at A is a right angle?

Chapter III: Extreme Values

Section 1: Increasing and Decreasing Functions and Relative Extrema

1. Find where the following functions are increasing and where they are decreasing. Then find all relative maxima and minima of these functions.

   a. $f(x) = 8x^9 - 9x^8$

   b. $g(x) = 7x^9 - 18x^7$

   c. $h(x) = 5x^6 + 6x^5 - 15x^4$

2. Sketch the graph of a function whose derivative satisfies the properties given in the following table.

x	$(-\infty, -1)$	-1	$(-1, 1)$	1	$(1, 3)$	3	$(3, \infty)$
$f'(x)$	positive	0	negative	0	negative	0	positive

3. a. Find where the function $f(x) = (1000 - x)^2 + x^2$ is increasing and where it is decreasing.

   b. Use part a to decide whether 1000^2 is bigger or smaller than $998^2 + 2^2$. Check your answer by using a calculator.

   c. Generalize what you have discovered in part a to the function $f(x) = (c - x)^n + x^n$, where c is any positive number and n is any even positive integer. Use this to decide whether 10000^{100} is bigger or smaller than $9000^{100} + 1000^{100}$.

4. Let f and g be the differentiable functions whose graphs are shown below. The point c is the point where the vertical distance between the curves is the greatest in the interval $[a, b]$. Show that the line tangent to f at $x = c$ is parallel to the line tangent to g at $x = c$.

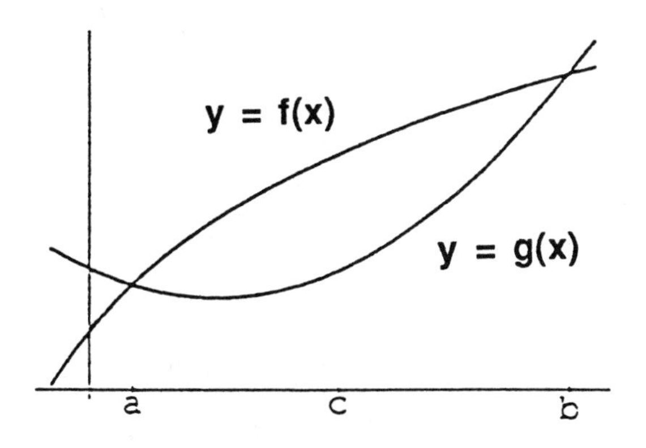

5. a. If $f'(x) \leq 2$ for all real numbers in the interval $[0, 3]$ and $f(0) = -1$, how large can $f(3)$ possibly be? Justify your answer.

   b. More generally, if $f'(x) \leq a$ for all real numbers and $f(0) = b$, how big can $f(c)$ be for a given positive number c?

6. Let $f(x) = |x - 1| + |x + 2|$. Find where f is increasing, where it is decreasing and where it is constant. Use this information to help sketch the graph of the function.

7. Let $f(x) = |x - 1| + |x| + |x + 2|$. Find where f is increasing, where it is decreasing and where it is constant. Use this information to help sketch the graph of the function.

8. Let $f(x) = |\cos^2 x - \frac{1}{3}|$. Find the absolute maximum and minimum values of f.

9. The graph below is the graph of the *derivative* of a function f.

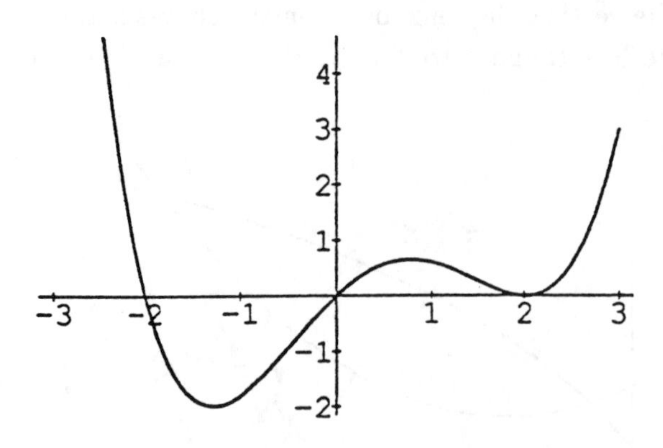

a. Find where f is increasing and where it is decreasing.

b. Find all relative maximum(s) and relative minimum(s) of f.

c. If $f(-3) = -2$, sketch the graph of f.

10. The graphs of the *derivatives* of two functions f and g are given below.

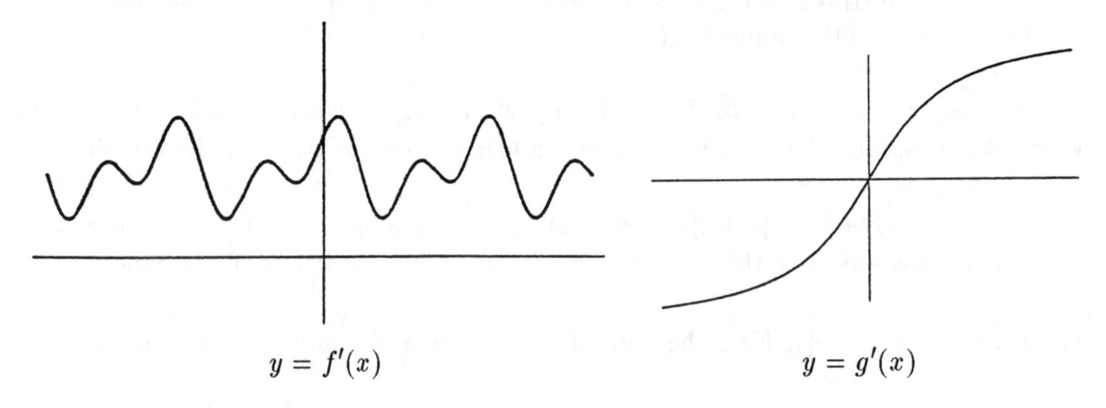

a. How many solutions can the equation $f(x) = 0$ have? Explain.

b. How many solutions can the equation $g(x) = 0$ have? Explain.

c. If $g(x) = 0$ has two solutions, what can you say about where these two solutions lie? Justify your answer.

11. a. Suppose f is a function for which $f'(x) > 0$ for all real numbers x and let $g(x) = f(f(x))$. Must g be increasing for all real numbers? Explain your answer.

b. Suppose f is a function for which $f'(x) < 0$ for all real numbers x and let $g(x) = f(f(x))$. Must g be decreasing for all real numbers? Explain your answer.

12. Determine whether the following statements are always true or are at least sometimes false and explain your answer. You should assume that f is differentiable on its domain.

    a. If f is increasing on an interval (a, b), then $f'(x) \geq 0$ for all x in (a, b).

    b. If f is (strictly) increasing on an interval (a, b), then $f'(x) > 0$ for all x in (a, b).

    c. If f is (strictly) increasing on an interval (a, b), then $f'(x) > 0$ for at least one number x in (a, b).

    d. If $f'(x) \geq 0$ for all x in an interval (a, b), then f is (strictly) increasing on this interval.

    e. If $f'(x) \neq 0$ for all x on an interval (a, b), then f has no relative extrema on this interval.

    f. If $f'(x) > 0$ for all x in an interval (a, b), and $f'(x) < 0$ in an interval (b, c), then f has a relative maximum at b.

    g. If f has an absolute maximum on an open interval (a, b), then it must be a relative maximum as well.

    h. If f has two relative maxima on an interval I, then it must have at least one relative minimum on that interval.

    i. There are functions for which $f(x) < 0$ for all real numbers x but $f'(x) > 0$ for all real numbers x.

13. Suppose f is a continuous function defined for all real numbers which has a maximum value of 5 and a minimum value of -7. Which of the following MUST be true, which MIGHT be true, and which can NEVER be true? Justify your answer.

    a. The maximum value of $f(|x|)$ is 7.

    b. The minimum value of $f(|x|)$ is 0.

    c. The maximum value of $|f(x)|$ is 7.

    d. The minimum value of $|f(x)|$ is 5.

14. a. Show that $f(x) = e^x$ is greater than $g(x) = 2x$ for all $x \geq 1$.

    b. Show that $f(x) = e^x$ is greater than $h(x) = x^2$ for all $x \geq 0$. Hint: Consider the cases when $0 \leq x < 1$ and $x \geq 1$ separately.

15. The graph below is that of a function $f(x) = ax^3 + bx^2 + cx + d$, where a, b, c, d are constants. Show that the x-coordinates of the two marked points are given by the formula

$$x = \frac{-b \pm \sqrt{b^2 - 3ac}}{3a}.$$

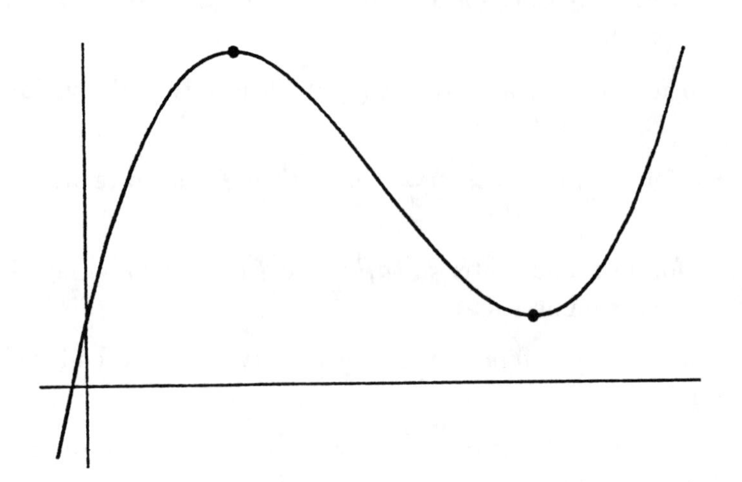

Section 2: Concavity and the Second Derivative

1. Sketch the graph of a continuous function which satisfies all the following conditions:
   $f'(x) < 0$ for all real numbers $x \neq 4$;
   $f'(4)$ does not exist;
   $f''(x) < 0$ for all $x < 4$; and
   $f''(x) > 0$ for all $x > 4$.

2. A function f is continuous on the interval $[-3, 3]$ and its first and second derivatives have the values given in the following table:

x	$(-3,-1)$	-1	$(-1,0)$	0	$(0,1)$	1	$(1,3)$
$f'(x)$	Positive	0	Negative	Negative	Negative	0	Negative
$f''(x)$	Negative	Negative	Negative	0	Positive	0	Negative

   a. What are the x-coordinates of all of the relative maxima and minima of f on $[-3, 3]$? Justify your answer.

   b. What are the x-coordinates of all points of inflections of f on the interval $[-3, 3]$? Justify your answer.

   c. Sketch a possible graph for f which satisfies all of the given properties.

3. The graph of a function f together with some points on the graph is given below.

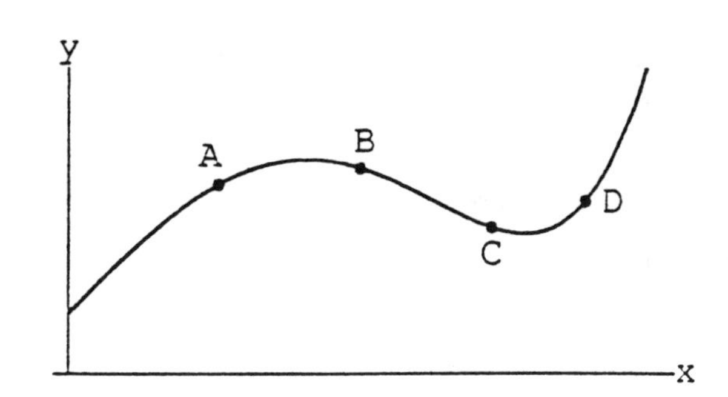

   a. At which point(s) is the first derivative of f positive?

   b. At which point(s) is the second derivative of f positive?

4. The graphs (i), (ii), and (iii) given below are the graphs of a function f and its first two derivatives f' and f'' (though not necessarily in that order). Identify which of these graphs is the graph of f, which is that of f' and which is that of f''. Justify your answer.

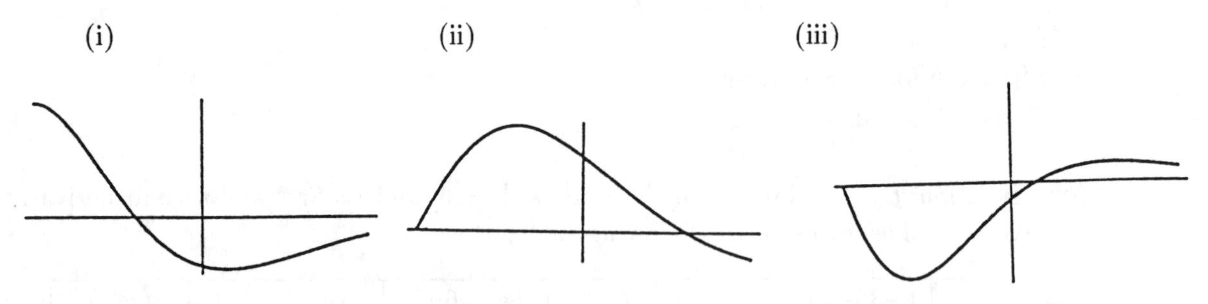

5. The graph below is the graph of the *derivative* of a function f. Use this graph to answer the following questions about f on the interval $(0, 10)$. In each case be sure to justify your answer.

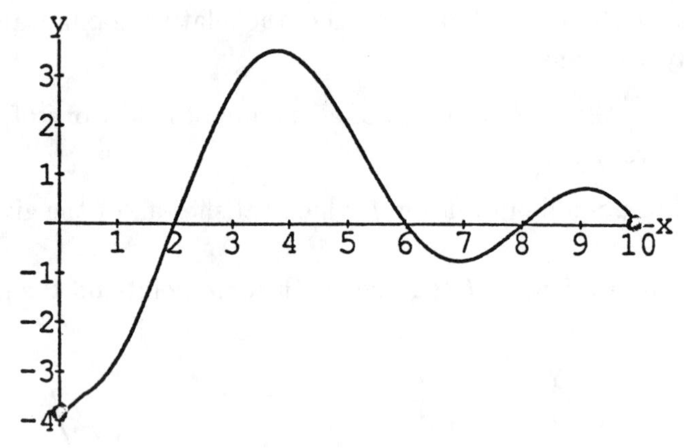

a. On what subinterval(s) is f increasing?

b. On what subinterval(s) is f decreasing?

c. Find the x-coordinates of all relative minima of f.

d. Find the x-coordinates of all relative maxima of f

e. On what subinterval(s) is f concave up?

f. On what subinterval(s) is f concave down?

g. Find the x-coordinates of all points of inflection of f.

6. Let f be a function which is twice differentiable for all real numbers and which satisfies the following properties:

    (i) $f(0) = 1$;

    (ii) $f'(x) > 0$ for all $x \neq 0$;

    (iii) f is concave down for all $x < 0$ and is concave up for all $x > 0$. Let $g(x) = f(x^2)$.

    a. Sketch a possible graph for f which takes into account its properties given above.

    b. Find the x-coordinates of all relative minimum point(s) of g. Justify your answer.

    c. Where is the graph of g concave up? Justify your answer.

    d. Use the information obtained in the three previous parts to sketch a possible graph of g.

7. Let f be a three times continuously differentiable function and suppose for some number c that $f'(c) = f''(c) = 0$ but $f'''(c) > 0$. Does f have a relative maximum, relative minimum, or a point of inflection at $x = c$? Explain your reasoning.

8. For what value(s) of x does the slope of the tangent line to the curve $y = -x^3 + 3x^2 + 1$ take on its largest value? Justify your answer.

9. For each part, determine whether it is possible for a twice differentiable function f to satisfy all of the properties listed in that part. Justify your answers.

    a. $f'(x) > 0$, $f''(x) > 0$ but $f(x) < 0$ for all real numbers x.

    b. $f''(x) > 0$ but $f'(x) < 0$ for all real numbers x.

    c. $f''(x) > 0$ but $f(x) < 0$ for all real numbers x.

Section 3: Max-Min Story Problems

One or both of the codes [CP] or [TA] may precede a problem. The code [CP] means "classical problem." It may appear in many textbooks, but has been included in this collection because it contains a noteworthy feature. The code [TA] means "trigonometric alert." It signifies that a trigonometric approach to the problem is possible and may even simplify matters.

Helpful Advice for Story Problems

1. Read the entire problem carefully.
2. Whenever possible, draw your own pictures.
3. Use mnemonically helpful letters for variables (e.g. A for area, h for height, etc.).
4. When in doubt, look for triangles. If there aren't any triangles, make some.
5. Consider using trigonometric variables (angles).

1. [TA] What is the area of the largest rectangle that can be inscribed in a semicircle of radius R so that one of the sides of the rectangle lies on the diameter of the semicircle? Hint: Express the area of the rectangle in terms of the angle θ in the figure below.

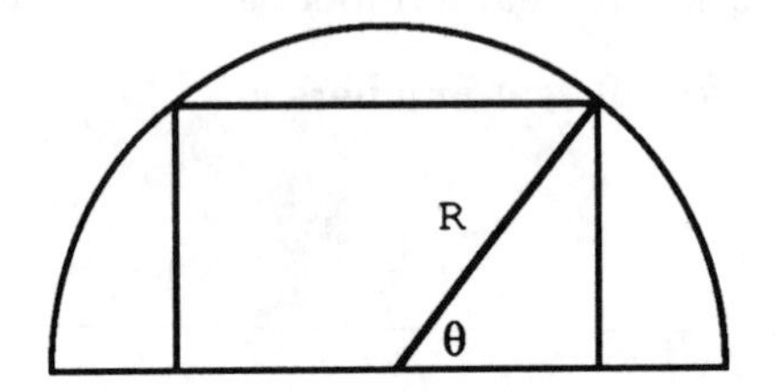

2. The number of bacteria in a culture at time t is given by $N = 1000(25 + te^{-t/20})$ for $0 \leq t \leq 100$.

   a. Find the largest and the smallest number of bacteria in the culture during the interval.

   b. At what time during the interval is the rate of change in the number of bacteria a minimum?

3. The concentration of a drug t minutes after it has been injected into the bloodstream is given by $C(t) = \dfrac{K(e^{-bt} - e^{-at})}{a - b}$ where $K > 0$ and $a > b > 0$. At what time does the maximum concentration occur?

4. In medicine, the reaction $R(x)$ to a dose x of a drug is given by $R(x) = Ax^2(B - x)$ where $A > 0$ and $B > 0$. The sensitivity $S(x)$ of the body to a dose of size x is defined to be $R'(x)$. Assume that a negative reaction is a bad thing.

   a. What seems to be the domain of R? What seems to be the physical meaning of the constant B? What seems to be the physical meaning of the constant A?

   b. For what value of x is R a maximum?

   c. What is the maximum value of R?

   d. For what value of x is the sensitivity a maximum?

   e. Why is it called sensitivity?

5. [CP] The range R of a projectile whose muzzle velocity in meters per second is v, and whose angle of elevation in radians is θ, is given by $R = \dfrac{v^2 \sin 2\theta}{g}$ where g is the acceleration of gravity. Which angle of elevation gives the maximum range of the projectile?

6. [TA] A billboard 54 feet wide is perpendicular to a straight road, and is 18 feet east of the nearest point on the road. A car approaches the billboard from the south.

   a. From what point does a passenger see the billboard at the widest angle?

   b. What assumptions are being made to simplify the problem?

   c. Under what circumstances are these assumptions reasonable, and under what circumstances might they be unreasonable?

   d. Can you generalize this result? Try to solve this problem for a billboard of width B at distance A from the road. Check that the answer you get is consistent with your answer for part a by letting $A = 18$ and $B = 54$.

7. [CP] A sailor in a rowboat finds himself at a point P which is 1 mile from the nearest point A on a straight shore line. His goal is to reach a point Q 1 mile directly inland from a point B which is on the shore 2 miles from A (see the figure below). He decides to row directly to a point X between A and B which is x miles from A. He then walks directly to Q. Assume that his rowing speed is R MPH, and that his walking speed is W MPH.

   a. Obtain an expression in terms of x for the total time taken.

   b. Let α denote the angle between PX and a line perpendicular to AB through X, and let β denote the angle between QX and a line perpendicular to AB through X (as shown). Show that minimizing the total time taken requires that $\dfrac{\sin\alpha}{\sin\beta} = \dfrac{R}{W}$.

   c. Suppose that $R = 2$ MPH and $W = 4$ MPH. Show that to minimize the total time taken requires x to satisfy the equation $3x^4 - 12x^3 + 15x^2 + 4x - 4 = 0$.

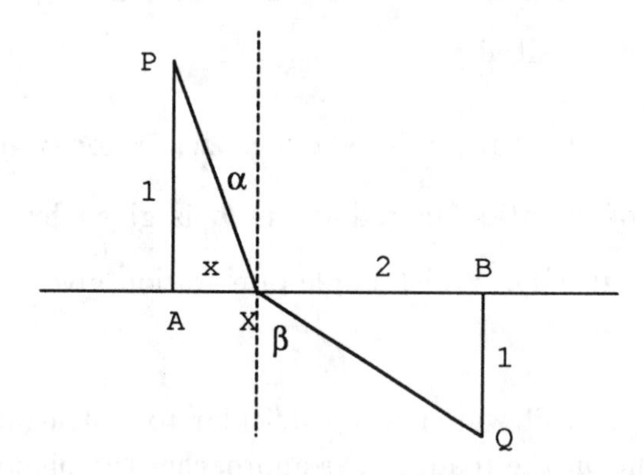

8. A boat is 4 miles from the nearest point on a straight shore line which is 6 miles from a shoreside restaurant. A woman plans to row to a point on shore, and then walk to the restaurant. If she can walk at 3 MPH, at what speed must she be able to row so that the quickest way to get to the restaurant is to row directly?

9. The depth of the water in Moon River x miles downstream from Hard Rock is $D(x) = 20x + 10$ feet, and the width of the river is $W(x) = 10(x^2 - 8x + 22)$ feet. To create Moon Lake, it is necessary to place a dam downstream from Hard Rock.

   a. If the dam cannot be more than 310 feet wide and 130 feet above the riverbed, and the top of the dam must be 20 feet above the present river surface, where can the dam be placed?

   b. What are the dimensions of the widest and narrowest dam that can be constructed in accordance with the above constraints?

   c. If cost is proportional to the product of the width and the height of the dam, where is the cheapest dam located?

10. [CP, TA] A square plot of ground 100 feet on a side has vertices labeled A, B, C, D clockwise. Pipe is to be laid in a straight line from A to a point P on BC (P may be one of the vertices), and from there to C. The cost of laying the pipe is \$20 a foot if it goes through the lot (since it must be laid underground), and \$10 a foot if it is laid along one of the sides of the square. What is the most economical way to lay the pipe?

11. An outfielder throws a ball toward home plate with an initial velocity of 100 ft/sec towards home plate, which is 300 feet away. An infielder positions himself in a direct line between the outfielder and home plate. This infielder, who can throw the ball with an initial velocity of 110 ft/sec, can position himself x feet from home plate, at which point he can, if he so chooses, catch the ball and relay it towards home plate. Let R be the amount of time necessary for the infielder to make the relay. Due to air resistance, the velocity $v(t)$ of a thrown ball satisfies the differential equation $\frac{dv}{dt} = -\frac{v}{10}$. Ignore vertical motion to answer the following.

    a. Express the time T for the ball to reach home as a function of R and x.

    b. Minimize T through an appropriate choice of x.

    c. How fast must the relay be made in order to make it advisable?

12. A truck driving over a flat interstate at a constant rate of 50 MPH gets 4 miles to the gallon. Fuel costs \$0.89 per gallon. For each mile per hour increase in speed, the truck loses a tenth of a mile per gallon in its mileage. Drivers get \$27.50 per hour in wages, and fixed costs for running the truck amount to \$11.33 per hour. What constant speed (between 50 MPH and the speed limit of 65 MPH) should a dispatcher require on a straight run through 260 miles of Kansas interstate to minimize the total cost of operating the truck?

13. [TA] Sue finds a cigarette burn in her new scarf. She decides the best solution is to cut off one corner of the scarf through the burn. If the scarf is a rectangle and the burn is located 2 cm from one edge and 5 cm from a perpendicular edge, what is the smallest area of the triangle she has to cut off to remove the burn, and how should she make the cut? If she decides to make the cut through a point 2.1 cm from one edge and 5.1 cm from the other in order to make sure that all of the burn is removed, how much extra area from the scarf will she remove?

14. In constructing the new Trump Colosseum, projected to occupy the entire state of Rhode Island, the builder estimates the initial costs (buying Rhode Island, etc.) as 450 times the cost of the first floor. The second floor is projected to cost twice as much as the first floor, the third floor three times as much as the first floor, etc. What number of floors in the building will give the cheapest average cost per floor?

15. [CP] A pumping station is to be built on the edge of a straight river to serve two towns, A and B, which lie on the same side of the river. The towns are 4 and 5 miles, respectively, from the nearest points C and D on the river. C and D are 12 miles apart. Let x be the distance from C tothe point P where the station is to be built. For what value of x is the sum of the pipe lengths from P to A and from P to B a minimum?

16. The hourly cost of fuel in operating a luxury liner is proportional to the square of the speed v, and is \$300 for a speed of 10 km/hr. Other fixed costs amount to \$1200/hr. Assuming that the ship is to make a trip of total distance D, find the speed at which it should be run to minimize the cost of making the trip, and show that this speed is independent of D.

17. [TA] Students are experimenting with pencils of various lengths to see which ones they can pass through an L-shaped channel without lifting them off the floor of the channel. If the width of the base of the channel is b inches and the width of the side of the channel is s inches, find the length of the longest pencil which can be passed through the channel. Hint: Model the pencil (as shown in the figure below), as a line segment PRD, and find a formula for the length L of the pencil in terms b, s, and the angle θ at R.

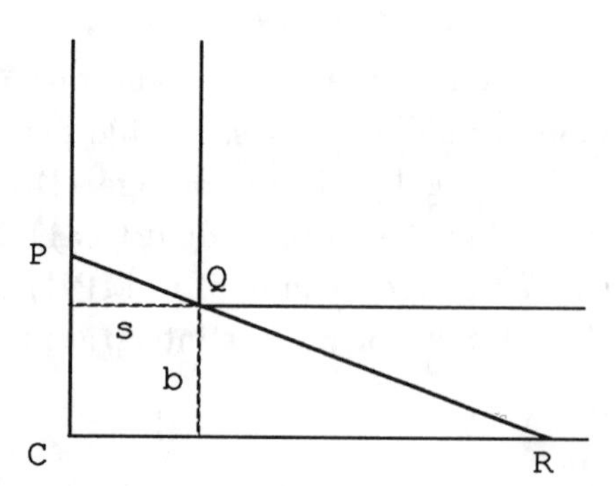

18. An electronics store needs to order a total of 2400 CD players over the course of a year. It will receive them in several shipments, each containing an equal number of CD players. The shipping costs are \$50 for each shipment, plus a yearly fee of \$2 for each CD player in a single shipment. What size should each shipment be in order to minimize yearly shipping costs?

19. What is the area of the largest triangle that can be formed in the first quadrant by the x-axis, the y-axis, and a tangent to the graph of $y = e^{-x}$?

20. [CP] Take a circle of radius R, and cut out a sector subtended by an angle θ, producing a Pac-Man-like figure. Join the cut ends together (the upper and lower jaws of Pac-Man) to form a cone. What angle θ produces the largest volume of such a cone?

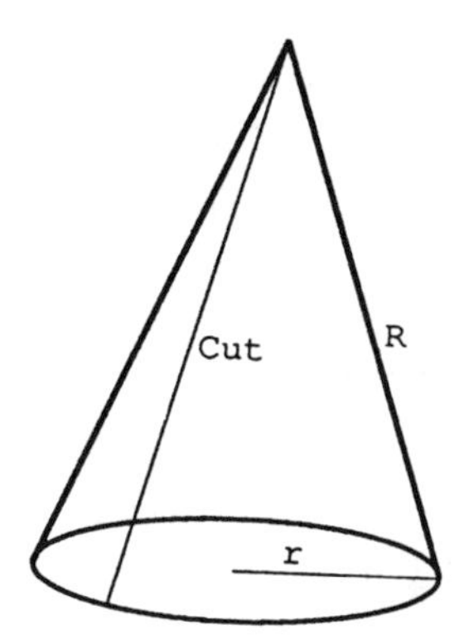

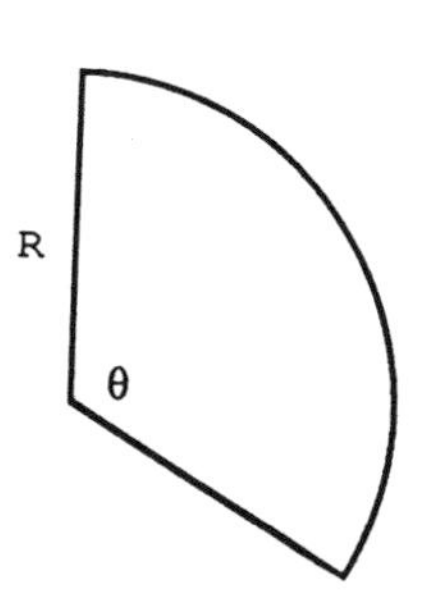

21. At Pizza Hut, they deliver for you, and in pizza boxes formed from 47 cm by 95 cm rectangular pieces of cardboard by cutting out eight squares of equal size, four from each of the 95 cm edges: In the figure below, one square from the corner on the left, one from the middle of the edge, and two (side-by-side) from the corner on the right, to make a "lip" for the box. The result is then folded in the obvious fashion. (If it isn't obvious, order one with sausage and mushrooms to investigate.) How does one obtain the box of largest volume?

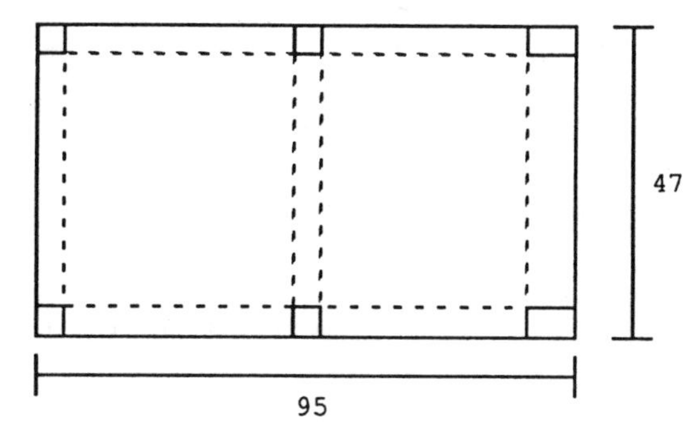

22. A plot of ground in the shape of a circular sector (a wedge of pie) is to have a border of roses along the straight lines and tulips along the circle. If the area of the plot is to be 100 square meters, and roses cost \$20 a meter and tulips \$15 a meter, what is the least the flowers will cost?

23. [TA] A fence of height H is D feet away from a vertical wall. At what angle θ should a ladder be leaned against the fence in order that the minimum length ladder be required to stretch from the ground to the wall?

24. A rectangular piece of metal is bent to form a gutter in the following manner: The left and right thirds are bent up so that each forms an angle θ as shown in the figure below. The cross-section of the gutter is therefore a trapezoid. What should θ be in order that the cross-sectional area be maximized?

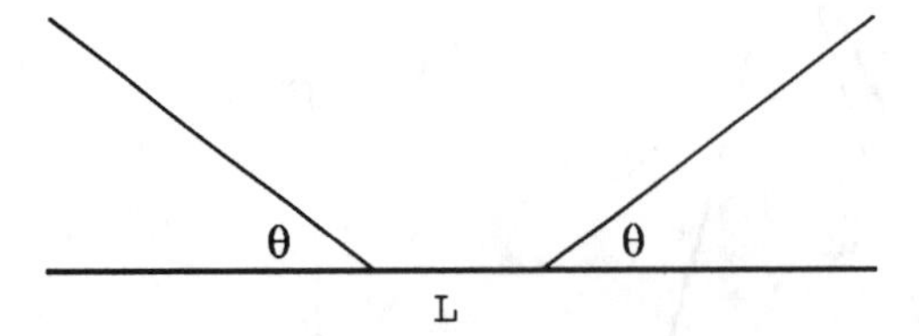

25. [TA] A rectangle has length L and width W. What is the area of the largest rectangle that can be circumscribed around it? Hint: Express the area A of the rectangle in terms of the angle θ in the figure below.

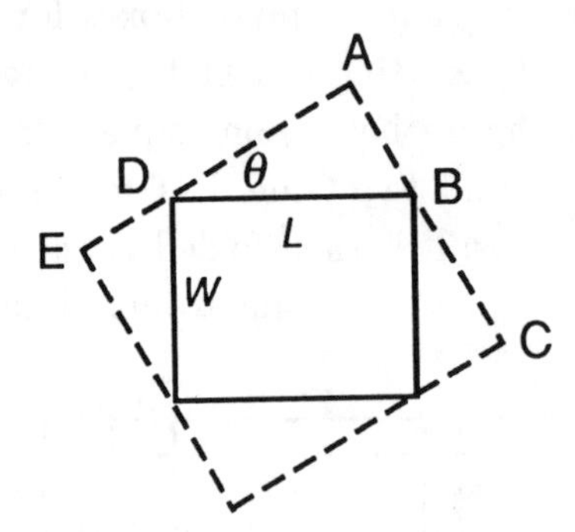

26. A piece of wire 100 cm long is to be cut into several pieces and used to construct the skeleton of a box with a square base.

    a. What is the largest possible volume that such a box can have?

    b. What is the largest possible surface area?

27. [CP] Given a cone with radius R and height H, what is the cone with largest volume that can be inscribed inside this cone such that the vertex of the inscribed cone is located on the center of the base of the original cone, and the axes of the two cones coincide?

28. [TA] The light in an offshore lighthouse is rotating at a constant rate. Show that, as the beam of light moves down the shoreline, it moves most slowly at the point on the shore directly opposite the lighthouse.

29. Financial reversals force Elms Lea Inns to convert a hotel into an apartment house with 65 rental units. At \$300, all the units can be rented. For each \$10 that the rent is raised, one of the units becomes vacant. Each occupied unit requires \$30 in service each month. How much rent should be charged to maximize cash flow?

    Hint: Let x be the rental price in dollars. Show that the number of vacant units is $(x - 300)/10$ for $x \geq 300$. Then the number of occupied units is $65 - (x - 300)/10$. Formulate an expression $C(x)$ for the total cash flow and maximize it.

30. According to Poiseuille's Law, if blood flows into a straight blood vessel of radius r branching off another straight blood vessel of radius R at an angle of θ, the total resistance T of the blood in the branching vessel is given by

$$T = C\left(\frac{a - b\cot\theta}{R^4} + \frac{b\csc\theta}{r^4}\right)$$

    where a, b, and C are constants and $r < R$. Show that the total resistance is minimized when $\cos\theta = \left(\dfrac{r}{R}\right)^4$.

31. A rectangular piece of metal is to be made into a trough by bending a side of length S into a circular arc. How should this be done so that the trough will have the most volume?

32. [TA] What is the largest perimeter of a rectangle that can be inscribed in an ellipse with semi-axes a and b? Hint: Find an expression for the perimeter in terms of an appropriate angle.

33. [TA] Two points are located diametrically opposite one another on a pond of radius 1 mile. A man wishing to go from one point to another plans to swim in a straight line to a point on the circumference, and walk from there along the circumference to the diametrically opposite point. If he can swim at 2 MPH and walk at 5 MPH, what is the minimum time that the trip will take?

34. [CP] Under the influence of a complicated collection of physical fields, the altitude h of a rocket from its moment of launch until its fall to Earth is given by the formula $h(t) = t^4 - 8t^3 + 375$ where the altitude h is measured in feet and the time from launch t is measured in seconds. Answer the following questions.

a. What is the altitude from which the projectile is initially launched?

b. When does the projectile impact Earth?

c. What is its maximum altitude?

d. What is its maximum upward velocity?

e. What is its maximum downward velocity?

f. What is its maximum upward acceleration?

g. What is its maximum downward acceleration?

Chapter IV: Antiderivatives and Differential Equations

Section 1: Antiderivatives

1. Consider the three graphs in the figure below. If the graph labeled A is the graph of a function f, then which graph is an antiderivative of f?

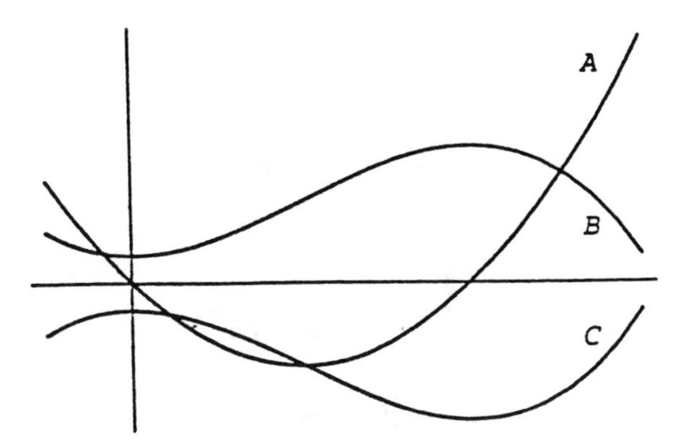

2. If $\dfrac{dy}{dx} = \cos 2x$, then $y =$

   A. $-\frac{1}{2}\cos 2x + C$ B. $-\frac{1}{2}\cos^2 x + C$ C. $\frac{1}{2}\sin 2x + C$

   D. $\frac{1}{2}\sin^2 2x + C$ E. $-\frac{1}{2}\sin 2x + C$

3. Consider the three graphs in the figure below. If the graph labeled C is the graph of a function f, then which graph is an antiderivative of f?

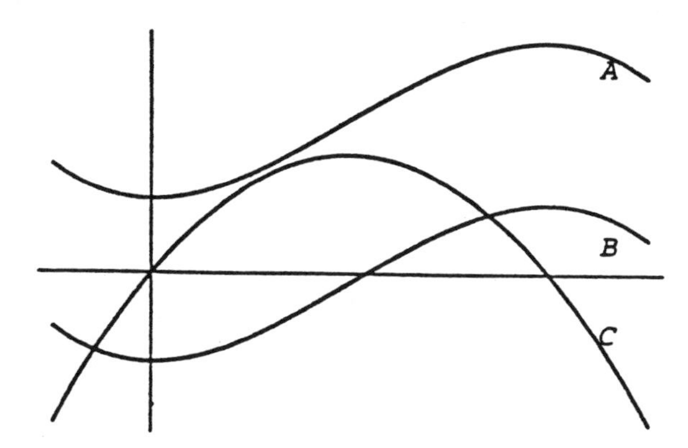

4. For each part of this problem you will be given two functions, f and g. Differentiate both functions. How are the derivatives related? How are f and g related? Is it possible for different functions to have the same derivative? What must be true of such functions?

   a. $f(x) = (x-1)^3$ and $g(x) = x^3 - 3x^2 + 3x$

   b. $f(x) = \tan^2 x$ and $g(x) = \sec^2 x$

5. An antiderivative of $f(x) = e^{x+e^x}$ is:

   A. $\dfrac{e^{x+e^x}}{1+e^x}$ B. $(1+e^x)e^{x+e^x}$ C. e^{1+e^x}

   D. e^{x+e^x} E. e^{e^x}

6. The graph of the derivative of f is shown in the figure below.

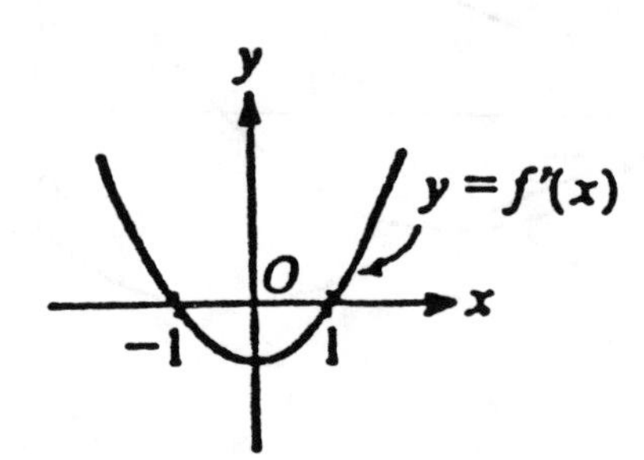

Explain why each graph below cannot be the graph of f.

a.

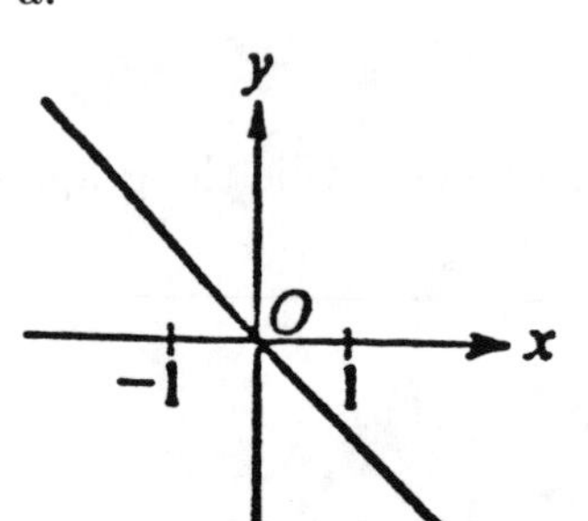

b.

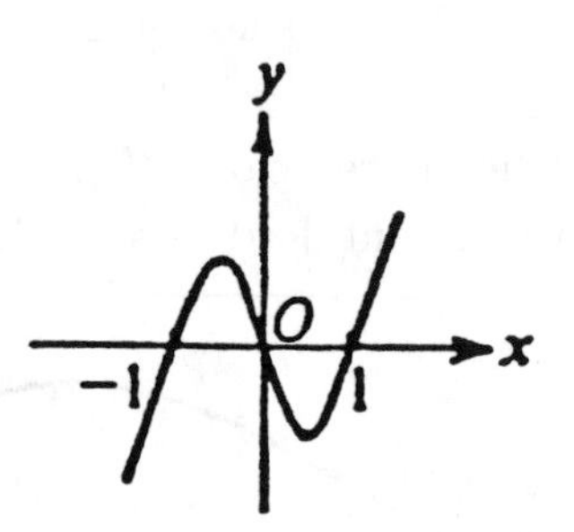

c.

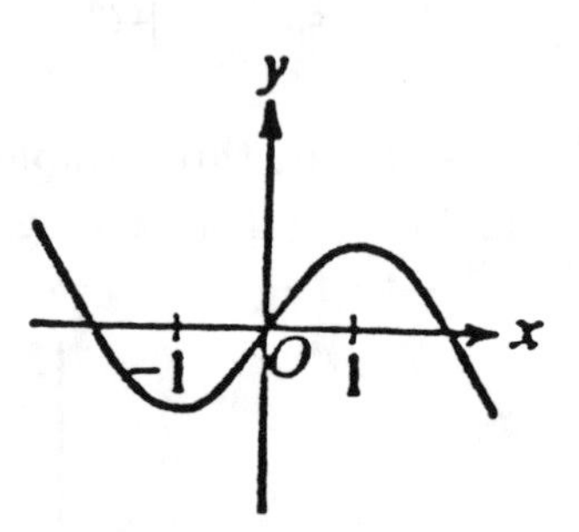

d. Sketch a graph that can be the graph of f.

7. Confirm that $\dfrac{d}{dx}\sin^2 x = 2\sin x\cos x = \sin 2x$. Therefore one antiderivative of $\sin 2x$ is $\sin^2 x$. Find another antiderivative of $\sin 2x$ and explain why this function can have two different antiderivatives.

8. a. To find an antiderivative of $f(x) = xe^x$, let the antiderivative be of the form: $F(x) = Axe^x + Be^x$. Find the constants A and B by differentiation.

   b. Repeat the procedure for $g(x) = 2xe^x$.

   c. What modifications are needed in the procedure for it to work with $h(x) = xe^{2x}$?

   d. Try to modify the procedure so that it will work for the function k defined by the rule $k(x) = x^2e^x$.

9. The product rule for derivatives states that if $F(x) = f(x)g(x)$, then $F'(x) = f'(x)g(x) + f(x)g'(x)$. Find $f(x)$ and $g(x)$ for the following and indicate what $F(x)$ is. (For example, if $F'(x) = 2x\cos x - x^2\sin x$, then $f(x) = x^2$ and $g(x) = \cos x$ and hence $F(x) = x^2\cos x$.)

   a. $F'(x) = 2xe^x + x^2e^x$

   b. $F'(x) = \dfrac{1}{x}\sin x + \ln x\cos x$

   c. $F'(x) = e^x\sin x + e^x\cos x$

   d. $F'(x) = e^x\sin x - e^x\cos x$

   e. $F'(x) = 1 + \ln x$

10. a. In the two figures below, you are given five line segments and five points. The points are indicated by the dot at the left end of each line segment. In each figure, draw a curve through the five points, by using the line segments as tangents to the curve at the indicated points.

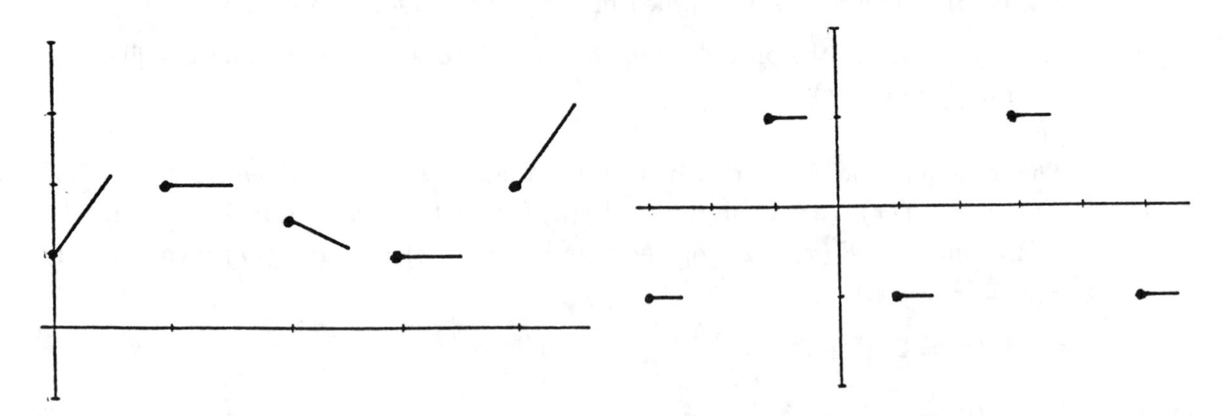

b. If the curve in the figure at the left is the graph of a polynomial, then what is the minimal degree of that polynomial?

c. Can you associate a function with the curve on the right?

d. Using a different color pencil, draw the graph of the derivative of each of the functions whose graph you drew in part a.

11. The figures below give what are called flow (direction) fields. At each point, the flow (direction) of a function is shown by the tangent to the graph of that function at that point. These look more complicated than the figures for the previous problem, but they are very much alike.

    a. For the flow field below, draw the graph of a function that follows the flow field and goes through the point labeled A.

    b. Do the same for the point labeled B and for the points labeled C and D on the other flow field.

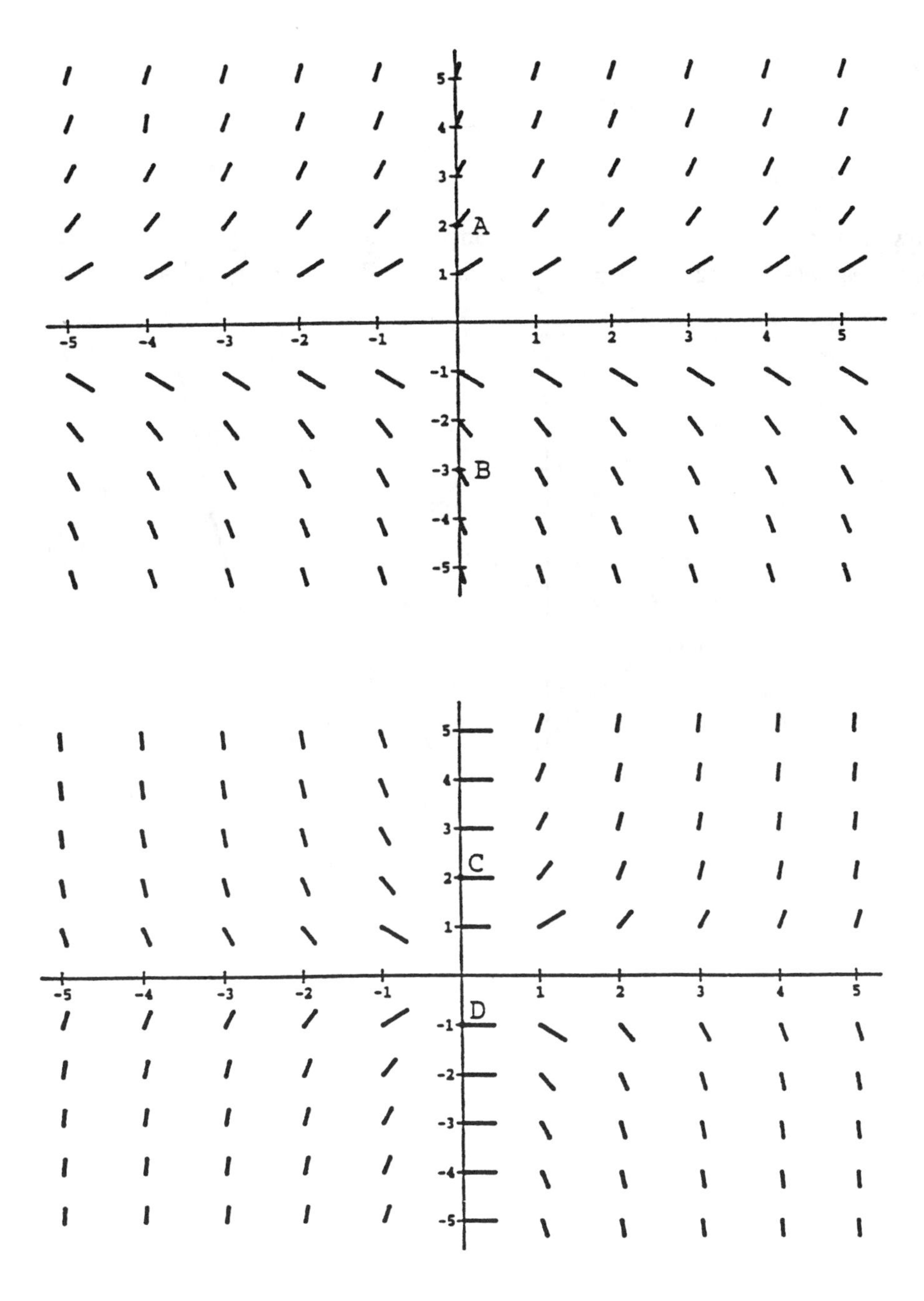

12. Sketch the graph of an antiderivative F of the function f whose graph appears below. Can you see any relation between the shapes of the graphs of f and F? What does this say about the functions f and F?

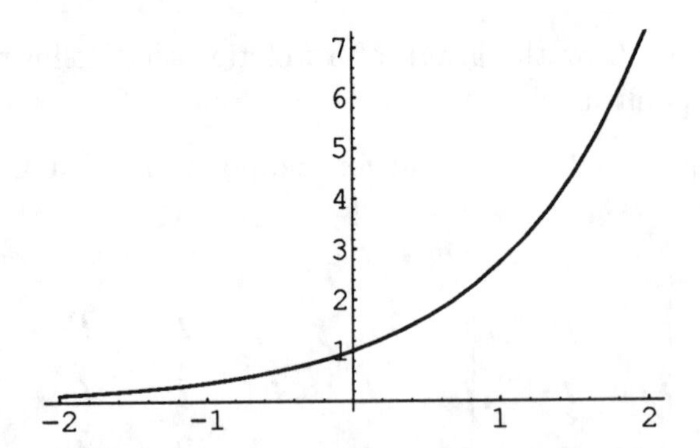

13. Below is the graph of a function f. Suppose another function g has the following properties: $g(-1) = 2$, $g'(x) = f'(x)$ for all x. Sketch the graph of g, using the same axes.

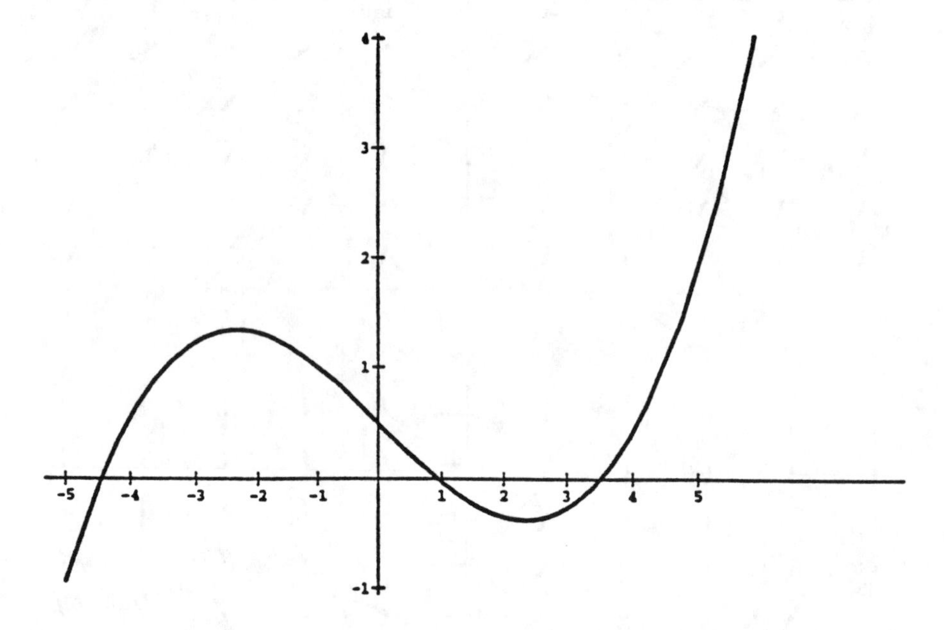

14. Let f and g be two differentiable functions such that $f'(x) = g'(x)$, for all x. What additional condition from the choices below is necessary in order to conclude that $f(x) = g(x)$ for all values of x?

   A. $f''(x) = g''(x)$ for all x.

   B. $f(0) = g(0)$.

   C. f and g are continuous functions.

   D. No additional condition will allow you to conclude that $f(x) = g(x)$.

   E. No additional condition is required.

Section 2: Introduction to Differential Equations

1. According to Newton's law of heat transfer, the rate at which the temperature of a body changes is proportional to the difference between the temperature T of the body and the temperature T_m of the surrounding medium. Express this law as a mathematical equation.

2. A biology student finds a large glass bottle which can be used to grow a bacterial culture. She has a bacterial culture that doubles in size every minute, and with the amount she currently has, she calculates that the bottle will be full in one hour. She sterilizes the bottle and places the culture in the bottle at 11 am.

   a. At what time will the bottle be half full?

   b. Let us say that the bottle is "almost empty" when the culture is contained in less than 1% of the volume of the bottle. Approximately what percentage of the time is the bottle "almost empty"?

   c. After placing the culture in the first bottle, the biologist finds a bottle four times as large (in volume) as the first bottle. If she places the same amount of bacteria in this bottle, when will it be full?

3. Using what you know about the derivatives of functions like e^x, $\ln x$, and $\sin x$, find functions which satisfy the following differential equations.

   a. $y' - y = 0$

   b. $y' + y = 0$

   c. $y'' + y = 0$ (Find two different functions.)

   d. $y'' - y = 0$ (Find two different functions.)

4. a. Given a differential equation of the form

   $$ay'' + by' + y = 0,$$

   find constants a and b so that $y = e^x$ and $y = e^{2x}$ are each a solution of this differential equation.

   b. Given a differential equation of the form

   $$a(x)y'' + b(x)y' + y = 0,$$

   find functions $a(x)$ and $b(x)$ so that $y = x$ and $y = x^2$ are each a solution of this differential equation.

5. At each point (x, y) on a certain curve, the slope of the curve is $3x^2y$. If the curve contains the point $(0, 8)$, then its equation is

   A. $y = 8e^{x^3}$ B. $y = x^3 + 8$ C. $y = e^{x^3} + 7$

   D. $y = \ln(x + 1) + 8$ E. $y^2 = x^2 + 8$

6. The estimated human population of the world (in millions) in various years is given in the table below.

year	1650	1750	1800	1850	1900	1920	1930	1940	1950	1960	1970	1975
pop.	508	711	912	1131	1590	1811	2015	2249	2509	3008	3610	3967

   a. Make a plot of world population versus time. Does it appear that the world population obeys the exponential law?

   b. Suppose the function f satisfies the following: $f'(t) = kf(t)$; $f(0) = A$, where $A \neq 0$. Let $L(t) = \ln(f(t))$. Show that the graph of $L(t)$ is a line with slope k.

   c. Make a plot of ln(world population) versus time. Do the points appear to lie on a line? Is the world population growing faster or more slowly than exponentially?

   d. Try to explain why the exponential growth law is not a completely accurate model for the growth of the world's population.

7. Oil is being pumped continuously from an Arabian oil well at a rate proportional to the amount of oil left in the well; that is, $\frac{dy}{dx} = ky$ where y is the number of gallons of oil left in the well at any time t in years. Initially there are 1,000,000 gallons of oil in the well, and 6 years later there are 500,000 gallons remaining. Assume that it is no longer profitable to pump oil when there are fewer than 50,000 gallons remaining.

   a. Write an equation for y in terms of t.

   b. At what rate is the amount of oil in the well decreasing when there are 600,000 gallons of oil remaining in the well?

   c. How long will it be profitable to pump oil?

8. a. Find the general solution of $y\frac{dy}{dx} = x$.

   b. Find the specific solution which satisfies the initial condition given in (i) below. Repeat for the different conditions given in (ii) and (iii).

   (i) $y(2) = 1$ (ii) $y(2) = -1$ (iii) $y(-2) = -1$

   c. Graph the answers to part *b*.

9. a. Find the general solution of $\frac{dy}{dx} = 2xy^2$.

   b. As in part b of the problem above, find the specific solution which satisfies each of the following initial conditions.

   (i) $y(1) = 1$ (ii) $y(1) = -1$ (iii) $y(0) = 0$

10. a. Find a solution of the differential equation below by the method of separation of variables

$$\frac{dy}{dx} = 3y^{\frac{2}{3}}, \quad y(0) = 0.$$

   b. Show that $f(x) = \begin{cases} (x-a)^3, & x \leq a, \\ 8, & a < x < b, \\ (x-b)^3, & x \geq b \end{cases}$
   is also a solution of the same differential equation.

11. A not uncommon error in calculus is to believe that the product rule for derivatives states that $(fg)' = f'g'$.

   a. If $f(x) = e^{3x}$, find a nonzero function g for which $(fg)'$ equals $f'g'$.

   b. Repeat part a for $f(x) = e^{x^2}$.

12. Suppose an amount A_0 is invested at time $t = 0$ at a rate $r\%$ per year, compounded continuously. Let $A(t)$ denote the amount of the investment after t years.

   a. Over the period from t to $t + \Delta t$, the amount of interest earned is $A(t + \Delta t) - A(t)$. Show that

$$A(t)\frac{r}{100}\Delta t \leq A(t + \Delta t) - A(t) \leq A(t + \Delta t)\frac{r}{100}\Delta t.$$

   (The division by 100 converts r to a decimal.)

   b. Show that $\frac{dA}{dt} = \frac{r}{100}A$.

   c. Solve the differential equation of part b.

13. a. A curve in the plane has the property that a normal line to the curve at the point $P = (x, y)$ always passes through the point $(2, 0)$. Find the equation of the curve if the curve passes through the point $(1, 1)$.

   b. Graph the curve in part a.

14. A curve has the property that each point P on the curve is the midpoint of the line segment RS, where R and S are on the coordinate axes and RS is tangent to the curve at P. See the drawing below. Find the equation of the curve if the curve passes through the point $(1, 1)$.

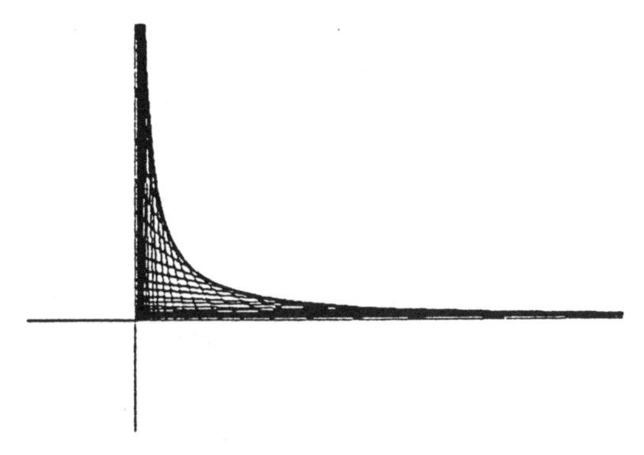

15. An important concept in business and banking is the time required for an investment of money to double. A common rule of thumb, called the "rule of 72," is that the doubling time in years is approximately 72 divided by the yearly interest rate r. When interest is compounded annually, money will grow according to the formula $A(t) = P(1 + \frac{r}{100})^t$, where P is the principal on deposit, r is the annual interest rate as a percentage, and t is time in years.

    a. Use the forumla for A given above to calculate the doubling time as a function of r.

    b. Compare the exact time required and the value $72/r$ for the following values of r: 2%, 4%, 8%, 12%, 18%.

    c. When interest is compounded continuously, then $A(t) = Pe^{\frac{r}{100}t}$. Find the exact doubling time in this case. Explain why a "rule of 69" is better than a "rule of 72" for doubling time in this case.

    d. Make a table showing doubling time in years when $r =$ 2%, 4%, 8%, 12%, and 18% when interest is compounded continuously. Compare the answer you get when using a "rule of 72," "rule of 69" and the exact doubling time. Why would business people use the "rule of 72" instead of the "rule of 69"?

16. Psychologists in learning theory study *learning curves*, the graphs of the "performance function" $P = P(t)$ of someone learning a skill as a function of the training time t. If M represents the maximal level of performance, it is noted that learning is at first rapid, and then it tapers off (the rate of learning decreases) as $P(t)$ approaches M.

    a. Explain why the differential equation $\dfrac{dP}{dt} = k(M - P)$, k a positive constant, describes this situation.

    b. Solve the differential equation under the initial condition $P(0) = P_0$ and sketch a graph of P.

    c. Suppose that for a specific learning activity, it is determined that $P_0 = .1M$ and $k = .05$ for t measured in hours. How long does it take to reach 90% of the maximal level of performance M?

17. When stated in the form of a differential equation, Newton's Law of Cooling (Warming) becomes $\dfrac{dT}{dt} = -k(T - T_a)$, where k is a positive constant and T_a is the ambient temperature.

    a. Find the general solution for T, satisfying the initial condition $T(0) = T_0$.

    b. What is the limiting temperature as $t \to \infty$? Explain the difference between what happens when $T_0 \leq T_a$, and when $T_0 \geq T_a$.

    c. A 5 lb roast, initially at $60°F$ is put into a $350°F$ oven. After two hours, the temperature of the roast is $120°F$. When will the roast be medium rare (a temperature of $150°F$)?

18. Suppose that rabbits are introduced onto an island where they have no natural enemies. Because of natural conditions, the island can support a maximum of 1000 rabbits. Let $P(t)$ denote the number of rabbits at time t (measured in months), and suppose that the population varies in size (due to births and deaths) at a rate proportional to both $P(t)$ and $1000 - P(t)$. That is, suppose that $P(t)$ satisfies the differential equation $\dfrac{dP}{dt} = kP(1000 - P)$, where k is a positive constant.

    a. Find the value of P when the *rate of change* of the rabbit population is maximal.

    b. When is the *rate of change* of the rabbit population a minimum? Discuss your answers.

    c. Assuming 50 rabbits were placed on the island, sketch of the graph that would show how t and $P(t)$ are related.

19. Consider a predator-prey population, e.g. wolf and deer population on a Lake Superior island. Assume the size of the predator population is $x = x(t)$ and the prey population is $y = y(t)$. This predator-prey situation can be modeled by the following differential equations.

$$\frac{dx}{dt} = Axy - Bx, \qquad \frac{dy}{dt} = Cy - Dxy,$$

where A, B, C, D are positive constants.

a. What does the xy term in both equations model?

b. Show that the predator population is increasing if and only if the prey population is greater than $\frac{B}{A}$.

c. When is the prey population increasing?

d. Interpret the meaning of the results of parts b and c in the figure below.

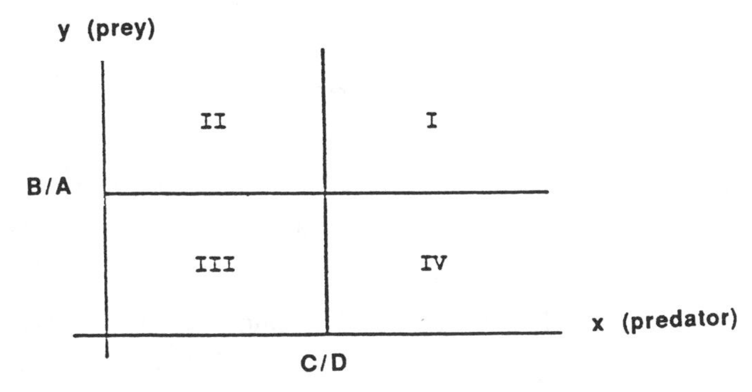

e. Eliminate the time parameter, compute $\frac{dy}{dx}$, and solve the resulting differential equation.

20. Consider the differential equation $\frac{dy}{dx} = \sqrt{1 - y^2}$.

a. Find the general solution of this differential equation.

b. Sketch a graph showing several representative curves from the family of solutions. Be sure to consider the *sign* of $\frac{dy}{dx}$ in plotting the solution curves. For solutions of this particular equation, should the curves be always increasing, always decreasing, or some of each?

c. Should solution curves from a first-order differential equation cross each other? If yours do, reconsider your answers to part b.

Chapter V: The Definite Integral

Section 1: Riemann Sums

1. Assume that the rate for postage is 29 cents for the first ounce and 23 cents for each additional ounce up to 11 ounces. Therefore, if C is the cost of mailing a letter weighing w ounces then C is of the form

$$C(w) = \begin{cases} 29, & \text{if } 0 \leq x < 1, \\ 52, & \text{if } 1 \leq w < 2, \\ 75, & \text{if } 2 \leq w < 3, \\ 98, & \text{if } 3 \leq w < 4, \\ \text{etc.} \end{cases}$$

   a. Draw a graph of C as a function of w.

   b. What is the total cost of mailing 11 letters, one for each of the following weights: .5, 1.5, 2.5, ..., 10.5 ounces. Describe your answer geometrically on the graph of part a.

   c. If the cost function continued on with the same pattern after 11 ounces, how much would it cost to mail 100 letters, one for each of the weights: .5, 1.5, 2.5, ..., 99.5? (Note: In fact, this is *not* how the Post Office currently charges for postage.)

2. Let $f(x) = x^2 + x$. Consider the region bounded by the graph of f, the x-axis, and the line $x = 2$. Divide the interval $[0, 2]$ into 8 equal subintervals. Draw a picture to help you answer the following.

   a. Obtain a lower estimate for the area of the region by using the left endpoint of each subinterval.

   b. Obtain an upper estimate for the area by using the right endpoint of each subinterval.

   c. Find an approximation for the area that is better than either of the answers obtained in parts a and b.

   d. Determine whether the answer in part c is larger or smaller than the true area, and supply a reason.

3. Let $g(x) = 4 - x^2$. Repeat the analysis of the previous question over the interval $[0, 2]$, divided as before into 8 equal subintervals.

4. Consider the problem of finding the front cross sectional area of your body when you are standing erect with your hands at your sides.

   a. In 60 seconds or less, make an estimate of this area. How do you arrive at your estimate?

   b. In the next twenty minutes, possibly with the help of others, get a better estimate of this area. What method do you use and how does it improve upon your previous estimate?

   c. As you reflect upon what you have done, are there any concepts from calculus that you have used in solving this problem?

5. a. Draw a grid of equally-spaced horizontal and vertical lines on the map of Lake Ontario given below. By counting squares and multiplying by the appropriate scaling factor, obtain a lower and an upper bound on the area of the lake.

   b. Explain how you might improve the bounds obtained in part a.

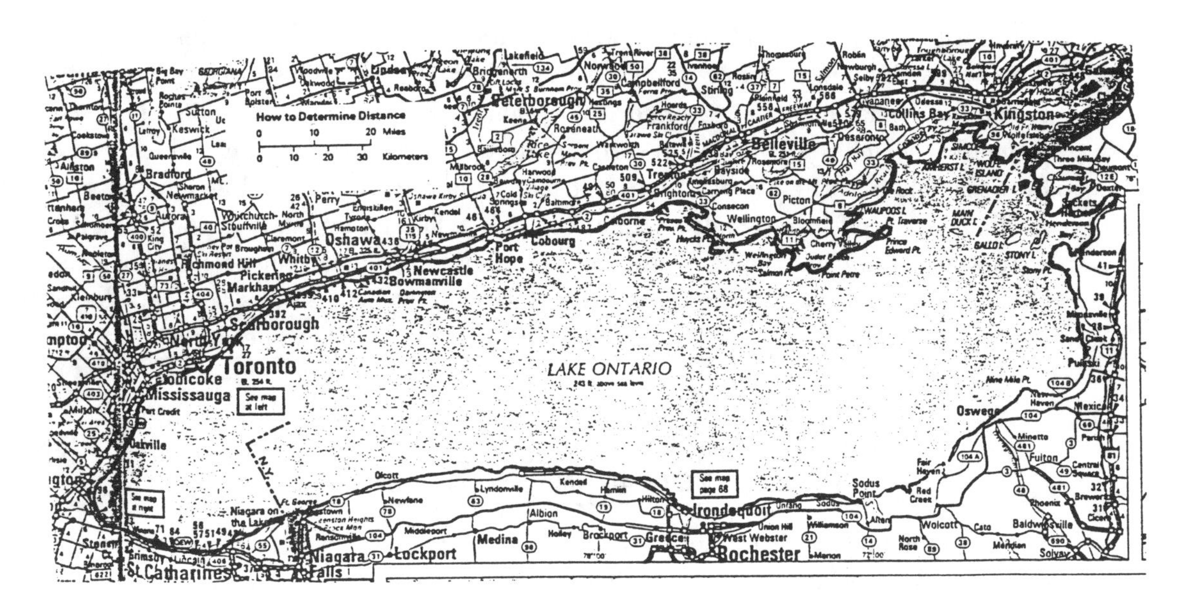

Lake Ontario

6. Brokers at a Wall Street firm have been following the share price of a stock , Roller Coaster, Inc., on the New York Stock Exchange. They accumulate the following data for a single day. Use upper and lower sums to obtain upper and lower bounds on the average value for the share price of Roller Coaster, Inc. for this day.

Time	9:00	10:12	11:05	12:00	1:33	2:45	3:15	3:50	4:00
Price	$10\frac{3}{8}$	12	$16\frac{3}{4}$	17	$15\frac{1}{4}$	14	$11\frac{1}{8}$	$10\frac{3}{4}$	$10\frac{1}{4}$

7. In order to determine the average temperature for the day, a meteorologist decides to record the temperature at eight times during the day. She further decides that these recordings do not have to be equally spaced during the day because she does not need to make several readings during those periods when the temperature is not changing much (as well as not wanting to get up in middle of the night). She decides to make one reading at some time during each of the intervals in the table below.

Time	12 AM–5 AM	5 AM–7 AM	7 AM–9 AM	9 AM–1 PM
Temp	42°	57°	72°	84°

Time	1 PM–4 PM	4 PM–7 PM	7 PM–9 PM	9 PM–12 AM
Temp	89°	75°	66°	52°

a. Using Riemannm sums, write a formula for the average temperature for the day.

b. Calculate the average temperature.

8. Bernhard Riemann V is planning to take a Sunday afternoon trip in his 1976 Mercedes Benz whose odometer broke when it turned over 200,000 miles. Being the descendant of the mathematician Georg Friedrich Bernhard Riemann (1826–1866), he decides to estimate the miles driven by recording his speed at various times during the trip. Because this is to be a restful trip and not a military drill, he does not plan to take readings at specified intervals of time, but only when he remembers to do so. He takes a stopwatch to record the time intervals (in minutes) to make the calculations easier for himself (and consequently for you). The following table contains the data he collects during his trip. Approximately how far does he travel? Each speedometer reading (in MPH) is made at some point during the time interval recorded.

time interval	15	25	30	15	20	35	40	20	10
speed	30	45	50	25	45	55	50	35	25

After completing the arithmetic to find the distance he has driven, Bernhard wonders about his average speed. Can you tell him what it is?

9. This problem concerns the region under the graph of $f(x) = \sqrt{16 - x^2} + 1$ from $x = 0$ to $x = 4$. You will be asked to find Riemann sums for the area of this region by picking sample points in various ways.

   a. Sketch the graph of f from $x = 0$ to $x = 4$.

   b. Divide the interval $[0, 4]$ into four subintervals each 1 unit wide. Pick x in the middle of each subinterval, find the corresponding values of $f(x)$, and then calculate the Riemann sum for these choices of x.

   c. Find a different Riemann sum by picking the value of x in each subinterval where $f(x)$ is the highest.

   d. Repeat part c, using the lowest value of $f(x)$ in each of the subintervals.

   e. Find an approximation for the area by averaging the two Riemann sums in parts *c* and *d*. Is this value the same as what you calculated in part *b*? Why or why not?

   f. A closer approximation to the area can be found by dividing the region into narrower strips. Divide the interval $[0, 4]$ into eight subintervals, each $\frac{1}{2}$ unit wide. Using the midpoint of each subinterval, find the corresponding values of $f(x)$, and use them to calculate a new Riemann sum.

   g. Using the concept of the limit, how might you define the exact area of this region?

   h. Calculate the exact area using familiar area formulas. How do the approximations you have calculated compare with the exact area?

10. Finding Survivors of a Plane Crash: A plane has just crashed six minutes after takeoff. There may be survivors, but you must locate the plane quickly without an extensive search. It was traveling due west and the pilot radioed its speed every minute from takeoff to the time it crashed. These data are given in the table below. Approximately how far west of the airport did the plane crash?

minutes	0	1	2	3	4	5	6
speed (mph)	90	110	125	135	120	100	70

11. The amount of money it takes to dig a tunnel equals the length of the tunnel times the cost per unit length. If the cost per unit length is a constant, calculating the cost to dig a tunnel poses no problem; however, the price per unit length increases as the tunnel gets longer because of the expense of carrying in workers and tools and carrying out dirt and rock. Assume that the dollar price per foot varies according to the table below.

length	0 to 100	100 to 200	200 to 300	300 to 400	400 to 500
cost per foot	500	820	1180	1580	2020

length	500 to 600	600 to 700	700 to 800	800 to 900	900 to 1000
cost per foot	2500	3020	3580	4180	4820

a. Find the total cost for digging a tunnel 1000 feet long.

b. How much money can be saved by starting the 1000-foot tunnel from both ends and making the two halves meet in the middle?

12. Let f be the function graphed below. Which of the following is the best estimate of $\int_1^6 f(x)\,dx$? Justify your answer.

A. -24 B. 9 C. 26 D. 38

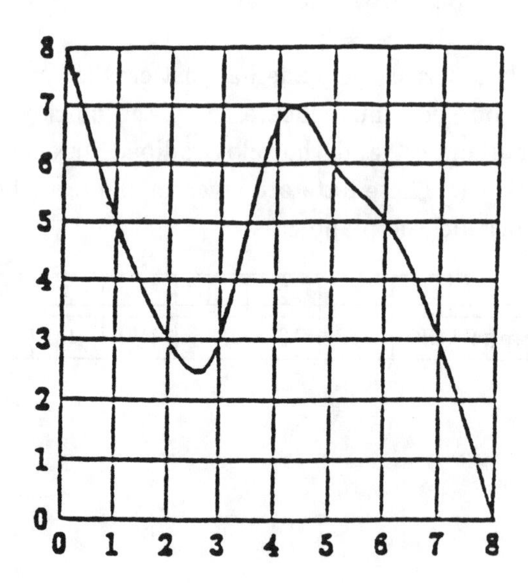

13. Assume that the function f is a *decreasing* function on the interval $[0,4]$ and that the following is a table showing some function values.

x	0	1	1.5	3	4
$f(x)$	4	3	2	1.5	1

Employ a Riemann sum to approximate $\int_0^4 f(x)\,dx$. Use a method so that your approximation will either be less than the value of the definite integral or will be greater than the definite integral. Finally, indicate whether your approximation is less than or greater than the value of the definite integral.

14. Consider the following table of values of a continuous function f at different values of x:

x	1	2	3	4	5	6	7	8	9	10
$f(x)$	.14	.21	.28	.36	.44	.54	.61	.70	.78	.85

a. From the data given, find *two* estimates of $\int_1^{10} f(x)\,dx$.

b. Obtain a different estimate for the integral by taking an average value of f over each subinterval.

c. Do you think that your estimates are too big or too little? Explain why.

15. The graph of a function f is given in the figure below.

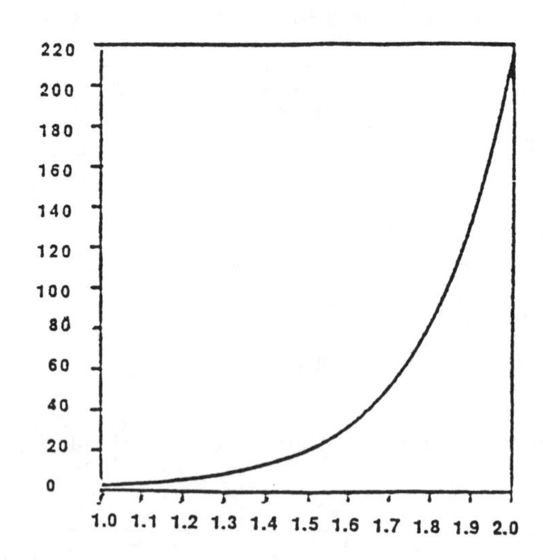

When asked to estimate $\int_1^2 f(x)\,dx$ to five decimal place accuracy, a group of calculus students submitted the following answers:

A. -4.57440 B. 4.57440 C. 45.74402 D. 457.44021

Although one of these responses is correct, the other three are "obviously" incorrect. Using arguments your classmates would understand, identify the correct answer and explain why each of the others cannot be correct.

16. a. Sketch the graph of the curve $y = \dfrac{1}{x}$ for $1 \le x \le 2$. Divide the interval into 5 subintervals of equal length to show that:

$$0.2\left[\tfrac{1}{1.2} + \tfrac{1}{1.4} + \tfrac{1}{1.6} + \tfrac{1}{1.8} + \tfrac{1}{2.0}\right] < \int_1^2 \frac{dx}{x} < 0.2\left[\tfrac{1}{1.0} + \tfrac{1}{1.2} + \tfrac{1}{1.4} + \tfrac{1}{1.6} + \tfrac{1}{1.8}\right].$$

b. Divide the interval into n equal parts instead of 5 to obtain the inequalities:

$$\sum_{r=1}^{n} \frac{1}{n+r} < \int_1^2 \frac{dx}{x} < \sum_{r=0}^{n-1} \frac{1}{n+r}.$$

c. Show that the difference between these two sums is $\dfrac{1}{2n}$.

d. How large should n be to evaluate $\ln 2$ correct to 4 decimal places using one of the sums above?

Section 2: Properties of Integrals

1. Given $\int_0^1 f(x)\,dx = \frac{4}{3}$, $\int_1^2 f(x)\,dx = \frac{8}{3}$, and $\int_0^3 f(x)\,dx = \frac{11}{3}$, find the values of:

   a. $\int_0^2 f(x)\,dx$ b. $\int_1^3 f(x)\,dx$ c. $\int_2^3 f(x)\,dx$

2. Suppose $\int_a^b f(x)\,dx = 18$, $\int_a^b g(x)\,dx = 5$, and $\int_a^b h(x)\,dx = -11$. Evaluate as many of the following as you can by using the properties of integrals. If you cannot give an answer, then admit it!

   a. $\int_a^b (f(x) + g(x))\,dx$

   b. $\int_a^b (f(x) - g(x))\,dx$

   c. $\int_a^b f(x)g(x)\,dx$

   d. $\int_a^b (g(x) + h(x))\,dx$

   e. $\int_a^b (g(x) - h(x))\,dx$

   f. $\int_a^b \frac{g(x)}{h(x)}\,dx$

   g. $\int_a^b (f(x) + g(x) + h(x))\,dx$

3. Decide whether the following statement is true or false. If f is continuous on R and $\int_a^b f(x)\,dx = 0$ for any choice of real numbers a and b, then $f(x) = 0$ for all x in R.

4. Let f be a continuous function on the interval $[-5, 10]$ and let $g(x) = f(x) + 2$. If $\int_{-5}^{10} f(t)\,dt = 4$, what is $\int_{-5}^{10} g(u)\,du$?

5. a. Let f be a function defined by the rule $f(t) = 2t$. Complete the following table to find the area under the graph of f and above the x-axis from $x = 0$ to the given value of x. Hint: The region is triangular.

x	1	2	3	4	5	6
Area						

The areas found above are values of a function F which is given by $F(x) = \int_0^x 2t\,dt$. Find a simple rule to define the function F which can be used to fill in the table in part (a).

b. Using the function $g(t) = 2t + 1$, answer the questions from part a. Note: The region is now trapezoidal.

x	1	2	3	4	5	6
Area						

6. Suppose that f is an integrable function and that $\int_0^1 f(x)\,dx = 2$, $\int_0^2 f(x)\,dx = 1$ and $\int_2^4 f(x)\,dx = 7$.

a. Find $\int_0^4 f(x)\,dx$.

b. Find $\int_1^0 f(x)\,dx$.

c. Find $\int_1^2 f(x)\,dx$.

d. Explain why $f(x)$ must be negative somewhere in the interval $[1, 2]$.

e. Explain why $f(x) \geq 3.5$ for at least one value of x in the interval $[2, 4]$.

7. For any real number b, $\int_0^b |2x|\,dx$ is

A. $-b\,|b|$ B. b^2 C. $-b^2$ D. $b\,|b|$

E. none of these.

8. Let f and g have continuous first and second derivatives everywhere. If $f(x) \leq g(x)$ for all real x, which of the following must be true?

I. $f'(x) \leq g'(x)$ for all real x

II. $f''(x) \leq g''(x)$ for all real x

III. $\int_0^1 f(x)\,dx \leq \int_0^1 g(x)\,dx$

A. none B. I only C. III only

D. I and II only E. I, II, and III

Section 3: Geometric Integrals

1. Let f be a continuous function on the closed interval $[0, 2]$. If $2 \le f(x) \le 4$, then the greatest possible value of $\int_0^2 f(x)\,dx$ is:

   A. 0 B. 2 C. 4 D. 8 E. 16

2. If f is the continuous, strictly increasing function on the interval $[a, b]$ as shown below, which of the following must be true?

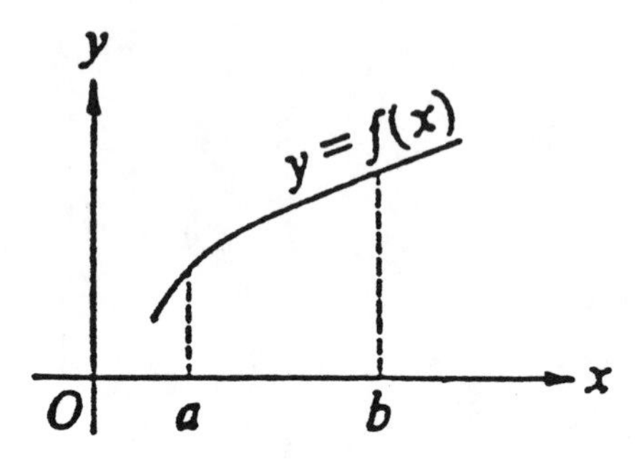

   I. $\int_a^b f(x)\,dx < f(b)(b-a)$

   II. $\int_a^b f(x)\,dx > f(a)(b-a)$

   III. $\int_a^b f(x)\,dx = f(c)(b-a)$ for some number c such that $a < c < b$.

   A. I only B. II only C. III only

   D. I and III only E. I, II, and III

3. Which of the following definite integrals is *not* equal to 0?

   A. $\int_{-\pi}^{\pi} \sin^3 x\,dx$ B. $\int_{-\pi}^{\pi} x^2 \sin x\,dx$ C. $\int_0^{\pi} \cos x\,dx$

   D. $\int_{\pi}^{\pi} \cos^3 x\,dx$ E. $\int_{-\pi}^{\pi} \cos^2 x\,dx$

4. Let f be the function graphed below. Note: The graph of f consists of two straight lines segments and two quarter-circles.

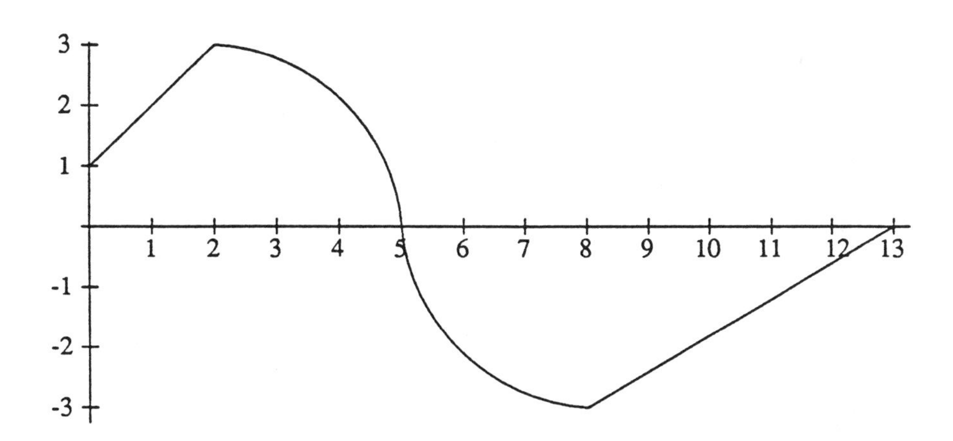

a. Evaluate $\displaystyle\int_0^{15} f(x)\,dx$.

b. Evaluate $\displaystyle\int_9^{12} f(x)\,dx$.

c. Evaluate $\displaystyle\int_0^{15} |f(x)|\,dx$.

5. If $[\![x]\!]$ denotes the greatest integer $\leq x$, let $f(x) = [\![x]\!]$ and $g(x) = [\![2x]\!]$.

a. Graph $f(x)$, $g(x)$, $f(x)+g(x)$, $3f(x)$, $f(x)g(x)$ on the interval $[0,5]$.

b. Using the graphs of part a, evaluate the following:

(i) $\displaystyle\int_0^5 f(x)\,dx$ (ii) $\displaystyle\int_0^5 g(x)\,dx$ (iii) $\displaystyle\int_0^5 3f(x)\,dx$

(iv) $\displaystyle\int_0^5 (f(x)+g(x))\,dx$ (v) $\displaystyle\int_0^5 f(x)g(x)\,dx$

c. Try to relate the integrals in iii), iv), and v) in part b to $\displaystyle\int_0^5 f(x)\,dx$ and/or $\displaystyle\int_0^5 g(x)\,dx$.

6. Let $f(x) = [\![x]\!]$, and $g(x) = x$.

   a. Graph $f(x)$, $g(x)$, $f(x) + g(x)$, $f(x)g(x)$, $g(x)/f(x)$ on the interval $[0, 5]$.

   b. Using the graphs of part a, evaluate the following:

   (i) $\int_0^5 f(x)\,dx$ (ii) $\int_0^5 g(x)\,dx$ (iii) $\int_0^5 (f(x) + g(x))\,dx$

   (iv) $\int_0^5 f(x)g(x)\,dx$ (v) $\int_1^5 \frac{g(x)}{f(x)}\,dx$ Note: Interval is $[1, 5]$.

   c. Try to relate the integrals in iii), iv), and v) in part b to appropriate integrals of f and/or g.

7. Calculate $\int_{-3}^{3} (x + 5)\sqrt{9 - x^2}\,dx$. Hint: $(x + 5)\sqrt{9 - x^2} = x\sqrt{9 - x^2} + 5\sqrt{9 - x^2}$.

   What do the graphs of $y = x\sqrt{9 - x^2}$ and $y = \sqrt{9 - x^2}$ for $x \in [-3, 3]$ look like? Think geometrically.

8. [C] Consider the function $f(x) = \sin(a\pi x)\sin(b\pi x)$ for x in $[0, 1]$.

   a. Using your CAS, plot the function for small integer values of a and b with $a \neq b$. Based upon what you see, make a conjecture about the value of $\int_0^1 f(x)\,dx$. Explain your reasoning.

   b. Plot the function f for small integer values of a and b when $a = b$. Make a conjecture about the value of $\int_0^1 f(x)\,dx$. Again, explain your reasoning. Hint: Draw the line $y = 1$ and turn the graphs upside down.

9. Four calculus students disagree as to the value of the integral $\int_0^{\pi} \sin^8 x\,dx$. Jack says that it is equal to π. Joan says it is equal to $\frac{35\pi}{128}$. Ed claims it is equal to $\frac{3\pi}{90} - 1$ while Lesley says its equal to $\frac{\pi}{2}$. One of them is right. Which one is it?

   Hint: Do *not* try to evaluate the integral; try instead to eliminate the three wrong answers.

10. By appealing to geometric evidence, show that the following hold true:

    a. $\int_0^1 x^n\,dx + \int_0^1 x^{1/n}\,dx = 1$ for n a positive integer.

    b. $\int_1^e \ln x\,dx + \int_0^1 e^x\,dx = e$.

11. Three regions are commonly associated with the graph of the function f between points P and Q in the figure below.

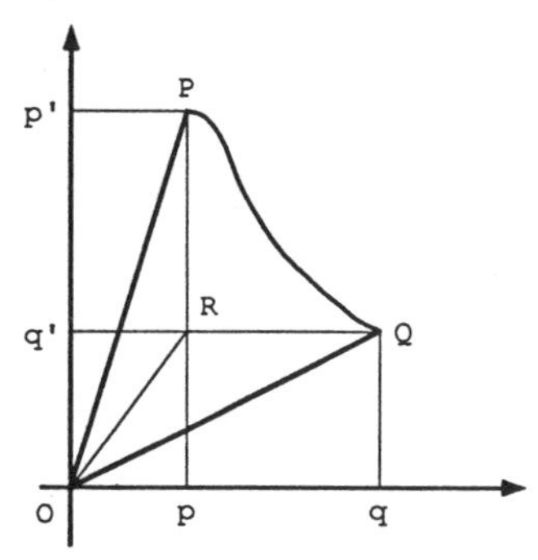

a. The "vertical region" with vertices P, Q, p, q where p and q are the projections of P and Q respectively on the x-axis.

b. The "horizontal region" with vertices P, Q, p', q' where p' and q' are the projections of P and Q, respectively, on the y-axis.

c. The "polar region" with verticies P, Q, O.

How are the areas of these three regions related? Hint: Forget integrals for a moment and think in terms of high school geometry. Draw the line segment between the origin O and the point R and consider the areas of the following subregions: The triangles ORP and ORQ, the rectangles $RQqp$ and $PRq'p'$, and the "common" subregion PRQ.

12. If you were asked to find $\int_1^2 x^2 e^{x^2}\, dx$, you could not do it analytically because you could not find an antiderivative of $x^2 e^{x^2}$. However, you should be able to estimate the size of the answer. Is it:

A. less than 0 B. 0 to 9.999 C. 10 to 99.99

D. 100 to 999.9 E. 1000 to 9,999 F. over 10,000

Section 4: The Fundamental Theorem of Calculus

1. Let f be a continuous function with antiderivative F on the interval $[a, b]$. Let c be any point in the interval. State whether the following are true or false. If false, then correct the statement or give an example to show why it is false.

   a. $\int_a^b f(x)\,dx = \int_a^c f(x)\,dx + \int_c^b f(x)\,dx$

   b. $\int_a^b F(x)\,dx = f(b) - f(a)$

   c. $\int_a^b f(x)\,dx \geq 0$

   d. $\int_a^b cf(x)\,dx = c(F(b) - F(a))$

   e. $\int_a^b f(x)\,dx = f(m)(b-a)$, for some $m \in [a, b]$

2. Suppose you know that a certain function f is twice differentiable and that its graph over $[0, 2]$ is given in the figure below. As you see, the printer was sloppy and spilled a lot of ink on the graph. Decide, if possible, whether each of the following definite integrals is positive, equal to 0, or negative.

   a. $\int_0^2 f''(x)\,dx$ b. $\int_0^2 f'(x)\,dx$ c. $\int_0^2 f(x)\,dx$

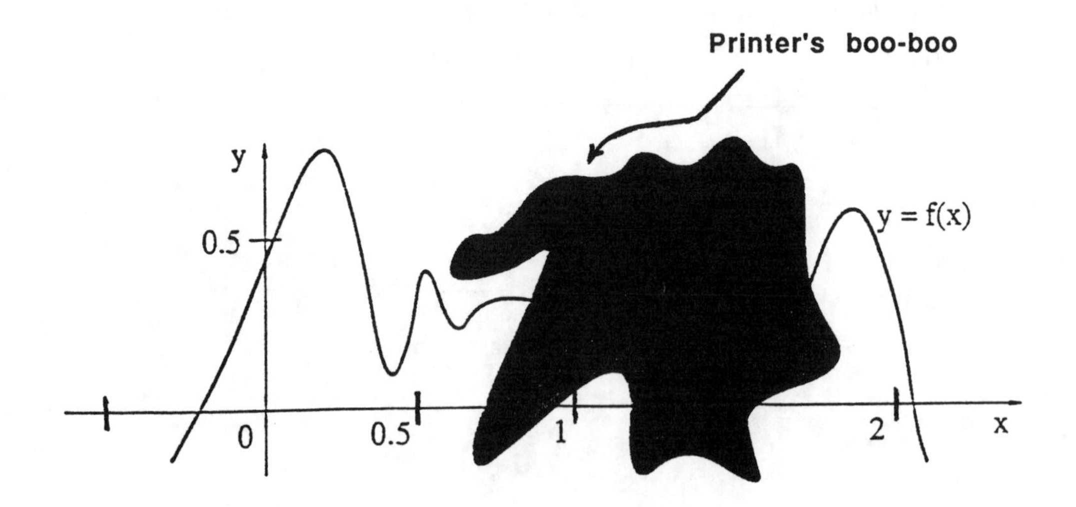

3. You are asked, in this question, to examine the region R under the graph of $f(x) = \sqrt{x}$ from $x = 1$ to $x = 4$.

   a. Draw a sketch of this region.

   b. Divide R into 3 vertical strips of equal width. Find an upper and a lower sum for the area of R. Keep all the accuracy your calculator will give you.

   c. Find an estimate for the area of R by averaging the upper and lower sums. Does this average overestimate or underestimate the actual area? Explain.

   d. (The secret choice) Find the Riemann sum for the area of R using the same three vertical strips of part b, but evaluating the function f at the following points in each subinterval.

subinterval	1	2	3
point	1.48584256	2.4916026	3.49402722

   How does this answer compare with the average found in part c?

   e. Calculate the exact area of R.

4. This continuation of the problem above shows you how the points are selected in part d. Let $g(x) = \frac{2}{3}x^{3/2}$.

   a. Show that the point 1.48584256 in the previous problem is the point in the interval $[1, 2]$ at which the conclusion of the Mean Value Theorem is true for the function g.

   b. How is the function g of this problem related to the function f of the previous problem?

   c. How are the points 2.4916026 and 3.49402722 calculated in the previous problem selected?

5. The average value of a function f over the interval $[a, b]$ is $\frac{1}{b-a}\int_a^b f(x)\,dx$.

   a. Find the average values of $f(x) = x$, $f(x) = x^2$, and $f(x) = x^3$ over the interval $[0, 1]$.

   b. From the pattern that is established in part a, what is the average value of $f(x) = x^n$, for an integer $n \geq 1$?

   c. What does the answer to part b imply about the average value of $f(x) = x^n$, as n gets larger and larger? Can you explain this from the graph of $f(x) = x^n$?

6. a. Find the average values of $f_1(x) = x$, $f_2(x) = x^{1/2}$, and $f_3(x) = x^{1/3}$ over the interval $[0, 1]$.

   b. From the pattern that is established in part a, what is the average value of $f_n(x) = x^{1/n}$, for integer $n \geq 1$?

   c. What does the answer to part b imply about the average value of $f_n(x) = x^{1/n}$, when n gets larger and larger? Can you explain what is happening from the graph of $f_n(x) = x^{1/n}$?

7. Find the average value of:

   a. $f(x) = x$ on $[-1, 1]$.

   b. $f(x) = x^3 - x$ on $[-1, 1]$.

   c. $f(x) = \sin x$ on $[-\pi, \pi]$.

   d. Why are all of these so easy?

8. Evaluate the following infinite limits by expressing each of them as a definite integral and then evaluating the integral.

   a. $\displaystyle\lim_{n\to\infty} \frac{1}{n}\left[\left(\frac{1}{n}\right)^3 + \left(\frac{2}{n}\right)^3 + \cdots + \left(\frac{n}{n}\right)^3\right]$

   b. $\displaystyle\lim_{n\to\infty} \frac{1}{n}\left[\left(\frac{1}{n}\right)^3 + \left(\frac{2}{n}\right)^3 + \cdots + \left(\frac{4n}{n}\right)^3\right]$

   c. $\displaystyle\lim_{n\to\infty} \frac{1^5 + 2^5 + 3^5 + \cdots + n^5}{n^6}$

   d. $\displaystyle\lim_{n\to\infty} \frac{1}{n}\left[\left(\frac{n+1}{n}\right)^3 + \left(\frac{n+2}{n}\right)^3 + \cdots + \left(\frac{2n}{n}\right)^3\right]$

9. Suppose that $5x^3 + 40 = \displaystyle\int_c^x f(t)\,dt$.

   a. What is $f(x)$?

   b. Find the value of c.

10. Let $G(x) = \displaystyle\int_0^x \sqrt{16 - t^2}\,dt$.

    a. Does $G(2) = -G(-2)$ or $G(2) = G(-2)$?

    b. What is $G(0)$?

    c. What is $G'(2)$?

    d. What are $G(4)$ and $G(-4)$?

11. If $F(x) = \int_0^3 t\sqrt{t+9}\,dt$, then $F'(1) = 0$. Why?

12. Phoebe Small is caught speeding. The fine is \$3.00 per minute for each mile per hour above the speed limit. Since she was clocked at speeds as much as 64 MPH over a 6-minute period, the judge fines her:

$$(\$3.00)(\text{number of minutes})(\text{MPH over } 55) = (\$3.00)(6)(64 - 55) = \$162.$$

Phoebe believes that the fine is too large since she was going 55 MPH at times $t = 0$ and $t = 6$ minutes, and was going 64 MPH only at $t = 3$. She reckons, in fact, that her speed v is given by $v = 55 + 6t - t^2$.

a. Show that Phoebe's equation does give the correct speed at times $t = 0$, 3 and 6.

b. Phoebe argues that since her speed varied, the fine should be determined by calculus rather than by arithmetic. What should she propose to the judge as a reasonable fine?

13. a. If f' is a continuous function on $[0, 2]$ and $f(0) = f(2)$, find $\int_0^2 f'(x)\,dx$.

b. Give an intuitive interpretation of your answer in terms of the average value of f' on $[0, 2]$.

14. [CP] There is a rule of thumb that the area of a parabolic arch is two thirds the product of its height h and the length of its base b:

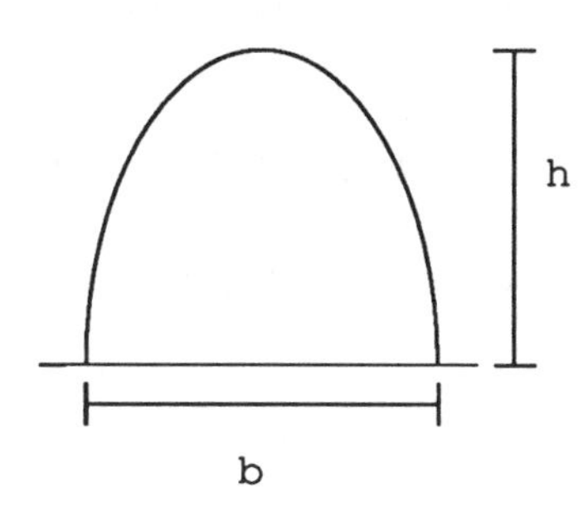

Confirm this formula by using the Fundamental Theorem to calculate the area in question.

Section 5: Functions Defined by Integrals

1. The Fundamental Theorem of Calculus says, in part, that:

$$\frac{d}{dx}\int_a^x f(t)\,dt = f(x), \quad \text{where } a \text{ is constant.}$$

Check this result by some examples that you already know. First find $\int_0^x f(t)\,dt$ for each of the following functions f, and then take the derivative of your answer with respect to x. For example, if $f(t) = t^2$, then $\int_0^x t^2\,dt = \frac{1}{3}x^3$ and $\frac{d}{dx}\left(\frac{1}{3}x^3\right) = x^2$. Do the same for the functions defined by the following rules:

a. $f(t) = t^3$

b. $f(t) = t^4 - 2t^3 + 1$

c. $f(t) = \cos t$

d. $f(t) = \sin t$

e. $f(t) = e^t + t$

2. Let us continue the problem above by trying to discover what happens when we change the form of the left side of the equation a little.

a. First consider $\int_x^a f(t)\,dt$ where a is still constant.

(i) Find $\frac{d}{dx}\int_x^0 t^2\,dt$ by the two-step method of Problem 1.

(ii) Find $\frac{d}{dx}\int_x^0 e^t\,dt$ using the same two steps.

(iii) Find $\frac{d}{dx}\int_x^0 \cos t\,dt$ again using the same two steps.

b. Can you now suggest a rule for these problems? What result from Section V.2 would be used in a proof of this rule?

3. We can now go one step further and try to find the result for $\dfrac{d}{dx}\displaystyle\int_0^{x^2} f(t)\,dt$.

   a. Find $\dfrac{d}{dx}\displaystyle\int_0^{x^2} e^t\,dt$ by the two-step method of Problem 1.

   b. Find $\dfrac{d}{dx}\displaystyle\int_0^{x^2} \cos(t)\,dt$ using the same two steps.

   c. Can you now suggest a rule which solves these problems in general? What result about derivatives are you using?

   d. If you made it to this part, can you go all the way and obtain a general rule for $\dfrac{d}{dx}\displaystyle\int_a^{u(x)} f(t)\,dt$?

4. $\displaystyle\lim_{h\to 0}\frac{\int_1^{1+h}\sqrt{x^5+8}\,dx}{h}$ is:

   A. 0 B. 1 C. 3 D. $2\sqrt{2}$

   E. nonexistent

5. Find the derivatives of the functions defined by the following integrals:

   a. $\displaystyle\int_0^x \frac{\sin t}{t}\,dt$ b. $\displaystyle\int_0^x e^{-t^2}\,dt$

   c. $\displaystyle\int_1^{\cos x} \frac{1}{t}\,dt$ d. $\displaystyle\int_0^1 e^{\tan^2 t}\,dt$

   e. $\displaystyle\int_x^{x^2} \frac{1}{2t}\,dt$ where $x > 0$

6. The graphs of three functions appear in the figure below. Identify which is f, which is f', and which is $\int_0^x f(t)\,dt$.

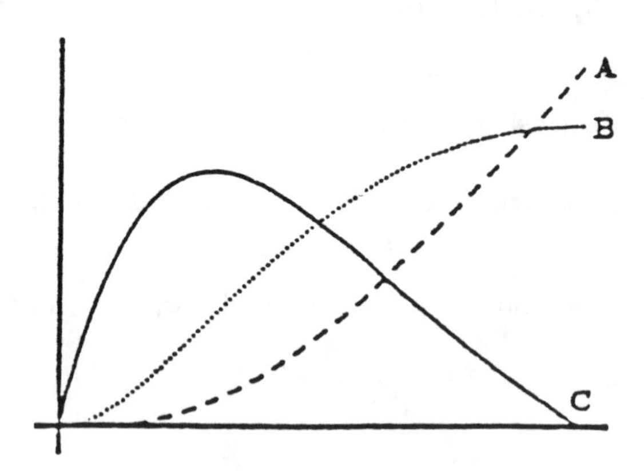

7. Let $F(x) = \int_0^x f(t)\,dt$ where f is the function graphed below. (The graph of f consists of straight lines and a semicircle.)

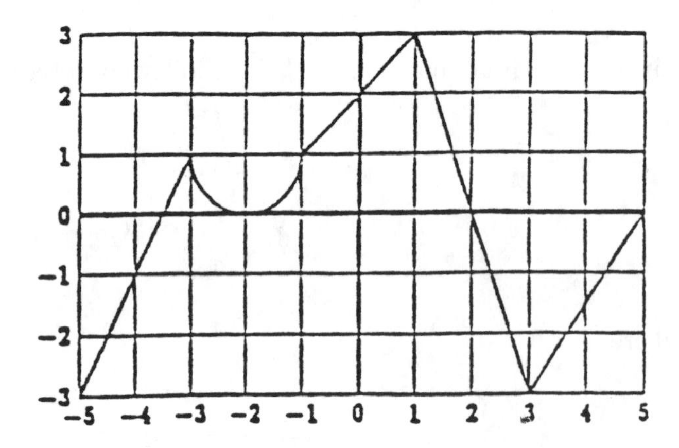

a. Evaluate $F(2)$, $F(0)$, and $F(-1)$.

b. Identify all the critical points of F in the interval $[-5, 5]$.

c. Identify all the inflection points of F in the interval $[-5, 5]$.

d. What is the average value of f on the interval $[-5, 5]$?

e. Let $G(x) = \int_0^x F(t)\,dt$. On what subintervals of $[-5, 5]$, if any, is G concave upward?

8. Find the equation of the tangent line to the curve $y = F(x)$, where $F(x) = \int_1^x \sqrt[3]{t^2+7}\,dt$, at the point on the curve where $x = 1$.

9. Let $F(x) = \int_0^x f(t)\,dt$ where f is the function graphed below.

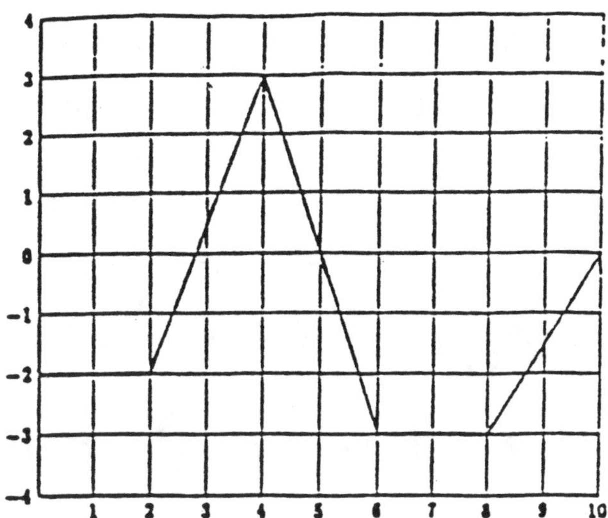

a. Evaluate $\int_0^2 f(t)\,dt$, $\int_0^4 f(t)\,dt$, $\int_2^4 f(t)\,dt$, $\int_5^{10} f(t)\,dt$, $\int_1^7 f(t)\,dt$.

b. Evaluate $F(0)$, $F(2)$, $F(5)$, and $F(7)$.

c. Find an analytic expression for $f(x)$.

d. Find an analytic expression for $F(x)$.

e. Sketch the graphs of f and F on the same axes over the interval $[0, 10]$.

f. Where does F have local maxima on the interval $[0, 10]$?

g. On which subintervals of $[0, 10]$, if any, is F decreasing?

h. On which subintervals of $[0, 10]$, if any, F increasing?

i. On which subintervals of $[0, 10]$, if any, is the graph of F concave up?

j. On which subintervals of $[0, 10]$, if any, is the graph of F concave down?

10. Let $F(x) = \int_1^x f(t)\,dt$ where f is the function graphed below.

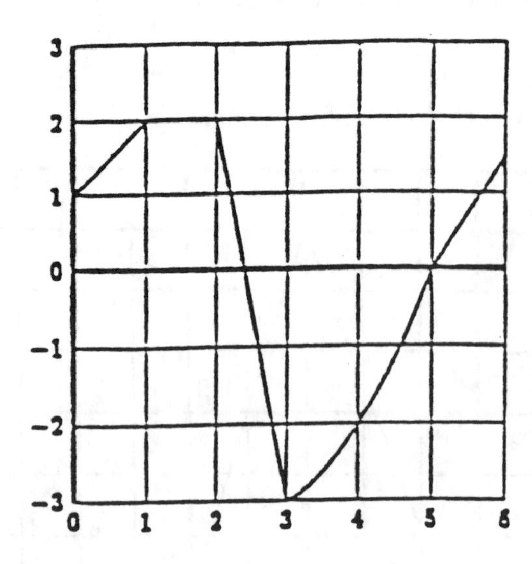

a. Suppose $\int_0^5 f(t)\,dt = -\frac{2}{3}$. What is $F(5)$?

b. Show that F has exactly one zero between 3 and 4.

c. Find the equation of the tangent line to the graph of F at the point $(3, F(3))$. Hint: What is $F'(3)$?

d. Use the equation found in part c to approximate the zero of F between 3 and 4.

11. Let $f(x) = \begin{cases} 1 & \text{if } t < 0, \\ 2 & \text{if } t \geq 0. \end{cases}$ Let $F(x) = \int_{-2}^x f(t)\,dt$.

a. Sketch the graphs of f and F on the same axes over $[-2, 2]$.

b. Find $F'(-1)$ and $F'(1)$.

c. Is F continuous at 0? Does $F'(0)$ exist?.

12. Let g be a differentiable function defined on $[-5, 5]$ such that $g(-1) = 2$. The graph of g' below consists of line segments and a half circle.

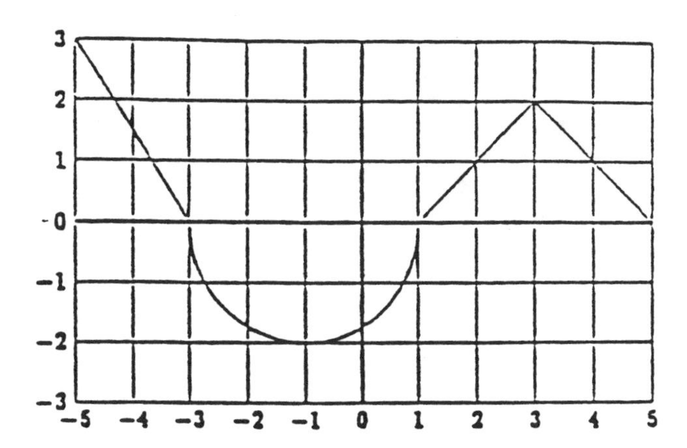

a. Find $g(5)$ and $g(-5)$. Explain your reasoning briefly.

b. Find $\int_{-5}^{5} g'(x)\,dx$.

13. Let $g(x) = \int_0^x f(t)\,dt$ where f is the function graphed below:

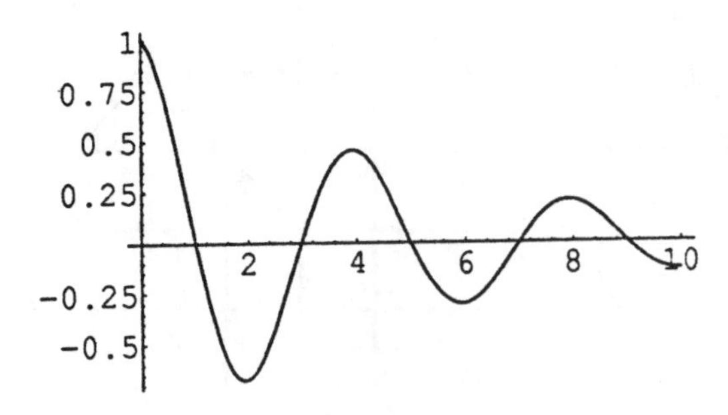

a. Does g have any local maxima within the interval $[0, 10]$? If so, where are they located?

b. Does g have any local minima within the interval $[0, 10]$? If so, where are they located?

c. At what value of x does g attain its absolute maximum value on the interval $[0, 10]$?

d. At what value of x does g attain its absolute minimum value on the interval $[0, 10]$?

e. On what subinterval(s) of $[0, 10]$, if any, is the graph of g concave up?

f. Sketch a graph of g.

14. The following graphs can be used in place of the graph of f in the previous problem. Answer all parts of that problem for each of them. Each graph presents a slightly different situation and therefore should be considered carefully.

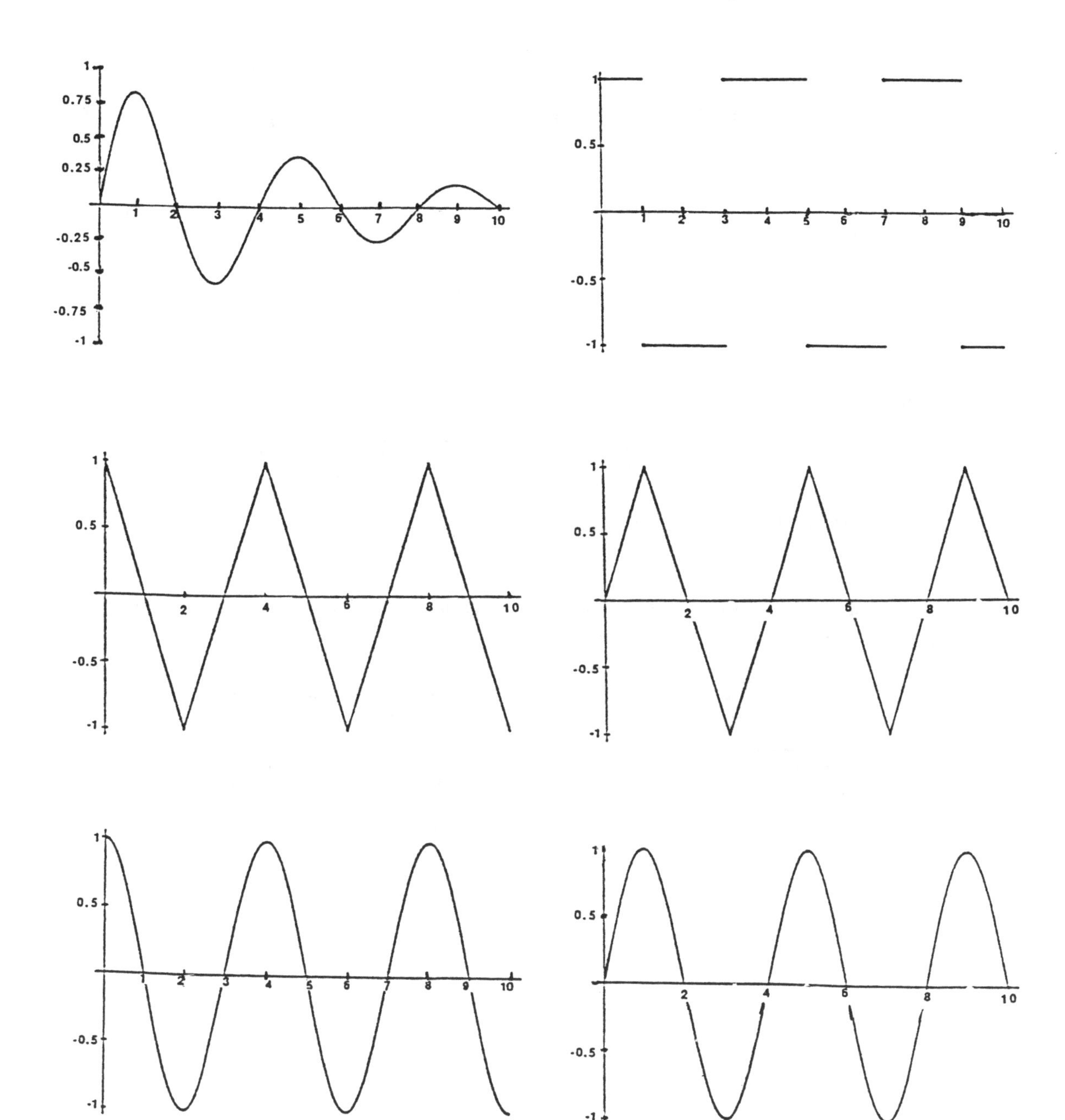

Chapter VI: The Definite Integral Revisited

Section 1: Exact Values from the Fundamental Theorem of Calculus

1. How can you test a conjecture that a given function g is an antiderivative of another given function f? Illustrate this by deciding which of the following equals $\int xe^{2x}\,dx$.

   $\frac{1}{4}x^2e^{2x} + C,\ \ \frac{1}{2}x^2e^{x^2} + C,\ \ \frac{1}{2}xe^{2x} - \frac{1}{4}e^{2x} + C.$

2. Put the idea of the problem above to work to establish the validity of the following formula:

$$\int \frac{dx}{(x^2+1)^{n+1}} = \frac{2n-1}{2n}\int \frac{dx}{(x^2+1)^n} + \frac{1}{2n}\frac{x}{(x^2+1)^n}.$$

   Use this result to find $\int \frac{dx}{(x^2+1)^3}$.

3. Which of the following integrals gives the length of the graph of $y = \sqrt{x}$ between $x = a$ and $x = b$ where $0 < a < b$?

   A. $\int_a^b \sqrt{x^2+x}\,dx$

   B. $\int_a^b \sqrt{x+\sqrt{x}}\,dx$

   C. $\int_a^b \sqrt{x+\frac{1}{2\sqrt{x}}}\,dx$

   D. $\int_a^b \sqrt{1+\frac{1}{2\sqrt{x}}}\,dx$

   E. $\int_a^b \sqrt{1+\frac{1}{4x}}\,dx$

4. A cable suspended from its two ends hangs in the shape of a "catenary", which is the graph of an equation of the form $y = a\cosh(x/a)$. ($\cosh x$ is $(e^x + e^{-x})/2$.)

   a. Sketch the graph of a catenary for the following values of a: 50, 100, 200. Note that the graph is symmetric about the y-axis, so that points on the graph with abscissas b and $-b$ have the same ordinate.

   b. Calculate the length of a cable suspended from two poles 100 meters apart. [Hint: Position the poles so that their x-coordinates are -50 and 50, respectively. Your answer will be in terms of a.]

   c. Evaluate the actual length of the cable for the following values of a: 50,100, 200.

   d. Compare the lengths calculated in part c to the length of the horizontal line segment joining the two points where the cable is attached. Note that the point $(0, a)$ lies on the graph of the catenary and so $a\cosh(50/a) - a$ measures the amount by which the cable "sags". How does the amount of "sag" compare with the length of the cable? Does this strike you as paradoxical?

5. Consider the triangle whose vertices in rectangular coordinates are $(0,0)$, $(1,0)$, and $(1,1)$.

   a. Find polar coordinate equations for each of the lines in which the triangle's sides lie.

   b. The area of this triangle is clearly $\frac{1}{2}$. Confirm this by setting up and computing a polar coordinate integral.

   c. The arc length of the side that runs from $(1,0)$ to $(1,1)$ is clearly 1. Confirm this by setting up and computing a polar coordinate integral.

6. [C] Define a curve by the parametric equations $\begin{cases} x = e^{-.1t}\cos t \\ y = e^{-.1t}\sin t \end{cases}$

   a. Sketch the curve over the interval $0 \le t \le 8\pi$.

   b. Find an equation for the tangent line to the curve at the point $(1,0)$.

   c. Find the arc length of the curve over the interval $0 \le t \le 8\pi$.

   d. Set up an integral for the area in the first quadrant bounded on the outside by this curve over the interval $0 \le t \le \frac{\pi}{2}$. Explain why $\pi/4$ is an upper bound for this integral. Compute the integral if you can;l approximate its value if you cannot.

7. a. *Guess* which of the following two integrals will be larger. Explain your reasoning.

$$\int_0^4 x\sqrt{16-x^2}\,dx \qquad \text{or} \qquad \int_0^4 \sqrt{16-x^2}\,dx.$$

   b. *Compute* which of the two integrals is actually larger. (In the case of the second integral, the exact value can be found without using antiderivatives.)

8. a. [CP] What is the probability that a needle of length L, dropped on a tongue-in-groove floor of plank width W, will land inside a single plank? Assume that $L < W$.

   b. A simulation is performed using a needle that is .8 as long as the width of a plank. In 2,000 drops, it lies inside a single plank 951 times. From this, estimate π.

9. A window is cut into the vertical side of a deep swimming pool. This window is rectangular in shape, has an area of $1\,\text{m}^2$, and the center of the window is exactly 3 meters beneath the surface of the water.

   a. Is this enough information for you to be able to compute the force of water on the window? If so, compute it, using the (metric) values of $9.8\,\text{m/s}^2$ for g, the acceleration due to gravity, and $1000\,\text{kg/m}^3$ for ρ, the density of water.

   b. Could you calculate the force on the window if it had any other shape? Justify your answer.

10. For each of the following functions f, find the value of c for which the line $x = c$ divides the area of the region under the graph of f from $x = a$ to $x = b$ in a $2:1$ ratio.

    a. $f(x) = (x-2)^2, \quad a = -1,\ b = 1$

    b. $f(x) = 1 - 1/x^2, \quad a = 1,\ b = 2$

11. Show that the region enclosed by the graph of the parabola $y = f(x) = \frac{2}{a^2}x - \frac{1}{a^3}x^2$, $a > 0$, and the x-axis has an area independent of a. How large is this area? What curve is determined by the vertices of all these parabolas?

12. Suppose that you had two spheres of the same kind of wood but of different radii. You make napkin rings out of them by drilling holes of different diameters through their middles. When you are done, you notice that the rings have exactly the same height h, as seen in the following sketch of vertical cross-sections through the centers of the rings.

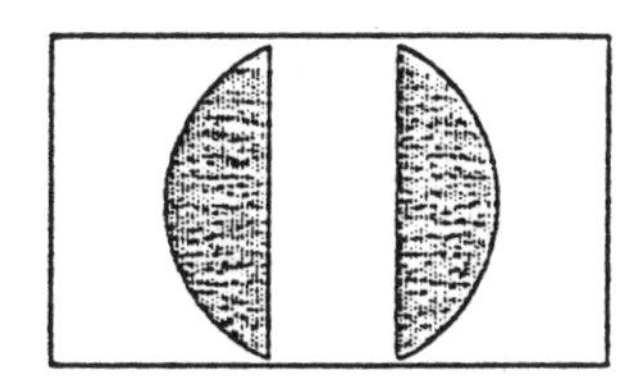

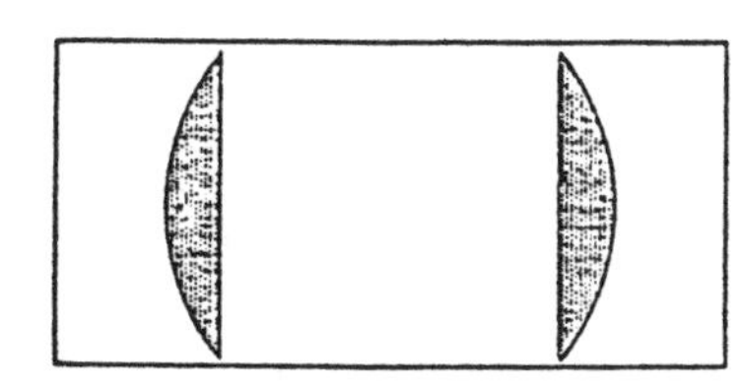

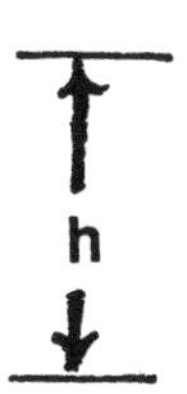

Napkin Ring A

Napkin Ring B

   a. Guess which ring has the greater volume, i.e. has more wood in it. Can you offer any evidence for your answer?

   b. Compute the volumes of the two rings to check your answer to part a.

13. Consider the region bounded below by the curve $y = x^m$ and above by the curve $y = x^n$, for $0 \leq x \leq 1$, and for $0 \leq n < m$.

   a. Sketch this region.

   b. Find general formulas depending on n and m for the x-coordinate and the y-coordinate of the center of mass of this region. Assume a constant density for the region.

   c. Can you find a combination of n and m such that the center of mass lies *outside* the region in question?

14. Calculate $\int e^{-x}x^5\,dx$. Hint: This problem may be more fun if you forego integration by parts and use a method which you will encounter again and again in your mathematics courses: The method of undetermined coefficients. Make an educated guess that

$$I(x) = e^{-x}[a_0 + a_1 x + \ldots + a_5 x^5],$$

and set $I'(x) = e^{-x}x^5$.

15. Consider a loaf of bread to be a hemisphere of radius r, sliced into n pieces of equal width by parallel planes.

    a. Ignoring the "bottom" horizontal crust, calculate the area of the crust of a typical slice.

    b. Which slice has the most crust?

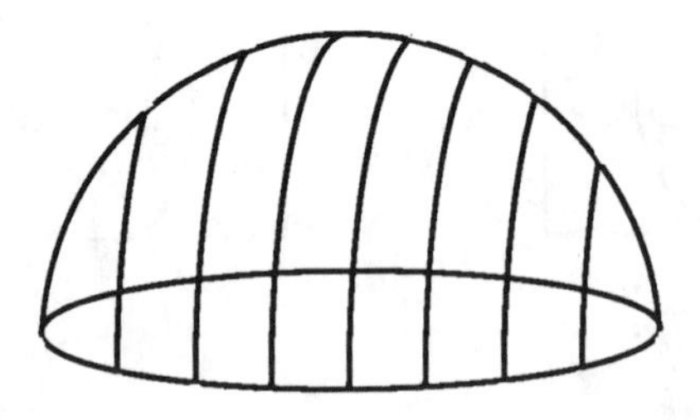

Section 2: Techniques of Integration

1. If the substitution $u = \frac{x}{2}$ is made, the integral $\displaystyle\int_2^4 \frac{1-(\frac{x}{2})^2}{x}\,dx =$

   A. $\displaystyle\int_1^2 \frac{1-u^2}{u}\,du$ B. $\displaystyle\int_1^2 \frac{1-u^2}{4u}\,du$ C. $\displaystyle\int_2^4 \frac{1-u^2}{u}\,du$

   D. $\displaystyle\int_2^4 \frac{1-u^2}{2u}\,du$ E. $\displaystyle\int_1^2 \frac{1-u^2}{2u}\,du$

2. a. Sketch a graph of the function f defined by the rule $y = f(x) = \dfrac{1}{x^2-4}$.

   b. Is the definite integral $\displaystyle\int_{-1}^1 \frac{dx}{x^2-4}$ negative or positive? Justify your answer.

   c. Compute the integral in part *b* by making a partial fraction decomposition of the integrand.

   d. A friend suggests that the integral can most easily be worked by making the substitution $x = 2\sec\theta$. What do you think?

3. If $\displaystyle\int f(x)\sin x\,dx = -f(x)\cos x + \int 3x^2\cos x\,dx$, then $f(x)$ could be

   A. $3x^2$ B. x^3 C. $-x^3$ D. $\sin x$ E. $\cos x$

4. Sketch the graph of the cardioid $r = a(1+\cos\theta)$ where r and θ are polar coordinates and a is a positive constant. Show that an element of arc length of the curve is

$$ds = a\sqrt{2(1+\cos\theta)}\,d\theta.$$

   Go on to find its total length. A table of integrals may prove useful.

5. a. Suppose you and two friends order a 12-inch diameter pizza and decide to split it up among you by sectioning it as shown in the accompanying sketch. If you are really hungry, which slice do you ask for? Justify your answer (although not in the presence of your friends).

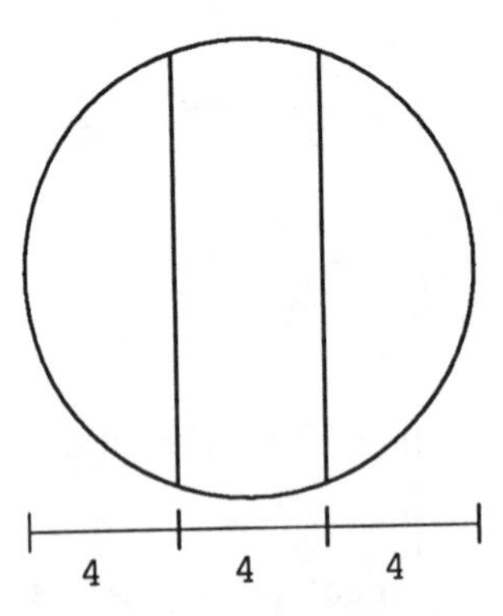

b. Suppose that, on another occasion, you order a bigger pizza and split it up as shown. Now which slice do you request?

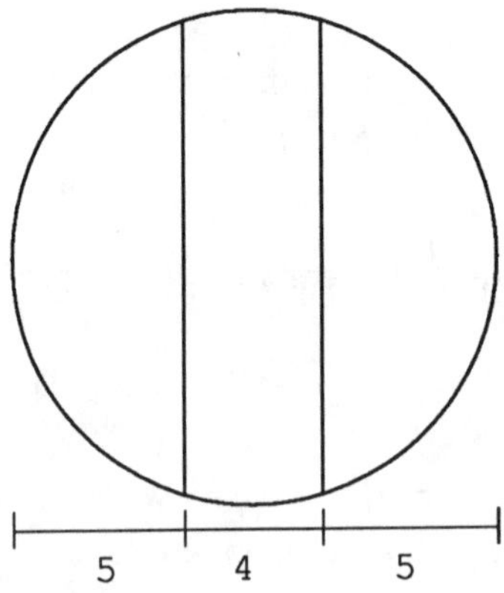

6. Here are some indefinite integrals for you to evaluate using a CAS. Observe the results to determine which, if any, could be done by a method of substitution. Can you make a general observation about the integrability, in closed form, of functions of the form $x^n \sin x^2$?

(i) $\int \sin x^2 \, dx$ (ii) $\int x \sin x^2 \, dx$

(iii) $\int x^2 \sin x^2 \, dx$ (iv) $\int x^3 \sin x^2 \, dx$

7. Consider this set of parametric equations: $\left\{ \begin{array}{l} x = \frac{1}{2}\cos t + \frac{1}{4}\cos 2t \\ y = \frac{1}{2}\sin t - \frac{1}{4}\sin 2t \end{array} \right\}$ for $0 \le t \le 2\pi$.

   a. Sketch the graph of this curve. The region contained by it is the smallest in area (among those without holes) in which a line segment of length 1 can be completely turned about in a rigid fashion.

   b. Set up an integral for the length of this curve. To simplify the integral, recall the trigonometric identity $\cos(\alpha + \beta) = \cos\alpha\cos\beta - \sin\alpha\sin\beta$.

   c. Use a CAS or a table of integrals to calculate the integral in part *b* or, if you wish, use Simpson's Rule to approximate it. Use the symmetry of the curve to find one third its length, then multiply by 3.

8. a. Show that $\int x^n e^x\,dx = x^n e^x - n\int x^{n-1}e^x\,dx$ where n is a non-zero number. This, by the way, is an example of a *reduction formula*.

   b. Derive a reduction formula for each of the following integrals

$$\int \ln^n x\,dx \qquad \text{and} \qquad \int x^n \sin x\,dx.$$

9. a. Now that you have had some experience with reduction formulas, find one for the indefinite integral $\int \sin^n x\,dx$. [Hint: Set $u = \sin^{n-1} x$, $dv = \sin x\,dx$.]

   b. Use the result in part a to show that $\int_0^{\pi/2} \sin^n x\,dx = \frac{n-1}{n}\int_0^{\pi/2} \sin^{n-2} x\,dx$. For n odd, go on to derive *Wallis' formula*:

$$\int_0^{\pi/2} \sin^n x\,dx = \frac{2\cdot 4\cdot 6\cdots(n-1)}{3\cdot 5\cdot 7\cdots n}.$$

10. Consider the function $f_n(x) = \cos(n \arccos x)$ for x in $[-1, 1]$, $n > 0$.

   a. Plot the function using your CAS for small integer values of n. Based upon what you see, make a conjecture about the value of $\int_{-1}^{1} f_n(x)\,dx$. For which values of n can you make a conjecture?

   b. Make a conjecture about the functions that were graphed in part a. Can you prove it?

   c. Use the substitution $u = \cos^{-1} x$ to change the integral in part a: $\int_{-1}^{1} f_n(x)\,dx = \int_0^{\pi} \cos nu \sin u\,du$. Then integrate by parts twice to get that

$$(1 - n^2) \int_0^{\pi} \cos nu \sin u\,du = (-1)^n + 1.$$

11. a. Using the picture below, give a geometric argument that the following identity is correct for $0 \leq a \leq 1$:

$$\int_0^a \arcsin x\,dx = a \arcsin a - \int_0^{\arcsin a} \sin y\,dy.$$

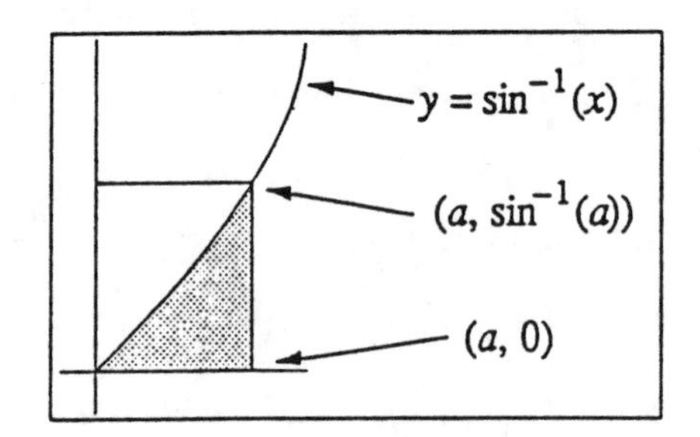

   b. Use integration by parts to compute $\int_0^a \arcsin x\,dx$.

12. Use a CAS to evaluate the following integrals. Check your answers and try to get a feel for what's going on by doing some by hand. Use integration by parts.

a. $\int_0^1 x(1-x)\,dx$ b. $\int_0^1 x^2(1-x)\,dx$

c. $\int_0^1 x(1-x)^2\,dx$ d. $\int_0^1 x^2(1-x)^2\,dx$

e. $\int_0^1 x(1-x)^3\,dx$ f. $\int_0^1 x^2(1-x)^3\,dx$

g. $\int_0^1 x^n\,dx$ h. $\int_0^1 (1-x)^n\,dx$

i. $\int_0^1 x(1-x)^n\,dx$ j. $\int_0^1 x^n(1-x)\,dx$

Can you find a pattern here? What happens to the last two integrals if the substitution $y = 1 - x$ is made? Does this shed some light on your attempt to find a pattern?

13. A projectile is shot from the top of a 100-meter high building with initial horizontal velocity of $x'(0) = 4\,\mathrm{m/sec}$ and initial velocity $y'(0) = 0\,\mathrm{m/sec}$. Ignore air friction to find the length of its trajectory until it hits the ground. Note: $y'' = -9.8\,\mathrm{m/sec}^2$.

Students sometimes fight about the answers that they find to questions. Analyze the following two arguments which hinge on different points.

14. Henry and Dorothy are having an argument as to the value of $\int \sec^2 x \tan x\,dx$. Henry makes the substitution $u = \sec x$ and gets the answer $\frac{1}{2}\sec^2 x$ whereas Dorothy makes the substitution $u = \tan x$ and gets the answer $\frac{1}{2}\tan^2 x$. Explain to Henry and Dorothy what is going on so that they will stop arguing.

15. Four calculus students disagree as to the value of the integral $\int_0^\pi \sin^8 x\,dx$. Jack says that it is equal to π, Joan says that it is equal to $\frac{35\pi}{128}$ while Ed claims that it is equal to $\frac{3\pi}{90} - 1$ and Lesley says that it is equal to $\frac{\pi}{2}$. One of them is right. Which one is it? Do not try to evaluate this integral yourself. Try instead to eliminate the three wrong answers.

16. Here is a "proof" that $1 = 0$. Your job is to find the bug. On the one hand,

$$\begin{aligned}\tfrac{2}{\pi}\int_0^\pi \cos^2\theta\,d\theta &= \tfrac{2}{\pi}\int_0^\pi \frac{1+\cos 2\theta}{2}\,d\theta \\ &= \tfrac{2}{\pi}\left[\tfrac{1}{2}\theta + \tfrac{1}{4}\sin 2\theta\right]\Big|_0^\pi \\ &= \tfrac{2}{\pi}\left(\tfrac{1}{2}\pi\right) = 1.\end{aligned}$$

On the other hand, let $u = \sin\theta$. Then $\cos\theta = \sqrt{1-\sin^2\theta} = \sqrt{1-u^2}$ and $du = \cos\theta\,d\theta$, so

$$\begin{aligned}\tfrac{2}{\pi}\int_0^\pi \cos^2\theta\,d\theta &= \tfrac{2}{\pi}\int_0^\pi (\cos\theta)(\cos\theta)\,d\theta \\ &= \tfrac{2}{\pi}\int_0^0 \sqrt{1-u^2}\,du = 0.\end{aligned}$$

17. What is the probability that a point in the unit square, with vertices $A = (0,0)$, $B = (1,0)$, $C = (1,1)$, and $D = (0,1)$ has a distance greater than or equal to one from the point $(-\frac{1}{2}, -\frac{1}{2})$?

18. Given two real numbers whose squares have a sum less than or equal to 9, what is the probability that their sum is greater than or equal to 2?

19. Find the second-degree polynomial $P(x)$ such that $P(0) = 1$, $P'(0) = 0$, and $\displaystyle\int \frac{P(x)}{x^3(x-1)^2}\,dx$ is a rational function.

Section 3: Approximation Techniques and Error Analysis

1. [C] For each of the following functions f, obtain a graph of the function for $x \in [a, b]$ and try to calculate $\int_a^b f(x)\,dx$. Do not be too discouraged if your attempts to integrate do not pan out. What is the difficulty here? Is there any way around it?

   a. $f(x) = e^x/x; \quad a = 1, b = 2.$

   b. $f(x) = \sqrt{\sin x}; \quad a = 0, b = \pi.$

   c. $f(x) = \dfrac{\sin x}{x}; \quad a = \frac{\pi}{4}, b = \frac{3\pi}{4}.$

   d. $f(x) = \dfrac{\ln x}{1 + x^2}; \quad a = 1, b = 4.$

2. [C] a. Using the Trapezoidal Rule, approximate the area under the curve $y = x \sin x$ from $x = 0$ to $x = \pi$, taking $n = 4$, $n = 8$, $n = 20$, and $n = 50$.

   b. Repeat this exercise, using Simpson's Rule.

   c. Calculate the value of the definite integral $\int_0^\pi x \sin x\,dx$ and compare it to the answers obtained in parts a and *b*. What does this exercise suggest about the relative accuracy of the trapezoidal and Simpson's rules?

3. [C] Since the function f defined by the rule $f(x) = x \sin x$ has been introduced, let's use it for one more calculation.

   a. Use both the Trapezoidal and Simpson's Rules to estimate the volume of the solid of revolution obtained by rotating the graph of f, from $x = 0$ to $x = \pi$, about the x-axis. As before, use $n = 20$, 50, 100 and give your answer in the form of a table:

n	Trap.	Simp.
20		
50		
100		

   b. Calculate the true volume approximated in part a and compare the accuracy attained by the two methods. Can you suggest a rule of thumb which can be used to tell you when to stop increasing the number of subdivisions in order to achieve good approximations, given the machinery at your disposal?

4. [C]

   a. Show that $\int_0^1 x^n \sin\left(\frac{\pi x}{2}\right)\, dx > 0$ for all $n \geq 0$.

   b. Show next that $\int_0^1 x^n \sin\left(\frac{\pi x}{2}\right)\, dx < \int_0^1 x^n\, dx = \frac{1}{n+1}$.

   c. Use a graphing calculator or a CAS to graph $x^n \sin\left(\frac{\pi x}{2}\right)$ for $x \in [0,1]$ and for $n = 0$, 1, 2, 3, and 4.

   d. Obtain approximations for $\int_0^1 x^n \sin\left(\frac{\pi x}{2}\right)\, dx$ for $n = 0, 1, 2, 3, 4$. Are your answers reasonable in light of parts a and *b*? Speculate about the answer for $n = 99$.

5. [C]

   a. Use the Trapezoidal Rule to approximate the distance traveled in the first 3 seconds by a particle with velocity given by $x'(t) = (t^2+3t)/(t+1)$. Use the following values of n : 4, 10, 20, 50, 100.

   b. Repeat the exercise using Simpson's Rule and the same values of n.

   c. Calculate $\int_0^3 x'(t)\, dt$ and comment on the apparent accuracy of the two rules.

6. Let $I = \int_1^3 \frac{dx}{x}$.

   a. Compute I.

   b. For 4 equal subdivisions of [1,3], approximate I by the Trapezoidal Rule. According to the error formula for the Trapezoidal Rule, what is the maximum possible error for this approximation? Compare it to the actual error. What is the least number of subdivisions for which the error formula guarantees accuracy to within .001?

   c. Repeat part *b* for Simpson's Rule.

7. a. Using two equal subdivisions of $[0,1]$ and the Trapezoidal Rule, approximate $\int_0^1 x^2\, dx = \frac{1}{3}$. Why is your answer too big? Compare the actual error made with the maximum error possible for this method.

   b. Re-do part a for $\int_0^1 x^4\, dx = \frac{1}{5}$, this time using Simpson's Rule.

8. [C] Consider the polar curve $r = 2\cos 3\theta$.

   a. Sketch this curve. Superimpose on it a sketch of the circle $r = 1$.

   b. Find the area of one lobe of the region outside the circle $r = 1$ but inside the curve $r = 2\cos 3\theta$.

   c. Set up the integral for the length of that part of the curve $r = 2\cos 3\theta$ which lies between the points used to calculate the area of part *b*.

   d. Use Simpson's Rule with $n = 10, 20, 50, 100$ to approximate the arc length in part *c*.

9. [C]

   a. Verify that $\displaystyle\int_0^1 4\sqrt{1-x^2}\,dx = \pi$.

   b. Use Simpson's Rule to approximate the value of the definite integral in part a, i.e. π, by choosing $n = 6, 10, 20, 50$.

   c. Confirm that $\displaystyle\int_0^1 \frac{4}{1+x^2}\,dx = \pi$.

   d. Use Simpson's Rule to approximate this definite integral by choosing $n = 6, 10, 20, 50$.

   e. Which method of approximating π in this problem converges faster? How do you explain this?

10. The Simpsons wish to install a liver-shaped swimming pool in their garden. To estimate the cost of the installation, it is necessary to know the pool's surface area. The family makes a sketch (given below) of the pool it wants, with width measurements in feet taken every three feet. Estimate the surface area of the pool.

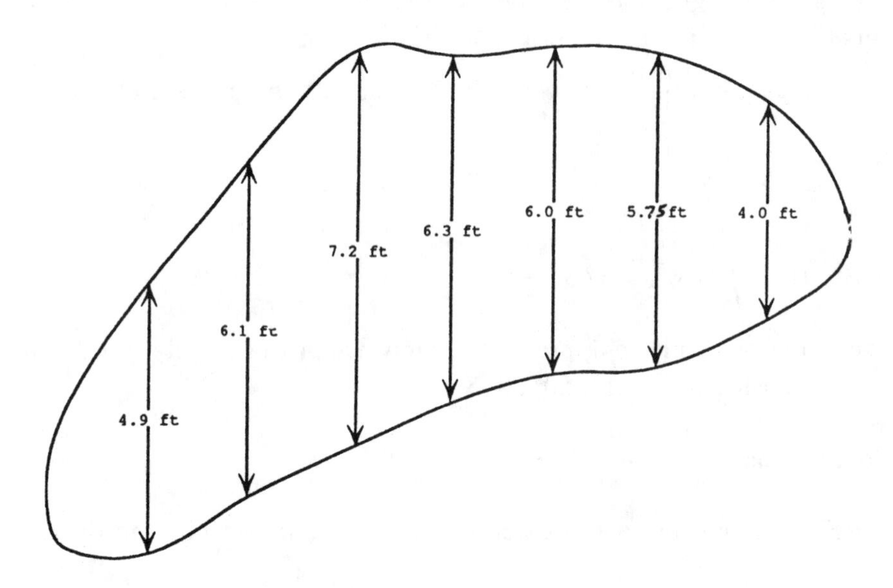

11. A plane has just crashed six minutes after takeoff. There may be survivors, but you must locate the plane quickly without an extensive search. It was traveling due west and the pilot radioed its speed every minute from takeoff to the time it crashed. These data are given in the table below. Approximately how far west of the airport did the plane crash?

Minutes	0	1	2	3	4	5	6
Speed (MPH)	90	110	125	135	120	100	70

12. [C] a. Use Simpson's rule, with 10 subdivisions, to estimate $\int_{-1}^{1} \frac{1}{\sqrt{2\pi}} e^{x^2/2}\, dx$.

    b. Where is such an integral encountered? [Hint: Look in the index of any statistics book under the entry "Normal Distribution."]

    c. After you have read about statisticians' use of the integral approximated in part a, explain the significance of your answer.

13. [C] There is evidence to support psychologists' claim that the intelligence quotient (IQ) of a population is normally distributed with a median of 100 and a standard deviation of 16. What this means is that the density function for IQ is given by the rule

$$f(x) = \frac{1}{16\sqrt{2\pi}} e^{-\frac{1}{2}\left(\frac{x-100}{16}\right)^2},$$

and that the proportion of people whose IQ's fall between α and β is $\int_{\alpha}^{\beta} f(x)\,dx$. Use Simpson's Rule to estimate the proportion of all people whose IQ's fall between 90 and 110.

14. Let $P = (0,0)$, $Q = (X,Y)$ and $R = (2X,Z)$ be three points where X, Y, and Z are positive. A quadratic curve passes through points P and Q and another quadratic curve passes through points Q and R. Assume that the two curves have slopes m and n respectively at Q.

   a. Calculate the area of the region bounded by the composite quadratic curves, the x-axis, and the vertical line $x = 2X$.

   Hint: The curve passing through P and Q is defined by a rule $f(x) = ax^2 + bx$. Why? Determine the values of a and b. Next, shift the y-axis to the right so that the coordinates of Q are $(0,Y)$ and the coordinates of R are (X,Z). The curve passing through Q and R is defined by a rule $g(x) = cx^2 + dx + e$. Find the values of the coefficients c, d, and e. The area of the region is given by $\int_0^X f(x)\,dx + \int_0^X g(x)\,dx$.

   b. For what conditions on the slopes m and n will Simpson's Rule give this area exactly? Explain your answer.

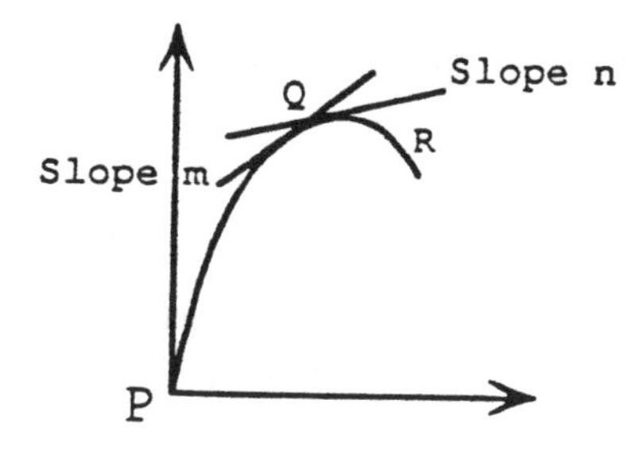

15. [C] The graph below gives the rate of production of oil in thousands of barrels per day by a large refinery. Use Simpson's Rule to estimate the total amount of oil produced through the end of the fourth day.

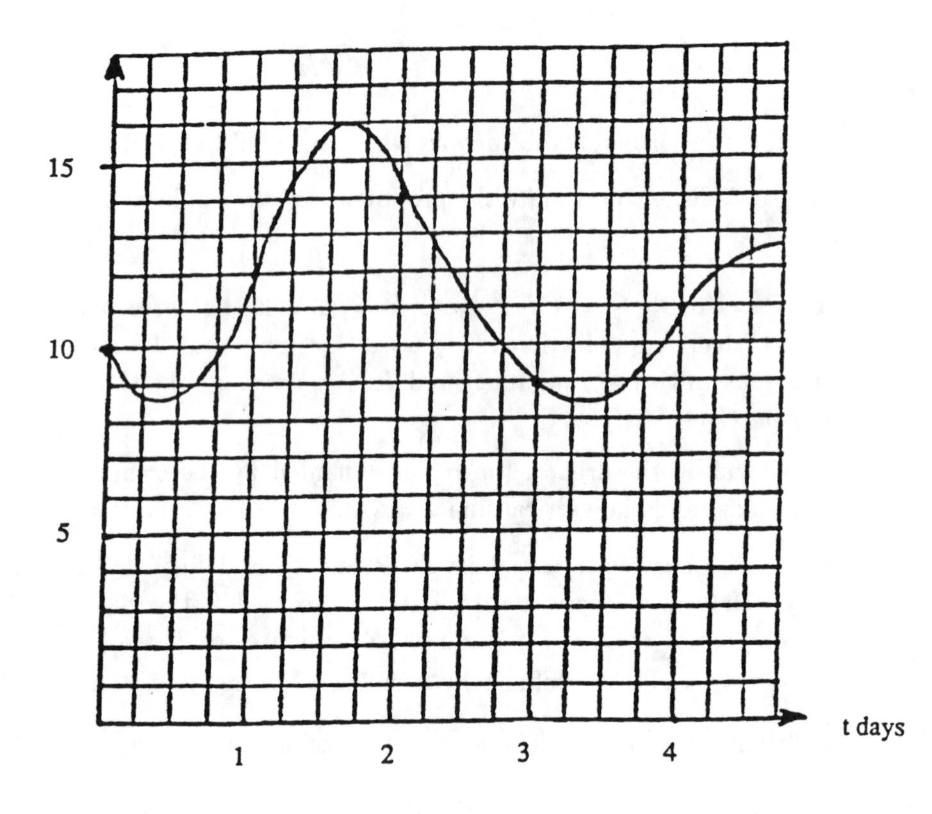

16. Let $f(x) = e^{\sin x}$ and $I = \int_{-50\pi}^{150\pi} f(x)\,dx$.

   a. Show that $I = 100 \int_0^{2\pi} f(x)\,dx$.

   b. Next show that $f^{(4)}(x) = \left(\cos^4 x - 6\cos^2 x \sin x - 4\cos^2 x + 3\sin^2 x + \sin x\right) e^{\sin x}$. It follows that $|f^{(4)}(x)| < 15e$.

   c. If you were to approximate the integral in part a by using Simpson's Rule, with n subintervals, then the error made, S_n, would be bounded as follows:

   $$|S_n| \leq \frac{(15e)(2\pi)^5}{180n^4}.$$

   What is the smallest even integer for which $|S_n|$ is less than .00001?

   d. With the value of n found in part c, approximate I, using Simpson's Rule. What is the accuracy of this approximation?

17. [C] What is the arc length (circumference) of the ellipse defined by the equation

$$\frac{x^2}{25}+\frac{y^2}{9}=1?$$

To answer this question, follow these steps:

a. Show that the ellipse can be parametrized by the equations

$$x=5\cos\theta, \quad y=3\sin\theta, \quad \theta\in[0,2\pi].$$

b. Show next that the arc length is given by $\int_0^{2\pi}\sqrt{16\sin^2\theta+9}\,d\theta$.

c. Use Simpson's Rule for n subintervals of $[0,2\pi]$ with $n=4,6,8,\ldots$. Stop when you seem to have achieved four decimal places of accuracy.

18. Let $S(f,I,n)$ be the Simpson Rule approximation, using n subdivisions, of $\int_a^b f(x)\,dx$, where $I=[a,b]$.

a. Show that if f is a polynomial of degree 2 or less, then $S(f,I,n)=\int_a^b f(x)\,dx$ for all even n.

b. Show that Simpson's Rule is linear in the sense that

$$S(f+g,I,n)=S(f,I,n)+S(g,I,n)$$

19. How can you approximate $\ln 2$ by Simpson's Rule?

Chapter VII: Sequences and Series of Numbers

Section 1: Sequences of Numbers

1. Explain how the geometric figures below can be used to produce the formula

$$1 + 2 + \cdots + n = \frac{n(n+1)}{2}.$$

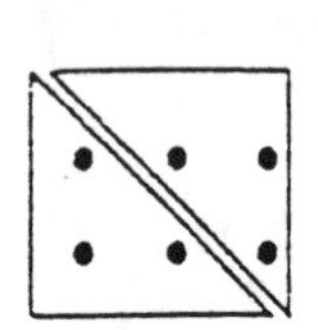

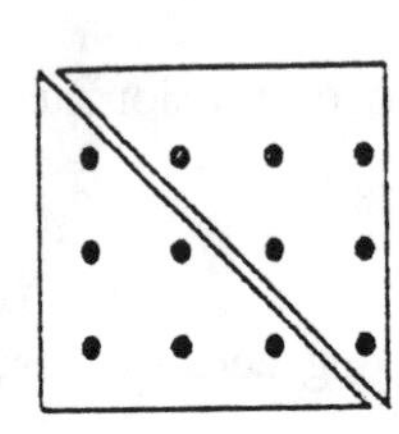

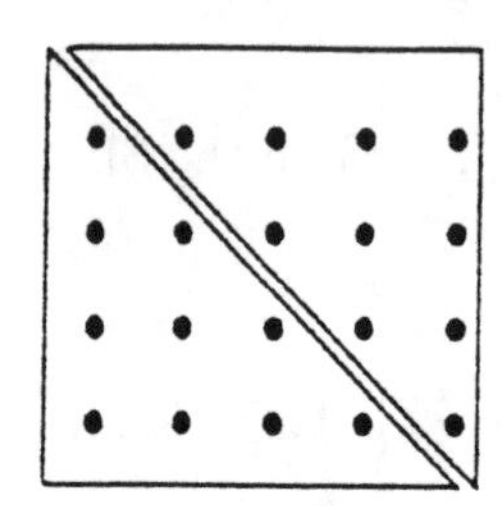

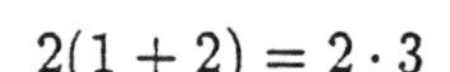

$2(1+2) = 2 \cdot 3$ $\quad$ $2(1+2+3) = 3 \cdot 4$ $\quad$ $2(1+2+3+4) = 4 \cdot 5$

2. On New Year's Day, you resolve to save one penny on the first day, two pennies on the second, three on the third, and so on. How many days will it take you to accumulate \$200?

3. Guess at the limits of these sequences (and guess whether the limits exist). You may use a calculator.

   a. $n^{1/n}$ $\quad$ b. $n \sin\left(\frac{1}{n}\right)$ $\quad$ c. $\left(1 + \frac{1}{n}\right)^n$ $\quad$ d. $\frac{\ln n}{n}$

   e. $\frac{n!}{n^n}$ $\quad$ f. $\frac{2^n}{n!}$ $\quad$ g. $\frac{n^2}{2^n}$

4. a. Find $\lim_{n\to\infty} a_n$ if $\{a_n\}$ converges, or explain why the sequence diverges, in each of:

(i) $a_n = \dfrac{4n^2 + 7n - 1}{n^2 + 5}$ (ii) $a_n = \dfrac{3n^5 + 13n^2 - 8n}{14n^9 - 7n^6 - 13}$ (iii) $a_n = \dfrac{2n^4 - 7n}{n^2 + 3}$

b. Based on the three problems above, find the general pattern. That is, if

$$a_n = \frac{cn^p + \text{(lower degree terms)}}{dn^q + \text{(lower degree terms)}}$$

where c and d are nonzero, under what conditions does $\{a_n\}$ converge and what is its limit? (If the three problems above are not enough for the pattern to be clear, make up more of the same type.)

5. Let $\{a_n\}$ be the sequence where

$$a_n = \begin{cases} n^2/(n^2 - 20) & \text{if } n \text{ is a multiple of 3,} \\ n/(n+1) & \text{if } n \text{ is one more than a multiple of 3,} \\ \sqrt{n}/\sqrt{n+3} & \text{if } n \text{ is two more than a multiple of 3.} \end{cases}$$

Evaluate $\lim_{n\to\infty} a_n$, or show that it doesn't exist.

6. Let $\{a_n\}$ be the sequence where

$$a_n = \begin{cases} \dfrac{2n}{n-1} & \text{if n is a multiple of 3,} \\ \dfrac{3n}{n+1} & \text{if n is one more than a multiple of 3,} \\ \dfrac{n^2}{n^2+10} & \text{if n is two more than a multiple of 3.} \end{cases}$$

Evaluate $\lim_{n\to\infty} a_n$, or show that it doesn't exist.

7. Find the limit of each of the following sequences with the aid of L'Hôpital's Rule. That is, replace n either by x (and let $x \to \infty$) or by $1/x$ (and let $x \to 0^+$).

a. $\dfrac{\ln n}{n}$ b. $n \sin\left(\dfrac{1}{n}\right)$ c. $\dfrac{n^2}{e^n}$

d. $n^{1/n}$ e. $\left(1 + \dfrac{1}{n}\right)^n$ f. $\dfrac{\ln\left(1 + \frac{2}{n}\right)}{\sin\left(\frac{3}{n}\right)}$

8. $\lim_{n\to\infty} \ln(\ln n) = \infty$, which means that, given any number M, we can find n so large that $\ln(\ln n) > M$ and the same can be said for all later terms of the sequence. How big must n be for $\ln(\ln n) > 5$?

9. Which of the following are always true, and which are at least sometimes false? Explain. ($\{a_n\}$ denotes an arbitrary sequence of real numbers.)

   a. If $a_n > 0$ for all n and $a_n \to L$, then $L > 0$.

   b. If $a_n \geq 0$ for all n and $a_n \to L$, then $L \geq 0$.

   c. If $\{a_n\}$ is bounded, then it converges.

   d. If $\{a_n\}$ is not bounded, then it diverges.

   e. If $\{a_n\}$ is decreasing, then it converges.

   f. If $\{a_n\}$ is decreasing and $a_n > 0$ for all n, then it converges.

   g. If $\{a_n\}$ is neither increasing nor decreasing, then it diverges.

   h. If $a_n > 0$ for all n and $(-1)^n a_n \to L$, then $L = 0$.

10. a. Show that $\left\{\frac{4^n}{n!}\right\}$ is eventually decreasing (starting at what term?), and therefore has a limit (why?).

    b. Find that limit.

11. In a course you are taking, your grade is based on the average of your exam scores and you are allowed to take the exams as often as you like. If each of your exam scores is no worse than the average of your previous scores, show that your average score approaches a limit. (Each exam has a maximum score of 100%.)

12. Define a sequence $\{a_n\}$ by $a_1 = 2$ and $a_{n+1} = \dfrac{1}{2}\left(a_n + \dfrac{2}{a_n}\right)$ for $n \geq 1$.

    a. Assuming the sequence converges, find its limit.

    b. Use a calculator to compute several terms. Does the sequence appear to converge to your answer in part a? Does the value of a_1 affect the limit?

13. Answer the same questions as in the preceding problem if the definition of the sequence is changed to: $a_{n+1} = \dfrac{1}{2}\left(a_n + \dfrac{B}{a_n}\right)$ for $n \geq 1$ where B is some other positive number.

14. Define a sequence $\{a_n\}$ by

$$a_1 = 1, \quad a_2 = 1 \cdot \tfrac{3}{4}, \quad a_3 = 1 \cdot \tfrac{3}{4} \cdot \tfrac{8}{9}, \quad a_4 = 1 \cdot \tfrac{3}{4} \cdot \tfrac{8}{9} \cdot \tfrac{15}{16}, \quad \ldots.$$

    In general $a_n = \left(1 - \dfrac{1}{n^2}\right) a_{n-1}$ for $n > 1$.

    a. Show that $\{a_n\}$ has a limit.

    b. Guess at the limit. You may use a calculator.

15. [C] Define a sequence $\{a_n\}$ by $a_{n+1} = \sqrt{a_n}$ for $n \geq 1$, where a_1 is any positive number.

   a. Assuming the sequence has a limit L, show that $L = 0$ or $L = 1$.

   b. By repeatedly pressing the square root key on your calculator (after entering some positive number a_1), confirm that $\{a_n\}$ approaches a limit. Which one (0 or 1)?

   c. On the graph below of $y = \sqrt{x}$ and $y = x$, use a "web diagram" to explain why $\{a_n\}$ converges. Choose at least two values of a_1, one larger than 1 and one less than 1.

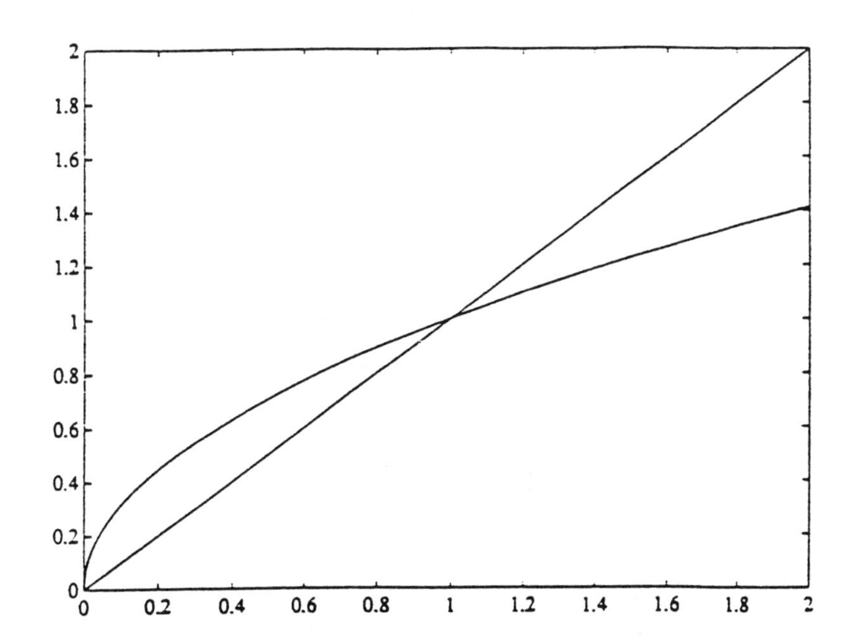

16. [C] Define a sequence $\{a_n\}$ by $a_{n+1} = 3a_n^2$ for $n \geq 1$, where a_1 is any number satisfying $0 < a_1 < 1$.

    a. Assuming the sequence has a limit L, show that $L = 0$ or $L = \frac{1}{3}$.

    b. With the aid of a calculator, compute several terms of the sequence. Does it always converge? To which of the two values of L? (Try at least two sequences, one where $a_1 = 0.2$ and one where $a_1 = 0.4$.)

    c. On the graph below of $y = 3x^2$ and $y = x$, use a web diagram to investigate more carefully those values of a_1 for which $\lim_{n\to\infty} a_n$ exists.

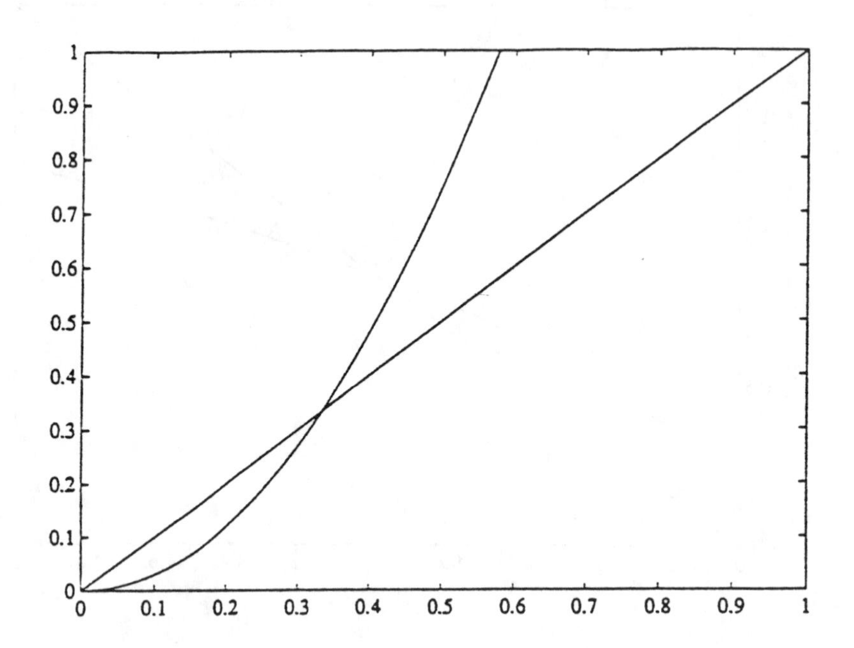

17. Define a sequence $\{a_n\}$ by $a_{n+1} = B^{a_n}$ for $n \geq 1$, where a_1 is any positive number and $B > 1$.

    a. Assuming the sequence has a positive limit L, show that $B = L^{1/L}$.

    b. On the graphs below of $y = B^x$ for $B = 1.1, 1.3, 1.5$ respectively, each superimposed on the graph of $y = x$, use a web diagram to discover the values of B for which the sequence converges for at least some values of a_1, the values of a_1 for which these sequences converge, and the value of the limit. (Approximate answers will suffice.)

    c. Find (exactly) the largest value of B for which the sequence converges for at least some values of a_1. Hint: For this value of B, $y = B^x$ intersects $y = x$ at exactly one point, so the two graphs are tangent there.

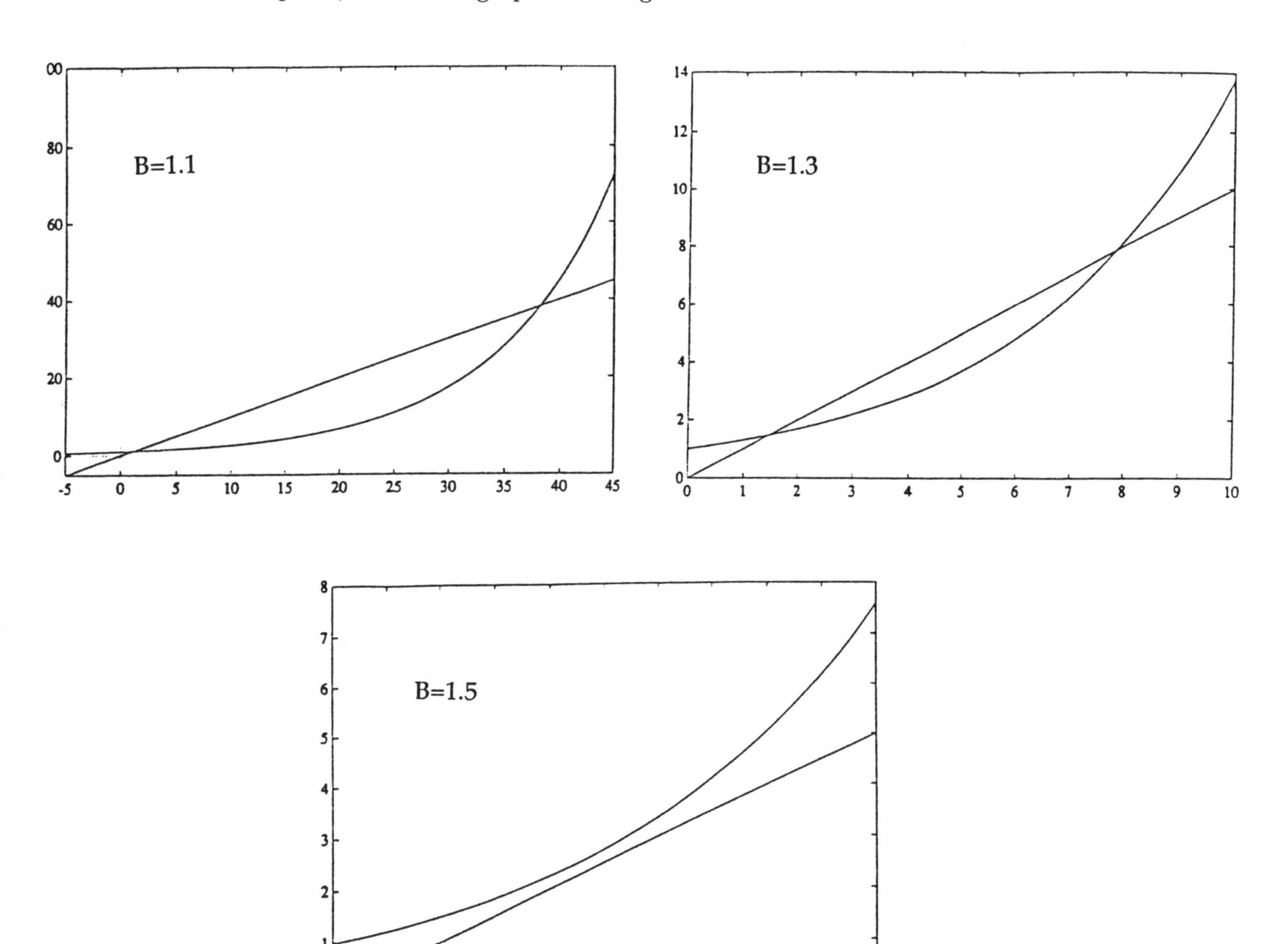

18. The Fibonacci sequence $\{F_n\}$ is defined by $F_0 = 0, \quad F_1 = 1$, and

$$F_{n+1} = F_n + F_{n-1} \qquad \text{for } n \geq 1.$$

The first few terms are 0, 1, 1, 2, 3, 5, 8, 13, 21, 34, 55, 89, ...

a. Assuming $\{F_{n+1}/F_n\}$ converges to a limit L, show that L is a root of $x^2 = x + 1$. Which root is it?

b. If x is either of the roots of $x^2 = x + 1$, use induction to prove that

$$x^n = xF_n + F_{n-1} \quad \text{for } n \geq 1.$$

c. Let y and z denote the two roots of $x^2 = x + 1$; that is, $y = \frac{1}{2}(1 + \sqrt{5})$ and $z = \frac{1}{2}(1 - \sqrt{5})$, so that $y^n = yF_n + F_{n-1}$ and $z^n = zF_n + F_{n-1}$. Subtract these equations to show that

$$F_n = \frac{1}{\sqrt{5}}\left[\left(\frac{1+\sqrt{5}}{2}\right)^n - \left(\frac{1-\sqrt{5}}{2}\right)^n\right].$$

d. Use part c to show that $\{F_{n+1}/F_n\}$ does indeed converge.

Section 2: Series of Numbers. Geometric Series

1. A series $\sum_{k=1}^{\infty} a_k$ has partial sums s_n given by $s_n = 5 - \frac{3}{n}$.

   a. Does $\sum_{k=1}^{\infty} a_k$ converge? To what?

   b. Find $\lim_{k\to\infty} a_k$.

   c. Find $\sum_{k=1}^{100} a_k$.

2. Does $\sum_{k=0}^{\infty}(-1)^k$ converge or diverge? In your explanation, use the definition of these terms.

3. Does $1-\frac{1}{2}+\frac{1}{2}-\frac{1}{3}+\frac{1}{3}+\cdots$ converge or diverge? In your explanation, use the definitions of these terms.

4. Find the sum of the series $\sum_{k=1}^{\infty}\left(\frac{1}{\sqrt{k+1}} - \frac{1}{\sqrt{k+3}}\right)$.

5. Find the sum of the series $\sum_{k=1}^{\infty}\frac{1}{k^2+2k}$. Hint: It can be made to telescope.

6. Do the following telescoping series converge?

   a. $\sum_{k=1}^{\infty}\left(\arctan(k+1) - \arctan k\right)$

   b. $\sum_{k=1}^{\infty}\left(\ln(k+1) - \ln k\right)$

7. For which values of x does each of the series below converge and what is its sum?

   a. $\sum_{k=0}^{\infty}(\sin x)^k$ b. $\sum_{k=1}^{\infty} x^{2k}$ c. $\sum_{k=2}^{\infty}\frac{1}{x^k}$ d. $\sum_{k=0}^{\infty}\frac{1}{(1-x)^k}$

8. a. Compute $\lim_{x\to 1^-} \sum_{k=0}^{\infty} (-1)^k x^k$.

   b. Does this mean that $\sum_{k=0}^{\infty} (-1)^k$ converges? Explain.

9. A rubber ball rebounds to two-thirds the height from which it falls. If it is dropped from a height of 4 feet and is allowed to continue bouncing indefinitely, what is the total distance it travels?

10. Two cyclists, 60 miles apart, approach each other, each pedaling at 10 MPH. A fly starts at one cyclist and flies back and forth between the cyclists at 15 MPH. When the cyclists come together, how far has the fly flown?

11. A hare runs a two-mile race with a tortoise who has a one-mile head start. Even though the hare runs ten times as fast as the tortoise, he seems unable to catch up for the following reason: When the hare gets to the spot where the tortoise started, the tortoise is one-tenth mile farther on; when the hare gets to this one-and-one-tenth-mile point, the tortoise is farther on yet; and so on. Explain this seeming paradox.

12. Evaluate the definite integral $\int_0^1 f(x)\,dx$, where f is the function whose graph is below.

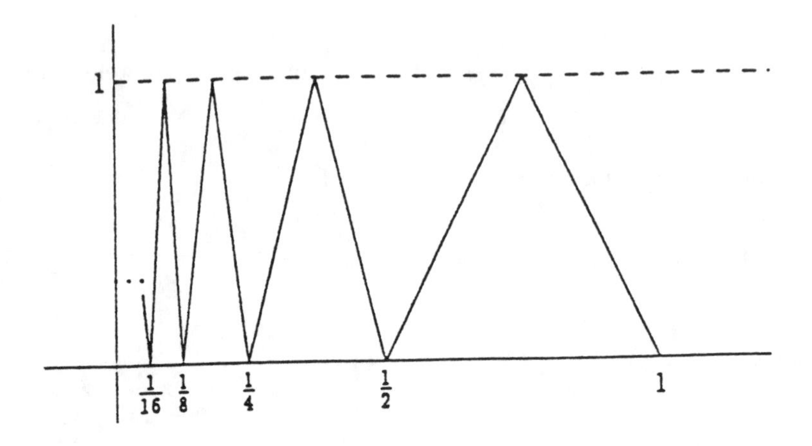

13. [C] Define the sequence $\{a_n\}$ by $a_n = \frac{1}{2}(a_{n-1} + a_{n-2})$ for $n \geq 2$ where a_0 and a_1 can have any value.

   a. Choose a_0 and a_1 arbitrarily and use a calculator to make a conjecture about $\lim_{n\to\infty} a_n$. Suggestion: Try $a_0 = 0$ and $a_1 = 1$; $a_0 = 1$ and $a_1 = 0$; then try a few values of your own.

   b. Evaluate $\lim_{n\to\infty} a_n$ analytically in terms of a_0 and a_1. Hint: Express $a_k - a_{k-1}$ in terms of a_1 and a_0, and then add these differences for $k = 1, \ldots, n$.

14. Construct a sequence of figures $T_0, T_1, T_2, \ldots, T_n, \ldots$ (see below) in the following way. T_0 is an equilateral triangle of side length one; T_1 is obtained from T_0 by replacing the middle third of each edge of T_0 by an outward equilateral triangle (of side length $\frac{1}{3}$); T_2 is obtained from T_1 by replacing the middle third of each edge of T_1 by an outward triangle (of side length $\frac{1}{9}$); and so on. (The limit of these figures is a strange curve called the *Koch Snowflake*.)

   a. Find S_n, the number of edges in the perimeter of T_n, and L_n, the length of each edge.

   b. Find P_n, the perimeter of T_n; find $\lim_{n\to\infty} P_n$.

   c. How much area is added in going from T_n to T_{n+1}?

   d. Find A_n, the area of T_n; find $\lim_{n\to\infty} A_n$. Does this answer contradict your answer in part *b*?

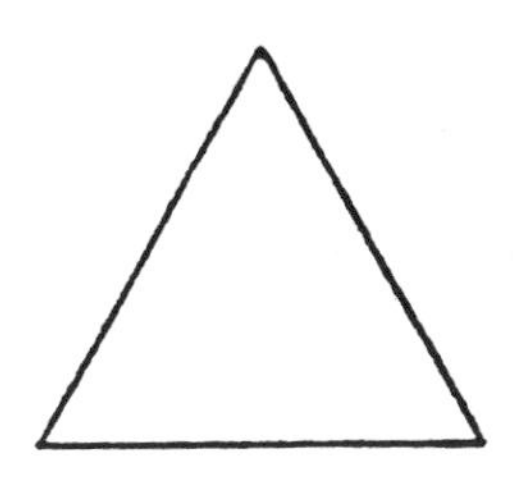

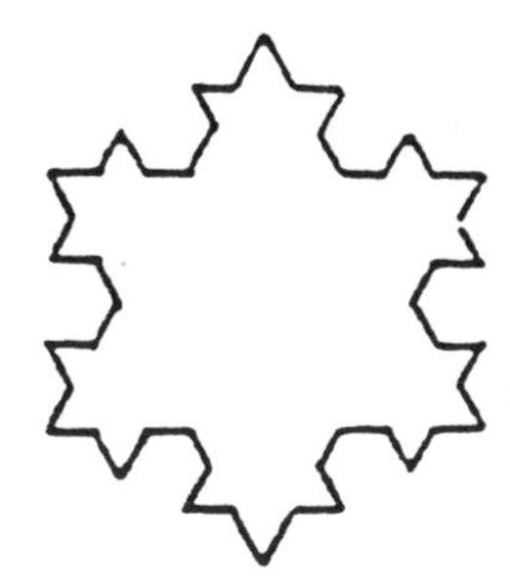

15. The curve $y = e^{-cx} \sin x$, $(c > 0,\ x \geq 0)$ when rotated about the x-axis, generates a solid that resembles a string of beads along the axis. The beads decrease in volume as x-increases.

    a. Using a CAS, compute the volume V_n of the nth bead. If you do not have access to such software or are having difficulty getting the answer to simplify, just write down the integral that must be evaluated and use the following answer in the remaining parts: $V_n = \dfrac{\pi e^{-2c\pi n}(e^{2c\pi} - 1)}{c(c^2 + 1)}$.

    b. Find the ratio V_{n+1}/V_n and show that it is independent of n.

    c. Find the total volume of the beads. Rather than computing an improper integral, express the answer in terms of V_1 and the ratio in part b.

16. Which of the following are always true and which are at least sometimes false? Explain.

    a. If $\sum a_n$ converges, then $a_n \to 0$.

    b. If $a_n \to 0$, then $\sum a_n$ converges.

    c. $\displaystyle\sum_{k=0}^{\infty} x^k = \frac{1}{1-x}$ whenever $x \neq 1$.

    d. If $\sum a_n$ and $\sum b_n$ diverge, then $\sum(a_n + b_n)$ diverges.

    e. If $\sum a_n$ converges and $\sum b_n$ diverges, then $\sum(a_n + b_n)$ diverges.

    f. If $\{a_n\}$ is monotonic and bounded, then $\sum a_n$ converges.

    g. If the partial sums of $\sum a_n$ are bounded, then $\sum a_n$ converges.

    h. If $a_n \geq 0$ and the partial sums of $\sum a_n$ are bounded, then $\sum a_n$ converges.

    i. If $\sum a_n$ and $\sum b_n$ converge, then $\sum(a_n b_n) = \left(\sum a_n\right)\left(\sum b_n\right)$.

Section 3: Convergence Tests: Positive Series

1. In the integral test, f is a decreasing positive function and $a_k = f(k)$. Which of the sums, $\sum_{k=1}^{n-1} a_k$ or $\sum_{k=2}^{n} a_k$, is larger than $\int_1^n f(x)\,dx$ and which is smaller? Illustrate your answer with a drawing.

2. Let $s_n = \sum_{k=1}^{n} \frac{1}{k}$, the nth partial sum of the harmonic series.

   a. Use drawings as in the integral test to show that $\ln(n+1) < s_n < 1 + \ln n$.

   b. Use part a to show that more than 8000 terms are needed to make $s_n > 10$.

   c. Use part a to show that $\lim_{n\to\infty} \frac{s_n}{\ln n} = 1$.

   d. Show that the sequence $\{s_n - \ln n\}$ is decreasing. Hint: Show that
   $$\frac{1}{n+1} < \int_n^{n+1} \frac{1}{x}\,dx = \ln(n+1) - \ln n.$$

   e. Use part d to show that the sequence $\{s_n - \ln n\}$ converges. The limit is called *Euler's constant*, and is denoted γ.

3. Show that it is possible to stack a pile of identical books face down so that the top book is as far as you like to one side of the bottom book (see below). Hint: Work from the top down. Find the center of mass of the top n books, and place the $(n+1)$st book below this stack so that the center of mass of the stack is just at the edge of this book.

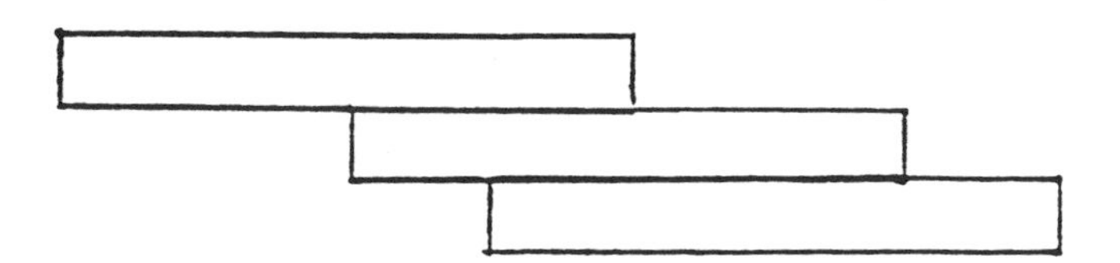

4. If $\sum a_n$ converges and $a_n \geq 0$, show that $\sum \sin(a_n)$ converges.

5. Let p_n denote the nth prime number. Use the fact that $\frac{p_n}{n \ln n} \longrightarrow 1$ to show that $\frac{1}{2} + \frac{1}{3} + \frac{1}{5} + \frac{1}{7} + \frac{1}{11} + \frac{1}{13} + \cdots$ diverges to ∞.

6. The result of the preceding problem can be interpreted to mean that there are a lot of primes; so many, in fact, that the sum of their reciprocals diverges. By the same criterion, are there a lot of integers that can be written in base 10 notation without using the digit 0? More precisely, if we sum the reciprocals of such positive integers, does the sum converge or diverge?

7. Every decimal can be written as an infinite series:

$$.d_1d_2d_3\cdots = \frac{d_1}{10} + \frac{d_2}{100} + \frac{d_3}{1000} + \cdots$$

where each d_k is a digit. Prove that every such series converges.

8. a. If f is a decreasing positive function and $a_k = f(k)$, use a picture as in the integral test to show that $\sum_{k=n+1}^{\infty} a_k \leq \int_n^{\infty} f(x)\,dx \leq \sum_{k=n}^{\infty} a_k$.

   b. Use part a to conclude that

$$-a_n \leq \sum_{k=1}^{\infty} a_k - \left[\sum_{k=1}^{n} a_k + \int_n^{\infty} f(x)\,dx\right] \leq 0.$$

9. If a series converges but we cannot find its sum exactly, we can approximate the sum with as much accuracy as we like by the following method. (We assume $a_k \geq 0$.)

$$S = \sum_{k=1}^{\infty} a_k = \sum_{k=1}^{n} a_k + \sum_{k=n+1}^{\infty} a_k = S_n + R_n.$$

Thus $|S - S_n| = |R_n|$. If we make $|R_n| < \varepsilon$ (where ε is a given small positive number), the partial sum S_n will be an approximation of the exact sum S with error less than ε. But to find n for a given ε, we need a way to compute an upper bound for $|R_n|$ that is easy to manipulate. So suppose we can find another series $\sum b_k$, where $a_k \leq b_k$ for all k, and where we either know how to sum $\sum_{k=n+1}^{\infty} b_k$ or we can use the integral test comparison to find an upper bound for $\sum_{k=n+1}^{\infty} b_k$ (see the preceding problem). Then we can apply the method.

Apply this method to approximate each of the following sums with error at most 0.0001. (If the number of terms in S_n is too great for your calculator and you do not have a computer program to compute the sum, simply find n and state what sum must be computed.)

a. $\displaystyle\sum_{k=1}^{\infty} \frac{1}{k^2}$ b. $\displaystyle\sum_{k=1}^{\infty} \frac{1}{k^3+2}$ c. $\displaystyle\sum_{k=1}^{\infty} \frac{2^k}{1+3^k}$

d. $\displaystyle\sum_{k=1}^{\infty} \frac{\ln k}{k^3}$ e. $\displaystyle\sum_{k=1}^{\infty} \frac{k^2}{2^k}$

10. a. Show that $\displaystyle\sum_{k=0}^{\infty} \frac{1}{k!}$ converges. Hint: $k! \geq 2^k$ for $k \geq 4$.

b. Define e to be the sum of this series, and show that $\frac{64}{24} < e < \frac{67}{24}$. (Use the hint in part a and the ideas of the preceding problem.)

11. To approximate the sum of a series, we can sometimes use a more rapidly converging variation of the method in the problem before last. Suppose $a_k = f(k)$, where f is a decreasing positive function, and suppose we can compute $\int_n^\infty f(x)\,dx$ by hand. Then (see two problems before the last)

$$-a_n \le \sum_{k=1}^{\infty} a_n - \left(\sum_{k=1}^{n} a_k + \int_n^\infty f(x)\,dx\right) \le 0,$$

which suggests that $\sum_{k=1}^{n} a_k + \int_n^\infty f(x)\,dx$ is a far more accurate approximation to the sum than the partial sum alone.

Apply this method to approximate each of the following sums with error at most 0.0001. (If the number of terms in S_n is too great for your calculator and you do not have a computer program to compute the sum, simply find n and state what sum must be computed.)

a. $\displaystyle\sum_{k=1}^{\infty} \frac{1}{k^2}$ b. $\displaystyle\sum_{k=1}^{\infty} \frac{\ln k}{k^3}$ c. $\displaystyle\sum_{k=2}^{\infty} \frac{1}{k(\ln k)^2}$

12. In *Kummer's acceleration method*, one writes $\sum a_k = \sum b_k + \sum (a_k - b_k)$ where $\sum a_k$ is the series whose sum is to be approximated and where $\sum b_k$ is a series whose sum can be computed exactly and whose terms are so close to the terms a_k that the series $\sum (a_k - b_k)$ converges much faster than $\sum a_k$ and so can be approximated with fewer terms.

Apply this method to find the sum of each of the following series $\sum_{k=1}^{\infty} a_k$ with error at most 0.0001 by using the suggested series $\sum_{k=1}^{\infty} b_k$ whose sum you must compute exactly. (If the number of terms in S_n is too great for your calculator and you do not have a computer program to compute the sum, simply find n and state what sum must be computed.)

a. $a_k = \dfrac{1}{k^2}$, $b_k = \dfrac{1}{k(k+1)}$.

b. $a_k = \dfrac{k}{k^3+1}$, $b_k = \dfrac{1}{k(k+1)}$.

c. $a_k = \dfrac{1}{k^3}$, $b_k = \dfrac{1}{k(k+1)(k+2)}$.

Section 4: Convergence Tests: All Series

1. Show that the ratio test fails for the series $\frac{1}{2}+\frac{1}{2}+\frac{1}{4}+\frac{1}{4}+\frac{1}{8}+\frac{1}{8}+\cdots$. Does the series converge?

2. Does the series $\sum 2^{-1/n}$ converge?

3. Let $a_n = \dfrac{1+2^n}{1+3\cdot 2^n}$.

   a. Does $\lim\limits_{n\to\infty} a_n$ exist? If so, find it.

   b. Does $\sum a_n$ converge? Explain.

   c. Does $\sum(-1)^n a_n$ converge? Explain.

4. Let $a_n = \dfrac{n2^n+1}{n^2 2^n+1}$.

   a. Does $\lim\limits_{n\to\infty} a_n$ exist? If so, find it.

   b. Does $\sum a_n$ converge? Explain.

   c. Does $\sum(-1)^n a_n$ converge? Explain.

5. Does $\frac{2}{3}-\frac{3}{5}+\frac{4}{7}-\frac{5}{9}+\cdots$ converge?

6. Show that the alternating series test fails for

$$\tfrac{1}{3}-\tfrac{1}{3}+\tfrac{1}{2}-\tfrac{1}{2}+\tfrac{1}{5}-\tfrac{1}{5}+\tfrac{1}{4}-\tfrac{1}{4}+\tfrac{1}{7}-\tfrac{1}{7}+\tfrac{1}{6}-\tfrac{1}{6}+\cdots.$$

   Does the series converge?

7. Which of the following are always true and which are at least sometimes false. Explain.

   a. If $a_n > 0$ for all n but $\{a_n\}$ is not eventually decreasing, then $\sum a_n$ diverges.

   b. If $\sum |a_n|$ converges, then $\sum a_n$ converges.

   c. If $\sum a_n$ converges, then $\sum |a_n|$ converges.

   d. If $a_n > 0$ for all n and $a_n \longrightarrow 0$, then $\sum (-1)^n a_n$ converges.

   e. If $a_n > 0$ for all n and $\sum a_n$ converges, then $\sum (-1)^n a_n$ converges.

   f. If $\sum a_n$ converges, then $\sum \frac{(-1)^n a_n}{n}$ converges.

   g. If $\sum a_n$ converges, then $\sum a_n^2$ converges.

   h. If $a_n > 0$ for all n and $\sum a_n$ converges, then $\sum a_n^2$ converges.

   i. If $a_n \le b_n$ and $\sum b_n$ converges, then $\sum a_n$ converges.

   j. If $0 \le a_n \le b_n$ and $\sum b_n$ converges, then $\sum a_n$ converges.

   k. If $\sum a_n$ converges and the sequence $\{b_n/a_n\}$ converges, then $\sum b_n$ converges. (Assume $a_n > 0$ and $b_n > 0$.)

   l. If $\sum a_n$ converges and the sequence $\{a_n/b_n\}$ converges, then $\sum b_n$ converges. (Assume $a_n > 0$ and $b_n > 0$.)

   m. If $\lim_{n \to \infty} (a_{n+1}/a_n) = L$ and $L < 1$, then $\sum a_n$ converges.

8. We shall later learn that $\ln(1+x) = \sum_{k=1}^{\infty} \frac{(-1)^{k+1} x^k}{k}$ when $-1 < x < 1$. Use this fact to deduce that the same equality holds when $x = 1$:

$$\ln 2 = 1 - \tfrac{1}{2} + \tfrac{1}{3} - \tfrac{1}{4} + \tfrac{1}{5} - \tfrac{1}{6} + \tfrac{1}{7} - \tfrac{1}{8} + \cdots$$

   Hint: When $0 < x < 1$, $\ln(1+x)$ is the sum of an alternating series; so the difference between $\ln(1+x)$ and an nth partial sum of this series is at most the absolute value of the $(n+1)$st term.

9. Show that the following rearrangement of the terms in the above series has the sum $\frac{3}{2}\ln 2$:

$$1+\tfrac{1}{3}-\tfrac{1}{2}+\tfrac{1}{5}+\tfrac{1}{7}-\tfrac{1}{4}+\tfrac{1}{9}+\tfrac{1}{11}-\tfrac{1}{6}+\cdots$$

(Hint: Multiply the original series by $\frac{1}{2}$, insert a term of 0 before each of these terms, and then add this series to the original series term by term.) Have we therefore proved that $\ln 2 = \frac{3}{2}\ln 2$?

10. a. Does $\sum \sin \frac{1}{n}$ converge? b. Does $\sum \frac{1}{n}\sin\frac{1}{n}$ converge?

11. If $\sum a_n$ converges and $|x| < 1$, prove that $\sum a_n x^n$ converges absolutely.

12. A convergent alternating series $\sum(-1)^k a_k$, where $a_k \geq a_{k+1}$, satisfies

$$\left|\sum_{k=1}^{\infty}(-1)^k a_k - \sum_{k=1}^{n}(-1)^k a_k\right| = \left|\sum_{k=n+1}^{\infty}(-1)^k a_k\right| \leq a_{n+1}.$$

Use this fact to approximate each of the following sums with error at most 0.0001. (If the number of terms in S_n is too great for your calculator and you do not have a computer program to compute the sum, simply find n and state what sum must be computed.)

a. $\displaystyle\sum_{k=0}^{\infty}\frac{(-1)^k}{2k+1}$ b. $\displaystyle\sum_{k=2}^{\infty}\frac{(-1)^k}{k\ln k}$

13. Suppose $\sum(-1)^k a_k$ is a convergent alternating series where not only is $\{a_k\}$ decreasing but so is $\{a_k - a_{k+1}\}$. Show that

$$\left|\sum_{k=1}^{\infty}(-1)^k a_k - \left(\sum_{k=1}^{n}(-1)^k a_k + \frac{(-1)^{n+1}a_{n+1}}{2}\right)\right| \leq \tfrac{1}{2}\left|a_{n+1} - a_{n+2}\right|.$$

14. Use the inequality of the preceding problem to approximate each of the following sums with error at most 0.0001.

a. $\displaystyle\sum_{k=0}^{\infty}\frac{(-1)^k}{2k+1}$ b. $\displaystyle\sum_{k=2}^{\infty}\frac{(-1)^k}{\ln k}$

Section 5: Newton's Method

1. Suppose the line $y = 2x - 3$ is tangent to the graph of a differentiable function f at $x = 2$. If Newton's method is used to find a root of $f(x) = 0$ and if the initial guess is $x_0 = 2$, what is x_1?

2. Apply Newton's method graphically on the graph below.

   a. Find x_1 and x_2 when x_0 has each of the following values:

   (i) -2.5 (ii) 3.5 (iii) 0

   b. For what values of x_0 does Newton's method fail immediately? Why?

   c. For what interval or intervals of values of x_0 will Newton's method clearly converge to the root? Why?

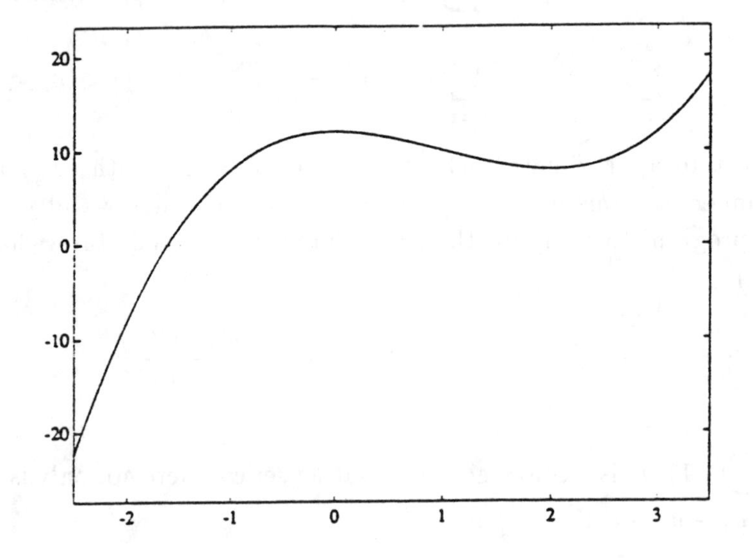

3. Newton's method fails rather spectacularly for $f(x) = x^{1/3}$.

   a. Compute and simplify Newton's recursion formula, and use it to explain what happens to the x_n.

   b. Sketch a graph of $y = x^{1/3}$, and use it to explain what happens to the x_n.

4. $x^2 - 2 = 0$ has two roots, $\sqrt{2}$ and $-\sqrt{2}$. Which values of x_0 lead to which root? Draw a graph to explain.

5. The function f defined by the rule $f(x) = x^3 - 3 + 1$ (graphed below) is an unremarkable cubic polynomial with three real roots, and yet an incautious choice of x_0 can lead to surprising results.

   a. For each of the following values of x_0, find out which root (if any) Newton's method converges to:

      (i) 1.05 (ii) 1 (iii) 0.95 (iv) .911

      (v) .91 (vi) .8893406567 (vii) 0.85

   b. Based on this experience, what precautions might you take in choosing x_0?

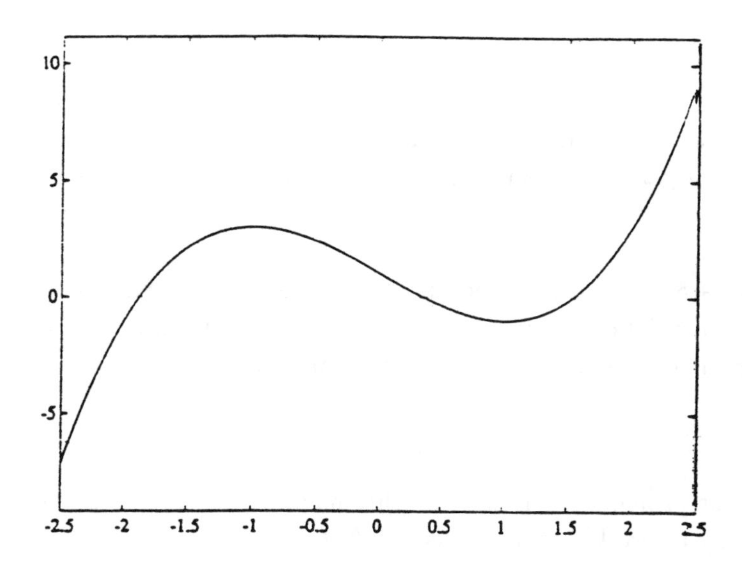

6. Use a symbolic calculator to find the "period-2 point" in the preceding problem. (That's the value of x_0 where $x_0 = x_2 = x_4 = \cdots$, and $x_1 = x_3 = x_5 = \cdots$.)

7. a. Apply Newton's method to $f(x) = x^2 + 1$ with

      (i) $x_0 = 1$,

      (ii) $x_0 = \frac{1}{\sqrt{3}}$ (compute the x_n's exactly, not with a calculator),

      (iii) $x_0 = 100$ (or any other large number).

   What results do you get?

   b. Draw a graph of $y = x^2 + 1$ and use it to explain the behavior of Newton's method in the three trials in part a. Is there any choice of x_0 that leads to a convergent sequence $\{x_n\}$?

8. $\sqrt{A}$ can be approximated by applying Newton's method to $f(x) = x^2 - A$. Show that this choice of f yields Newton's recursion formula

$$x_{n+1} = \frac{1}{2}\left(x_n + \frac{A}{x_n}\right).$$

9. $\frac{1}{A}$ $(A \neq 0)$ can be approximated without any divisions by applying Newton's method to $f(x) = \frac{1}{x} - A$. Show that this choice of f yields Newton's recursion formula

$$x_{n+1} = x_n(2 - Ax_n).$$

10. The *secant method* follows the secant line instead of the tangent line. Find the secant line through the two points on the graph of f whose x-coordinates are x_0 and x_1. Then find a formula for x_2, the point where this secant line crosses the x-axis, in terms of x_0 and x_1.

11. Anne borrows $6000 to help her buy a new car. She pays off her loan in monthly installments of $150 over four years. Find her monthly interest rate, accurate to four places after the decimal. Use the following formula:

$$\left(P - \frac{Y}{r}\right)(1+r)^n + \frac{Y}{r} = 0$$

where P is the principal, Y is the monthly payment, n is the number of months, and r is the monthly interest rate as a decimal.

12. A cable suspended from its two ends hangs in the shape of a "catenary," which is a graph of an equation of the form $y = a\cosh\left(\frac{x}{a}\right)$ for some fixed $a > 0$. ($\cosh x$ is $\frac{1}{2}(e^x + e^{-x})$.) Suppose a telephone wire is suspended between two poles 100 meters apart and that the wire sags 12 meters at its lowest point. Find the value of a, accurate to four digits. See graph below.

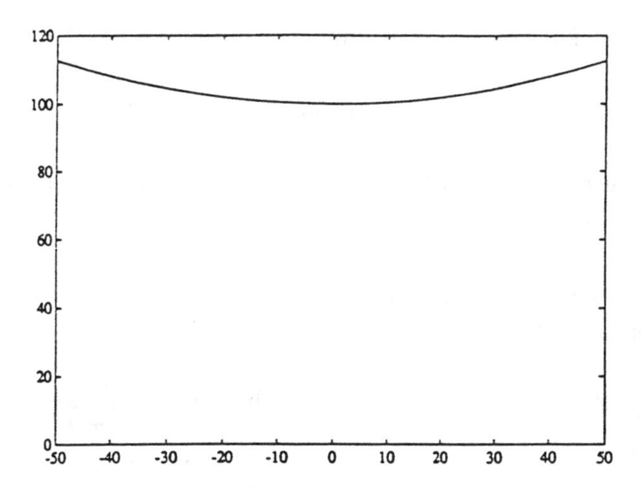

13. A cylindrical can with a bottom but no top is to be made from metal that costs 60 cents/ft^2. The vertical side seam and the base seam each cost 10 cents/ft. If the volume of the can is to be 1 ft^3, find the dimensions of the can that minimize its total cost. (Hint: Find the function to be minimized, and locate its critical point by calculus.)

Section 6: Improper Integrals

1. When an improper integral cannot be evaluated by hand, we may be able to decide whether it converges or diverges by applying a test much like the comparison test for series.

   a. State (invent) such a test for improper integrals of the form $\int_a^\infty f(x)\,dx$ and $\int_a^\infty g(x)\,dx$ where f and g are continuous on $[a, \infty)$.

   b. Give a geometric argument to justify the validity of this test, assuming f and g are positive everywhere.

2. Apply the comparison test for improper integrals to decide the convergence or divergence of each of the following:

   a. $\displaystyle\int_1^\infty \frac{1}{1+x^4}\,dx$

   b. $\displaystyle\int_1^\infty \frac{x}{\sqrt{1+x^3}}\,dx$

   c. $\displaystyle\int_0^\infty e^{-x^2}\,dx$

   d. $\displaystyle\int_1^\infty \frac{\sin x}{x^2}\,dx$

3. The *gamma function* $\Gamma(x)$ is defined for all $x > 0$ by $\Gamma(x) = \int_0^\infty e^{-t}t^{x-1}\,dt$.

   a. Evaluate $\Gamma(1)$.

   b. For $x > 1$, show that $\Gamma(x) = (x-1)\Gamma(x-1)$. Assume that all these improper integrals exist. Hint: Use integration by parts.

   c. Use parts a and b to find $\Gamma(2)$, $\Gamma(3)$, $\Gamma(4)$. What is the pattern?

   One of the few values of $\Gamma(x)$ for noninteger x than can be evaluated exactly is

   $$\Gamma(\tfrac{1}{2}) = \int_0^\infty e^{-t}t^{-1/2}\,dt.$$

   whose value is $\sqrt{\pi}$.

   d. Explain why $\Gamma(\frac{1}{2})$ converges. Hint: Break the integral up into an integral over $[0, 1]$ and an integral over $[1, \infty)$ so you can consider the improper behavior at 0 and at ∞ separately. Use a comparison test for each.

   e. Use the fact that $\int_0^\infty e^{-x^2}\,dx = \sqrt{\pi}/2$ (which is sometimes proved in a chapter on double integrals) to deduce that $\Gamma(\frac{1}{2}) = \sqrt{\pi}$. Hint: Use a change of variables.

   f. Use a computer package to confirm the value of $\Gamma(\frac{1}{2})$.

4. Another well-known integral is $\int_0^\infty \frac{\sin x}{x}\, dx$, which can be shown to have the value $\frac{\pi}{2}$.

   a. Despite the x in the denominator, this integral is not improper at $x = 0$. Why not?

   b. Use a computer package to confirm the value $\frac{\pi}{2}$.

   c. Show that the integral converges. Hint: Write the integral as a sum of integrals over the intervals $[0, \pi]$, $[\pi, 2\pi]$, $\ldots$.

Chapter VIII: Sequences and Series of Functions

Section 1: Sequences of Functions. Taylor Polynomials

1. a. Find the fourth-degree Taylor polynomial at $x = 0$ for $f(x) = 5x^3 - 3x^2 + 4x - 7$.

   b. What can you say in general about the mth degree Taylor polynomial at $x = 0$ for a polynomial of degree n?

2. At a given instant, a car is traveling 50 ft/sec and accelerating 6 ft/sec^2. Use a second-degree Taylor polynomial to estimate how far the car travels in the next second. In the next 3 seconds. In the next hour.

3. The tangent line to the graph of a function f at $x = a$ is the first-degree Taylor polynomial of f at $x = a$. Similarly, a certain close-fitting parabola is the second-degree Taylor polynomial of f at $x = a$.

   a. Draw a graph of $f(x) = \sqrt{25 - x^2}$, and (without doing any computations) draw a plausible graph of the second-degree Taylor polynomial of f at $x = 0$.

   b. Now compute this Taylor polynomial, and use it to check the accuracy of your parabolic graph in part a. Note where the vertex and x-intercepts of the parabola are located.

   c. Repeat the instructions for part b but for $x = 4$ instead of $x = 0$. Does the vertex of this parabola lie below or above the graph of f?

4. Find $\lim\limits_{x \to 0} \dfrac{(\sin x - x)^3}{x(1 - \cos x)^4}$ by approximating $\sin x$ and $\cos x$ with suitable Taylor polynomials, not by using L'Hôpital's Rule.

5. Why is it correct to say that $\sin x \approx x$ is a quadratic approximation of $\sin x$ near $x = 0$?

6. The error in approximating f by its first-degree Taylor polynomial at $x = a$, $f(x) - [f(a) + f'(a)(x - a)]$, is approximately $\frac{1}{2}f''(a)(x - a)^2$ for x near a. This error is positive when f has what property? Is the tangent line to the graph of f at $x = a$ then above or below the graph near $x = a$?

7. Taylor's remainder formula can be used to derive some handy inequalities. Derive the following:

   a. $\left|\ln(1+x) - x\right| \le \frac{1}{2}x^2$ if $x \ge 0$.

   b. $\left|\sin x - x\right| \le \frac{1}{6}|x|^3$ for all x.

8. a. Estimate graphically the maximum error in approximating e^x by $1 + x + \frac{1}{2}x^2$, its second-degree Taylor polynomial at $x = 0$, on the interval $[-1, 1]$.

   b. Use Taylor's remainder formula to compute a bound for this same error.

   c. How do you explain the discrepancy between your answers in parts a and *b*?

9. [C] On the interval $[0, \frac{\pi}{4}]$, use the approximation $\sin x \approx x - \dfrac{x^3}{3!} + \dfrac{x^5}{5!}$.

   a. Show that this approximates $\sin x$ to at least 3 places on $[0, \frac{\pi}{4}]$.

   b. Use this approximation and a computer or calculator to prepare a table of values of $\sin\theta$ for $\theta = 0°, 5°, 10°, \ldots, 45°$ (increments of 5°).

10. Approximate $\displaystyle\int_0^1 \frac{\sin x}{x}\,dx$ with error at most 0.0002. Hint: Use a Taylor polynomial for $\sin x$, not $\dfrac{\sin x}{x}$, and divide by x.

11. [C] Graph $\sin x$ and $x\sqrt[3]{\cos x}$ on $[-3, 3]$. Note that they agree closely on $[-1, 1]$ but less so when x is further away from 0. Use a symbolic calculator to find Taylor polynomials of the two functions at $x = 0$, and use these polynomials to help explain this graphical observation.

12. Approximate e with an error of at most 0.0001 by using a Taylor polynomial for e^x at $x = 0$.

13. [C] Approximate π to the limits of your calculator as follows.

a. Show that $\frac{\pi}{4} = \arctan\frac{1}{2} + \arctan\frac{1}{3}$ by using the addition formula

$$\tan(x+y) = \frac{\tan x + \tan y}{1 - \tan x \tan y}.$$

b. The Taylor polynomials of $\arctan x$ at $x = 0$ have the form $x - \dfrac{x^3}{3} + \dfrac{x^5}{5} - \dfrac{x^7}{7} + \cdots$. Confirm this by using a symbolic calculator if you have access to one.

c. Use parts *a* and *b* to express π in terms of sums of powers of $\frac{1}{2}$ and powers of $\frac{1}{3}$. Add terms in this sum until you get an approximation that equals π to 5 places (i.e., 4 to the right of the decimal). How many terms do you need?

Section 2: Series of Functions. Taylor Series

1. Why doesn't $x^{1/3}$ have a Taylor series about $x = 0$?

2. a. The Taylor series about $x = 0$ for $f(x) = (1+x)^p$ (p any real number) is called the *binomial series*. Derive it: $1 + \sum_{k=1}^{\infty} \frac{p(p-1)\cdots(p-k+1)}{k!} x^k$.

   b. If p is a positive integer, use part a to derive the binomial formula:

$$(1+x)^p = \sum_{k=0}^{p} \binom{p}{k} x^k \qquad \text{where} \qquad \binom{p}{k} = \frac{p!}{k!(p-k)!}.$$

3. When put into motion with veolocity v, an object with mass m_0 at rest has kinetic energy $\frac{1}{2}m_0v^2$. This, at least, is the prediction from Newtonian physics. For an object moving very fast, Einsteinian physics predicts that it has energy $E = mc^2$, where c is the speed of light in a vacuum. This formula also depends on the speed v since the mass increases with speed according to the formula

$$m = \frac{m_0}{\sqrt{1 - v^2/c^2}}.$$

   $E = mc^2$ is the total energy, not just the kinetic energy. Its energy at rest (when $v = 0$) is clearly $E_0 = m_0c^2$. So its kinetic energy K is the difference:

$$K = E - E_0 = mc^2 - m_0c^2 = m_0c^2\left[\left(1 - \frac{v^2}{c^2}\right)^{-1/2} - 1\right].$$

   Use the binomial series to expand this expression in powers of v (the first few terms will do). Then use your result to explain why the Newtonian expression for kinetic energy agrees with the Einsteinian expression when v is much less than c (ordinary speeds).

4. The electric potential generated by a uniformly charged disk of total charge Q and radius R is given by $F(y) = \frac{2Q}{R^2}\left(\sqrt{y^2 + R^2} - y\right)$ where y is the distance from the disk and $F(y)$ is the potential at distance y. Use the binomial series to expand this in powers of $1/y$ (the first few terms will do). Then use your result to explain why Q/y is a good approximation for the potential at points far from the disk.

5. When $x < 0$, the Taylor series for e^x about $x = 0$ is an alternating series. Use this fact to approximate e^{-1} with error at most $5 \cdot 10^{-7}$. Using your calculator, take the reciprocal of this approximation. How close is it to e?

6. Using the Taylor series for $\sin x$ about $x = 0$, express $\int_0^1 \frac{\sin x}{x}\,dx$ as an alternating series. Use this series to approximate the integral with error at most $5 \cdot 10^{-7}$.

7. Use the following ideas to prove that e is irrational. If $e = \frac{m}{n}$, where m and n are positive integers, then $N = m!\left[\frac{1}{e} - \left(1 - \frac{1}{1!} + \frac{1}{2!} + \cdots + \frac{(-1)^m}{m!}\right)\right]$ is an integer. (Why?) The number in brackets is less than $\frac{1}{(m+1)!}$ in absolute value. (Why?) These two sentences contradict one another. (Why?)

8. [C] A wrecking ball attached to a crane may be thought of as a pendulum. The period of such a pendulum is given (in seconds) by $P = 4k\int_0^{\pi/2} \frac{dx}{\sqrt{1 - a^2 \sin^2 x}}$ where $a = \sin\frac{\theta}{2}$, θ is the angle of the swing (measured in radians from the rest position), $k = \sqrt{\frac{r}{g}}$, r is the length of the cable holding the ball, and g is the acceleration due to gravity. Suppose the construction engineer determines that a period of $P = 5$ seconds is required for the wrecking ball to have its maximum impact, and suppose that $k = \frac{3}{4}$. Using a symbolic calculator, compute the angle θ required for the swing as follows: Since the integral cannot be computed exactly, approximate the integrand by its fifth-degree Taylor polynomial at $x = 0$. Then integrate, solve the equation for a, and compute θ from a.

Section 3: Power Series

1. What is wrong with the following computation (where $x > 0$)?

$$2 < (1 + x + x^2 + \cdots + x^n + \cdots) + \left(1 + \tfrac{1}{x} + \left(\tfrac{1}{x}\right)^2 + \cdots + \left(\tfrac{1}{x}\right)^n + \cdots\right)$$
$$= \frac{1}{1-x} + \frac{1}{1-\frac{1}{x}} = \frac{1}{1-x} - \frac{x}{1-x} = 1$$

2.

   a. What is the coefficient of x^{100} in the power series for e^{2x} about $x = 0$?

   b. Evaluate $f^{(100)}(0)$ for the function f, where

$$f(x) = \begin{cases} \dfrac{1-\cos x}{x^2} & \text{for } x \neq 0, \\ \dfrac{1}{2} & \text{for } x = 0. \end{cases}$$

3. For each of the following functions, find its Taylor series about $x = 0$ without computing $f^{(k)}(0)$. Instead, modify a familiar Taylor series by integrating it, differentiating it, making a substitution in it, or some combination of these.

   a. $\dfrac{1}{(1-x)^2}$ b. e^{-x^2} c. $\arctan x$

   d. $\dfrac{\ln(1+x)}{x}$ e. $\cos\sqrt{x}$, $(x \geq 0)$ f. $\arcsin x$

4. a. Compute, by hand, the first couple of terms of the Taylor series about $x = 0$ for $\tan x$. Either compute $f^{(k)}(0)$ or divide the series for $\cos x$ into the series for $\sin x$.

   b. Check your answer against the result produced by a symbolic calculator.

   c. Guess at the radius of convergence of the series. Explain. (You will probably not be able to prove you are right until you study more advanced mathematics.) Evidence that you might use are the graph of $\tan x$ compared to some partial sums of its Taylor series and/or a closer look at the definition $\tan x = \dfrac{\sin x}{\cos x}$.

5. Find the sum of each of the following series. Hint: Find a familiar Taylor series similar to it. Then convert one to the other by integrating, differentiating, making a substitution, or some combination of these.

   a. $x + x^3 + \frac{x^5}{2!} + \frac{x^7}{3!} + \cdots$

   b. $2 - 3 \cdot 2x + 4 \cdot 3x^2 - 5 \cdot 4x^3 + \cdots$

   c. $\frac{x^2}{2} + \frac{x^4}{4} + \frac{x^6}{6} + \frac{x^8}{8} + \cdots$

   d. $1 - \frac{3x^2}{2!} + \frac{5x^4}{4!} - \frac{7x^6}{6!} + \cdots$

6. Find the sums of the following series. Hint: Find a power series in x for which some specific value of x yields the series. Sum the power series as in the preceding problem.

   a. $\sum_{k=1}^{\infty} \frac{k}{2^k}$ b. $\sum_{k=1}^{\infty} \frac{1}{k2^k}$ c. $\sum_{k=0}^{\infty} \frac{2^k}{(k+1)!}$

7. a. Approximate $\int_0^1 e^{-x^2}\,dx$ with error at most $5 \cdot 10^{-7}$ as follows. Express e^{-x^2} as a power series about $x = 0$, integrate term by term, and note that the result is an alternating series of numbers.

   b. Do the same for $\int_0^1 \frac{1}{\sqrt{2\pi}} e^{-x^2/2}\,dx$.

8. Let $f(x) = e^{-1/x^2}$ when $x \neq 0$ and $f(0) = 0$. It can be shown that $f^{(k)}(0) = 0$ for all $k \geq 0$.

   a. What is the Taylor series for f about $x = 0$? What is its radius of convergence? For what values of x does $f(x)$ equal the sum of this series?

   b. Show that $f'(0) = 0$ by evaluating $\lim_{x \to 0} \frac{f(x) - f(0)}{x}$.

9. a. Consider the integral $\int_0^\infty \frac{xe^{-x}}{1-e^{-x}}\,dx$. (The antiderivative for the integrand cannot be computed in closed form using functions that you are familiar with.) Explain why this improper integral converges. (This question has two parts: Investigate what happens as $x \to \infty$ and what happens as $x \to 0$?)

   b. Demonstrate your skill in making substitutions in integrals by making the substitution $u = 1 - e^{-x}$ in this integral. (You will have to solve for x in terms of u.)

   c. Continue working with the integral in u that you have created in *b*. Use your knowledge of a power series for the logarithm to write the integrand as a power series. Integrate this series to get a series of numbers which adds up to this integral.

   d. About how many terms of this series would you have to add up to get a sum that is within .0001 of the correct value of the integral?

Chapter IX: The Integral in $\mathbb{R}^2$ and $\mathbb{R}^3$

Section 1: Real-valued Functions of Two and Three Variables

1. Match each of the following surfaces to the pattern of level curves which corresponds to it.

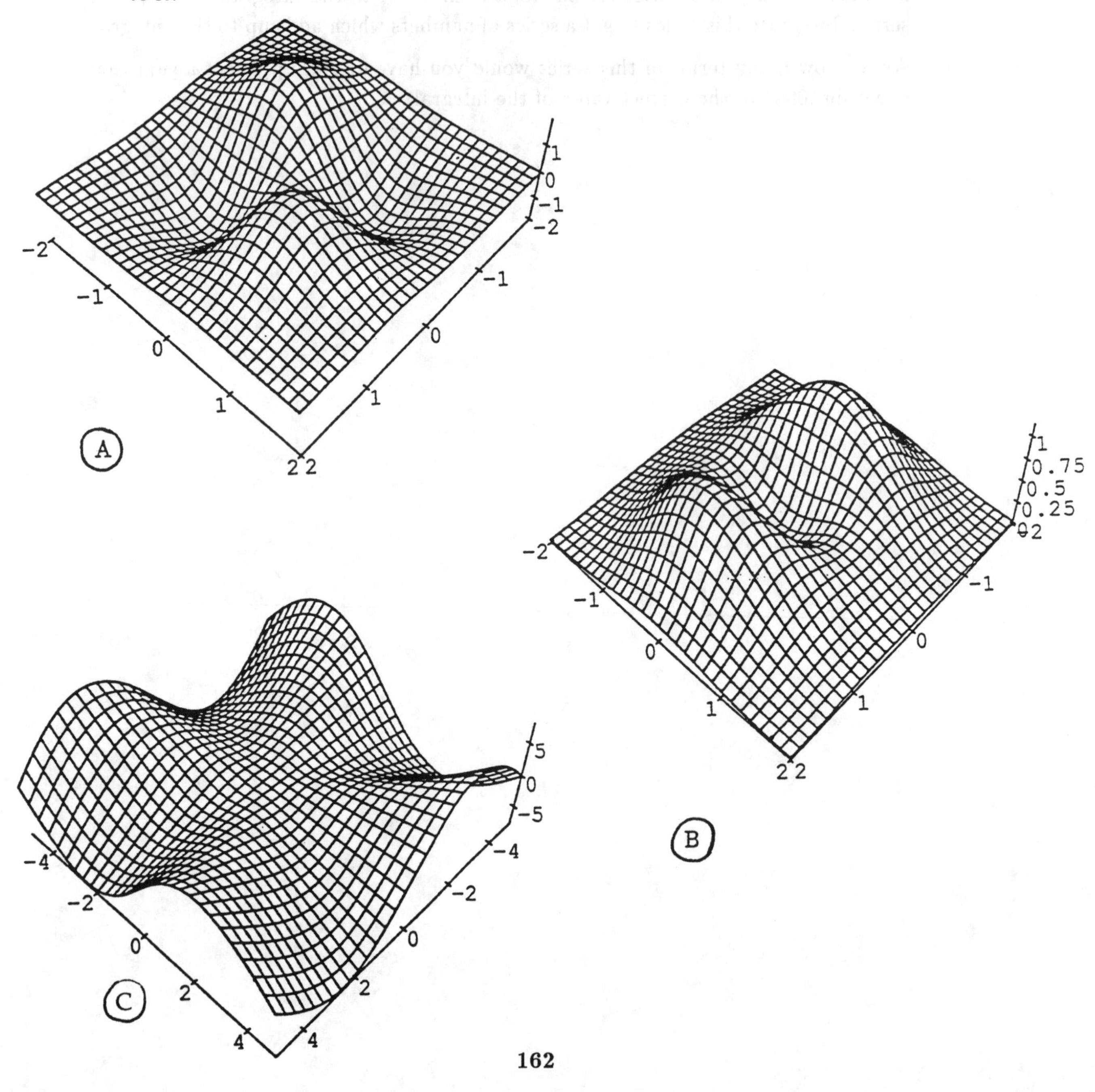

1. (continued)

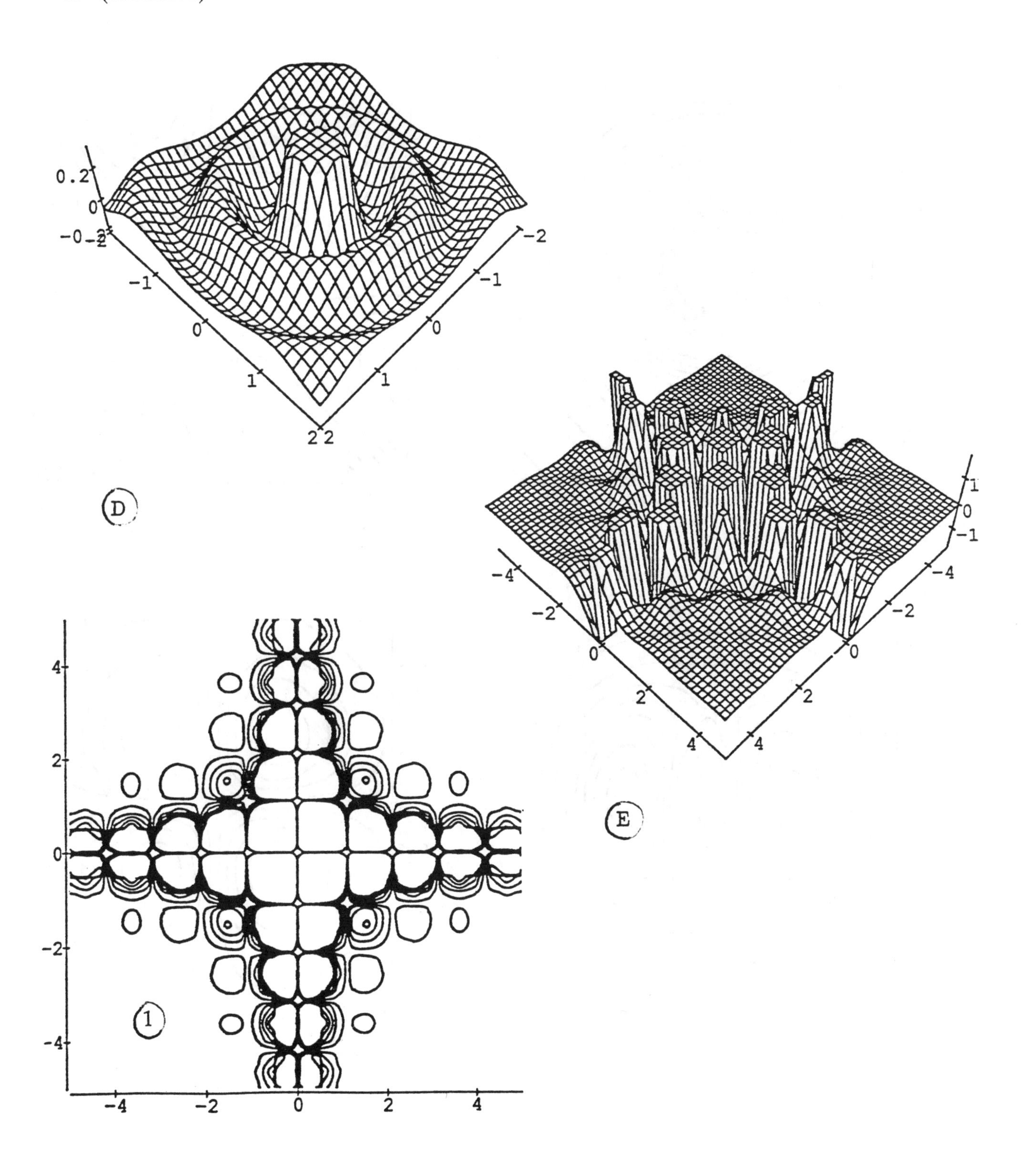

1. (continued)

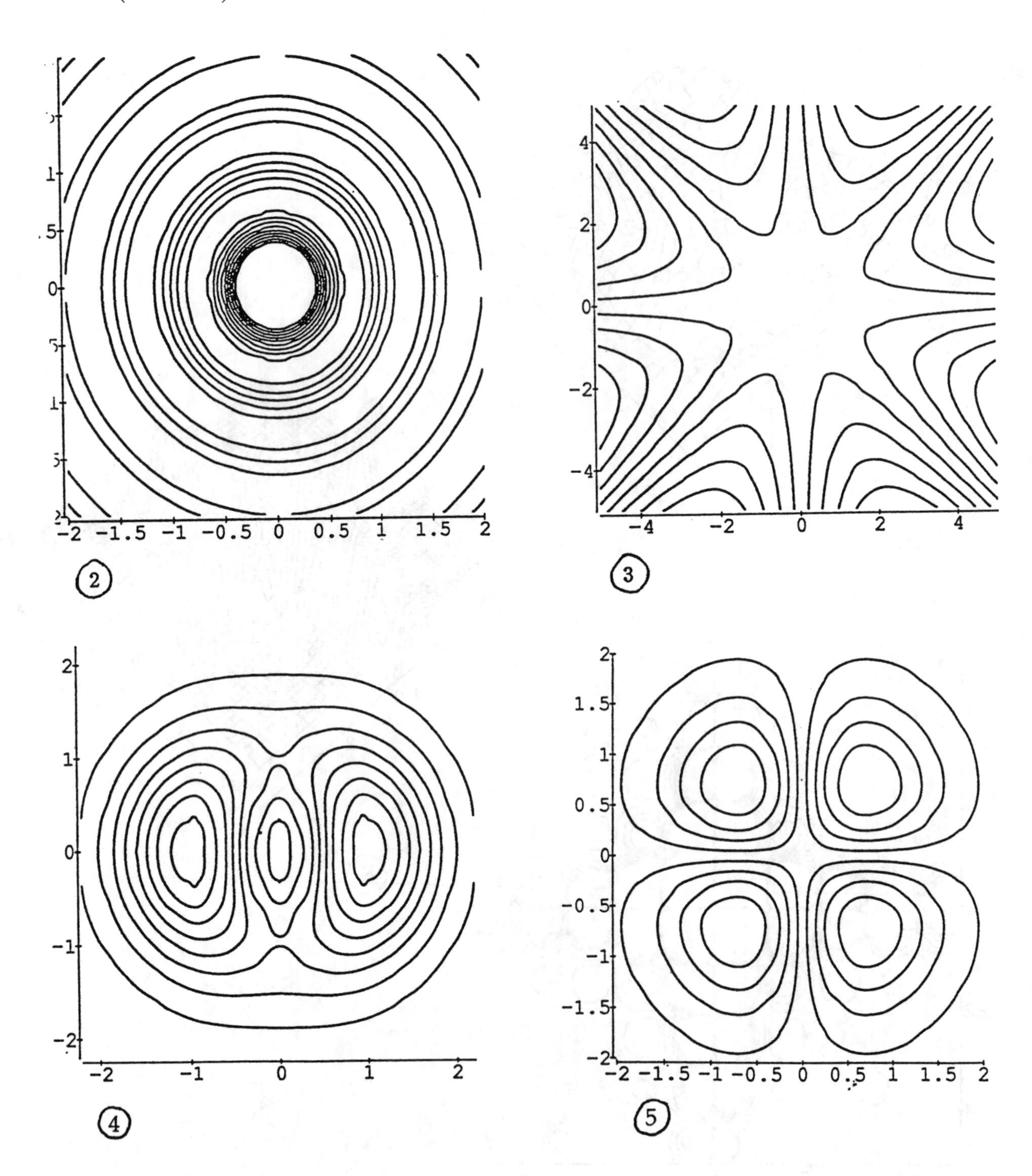

2. Here are two surfaces. For each of them, sketch the corresponding system of level curves.

a.

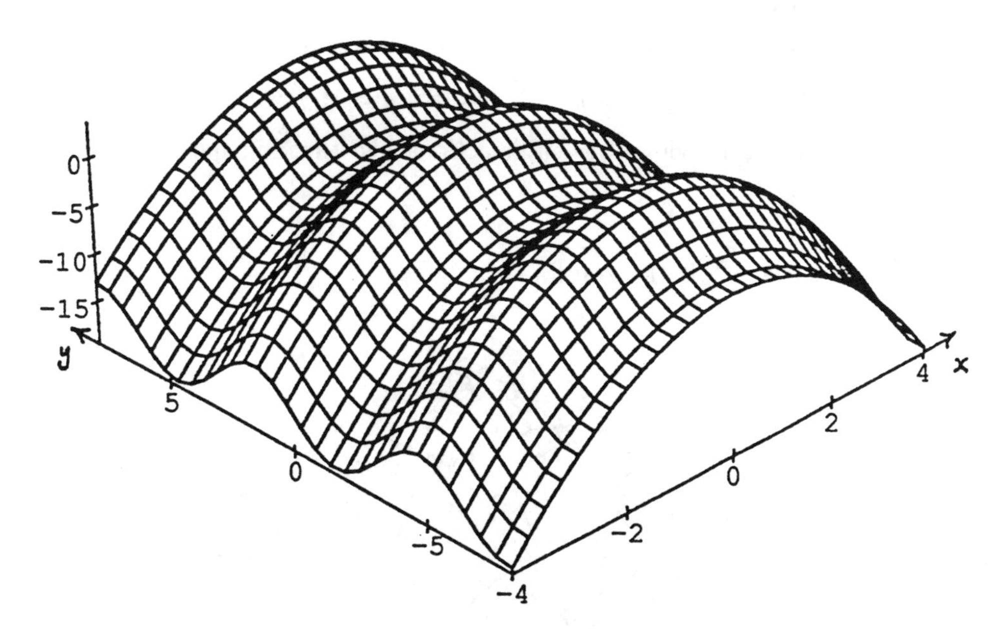

b.

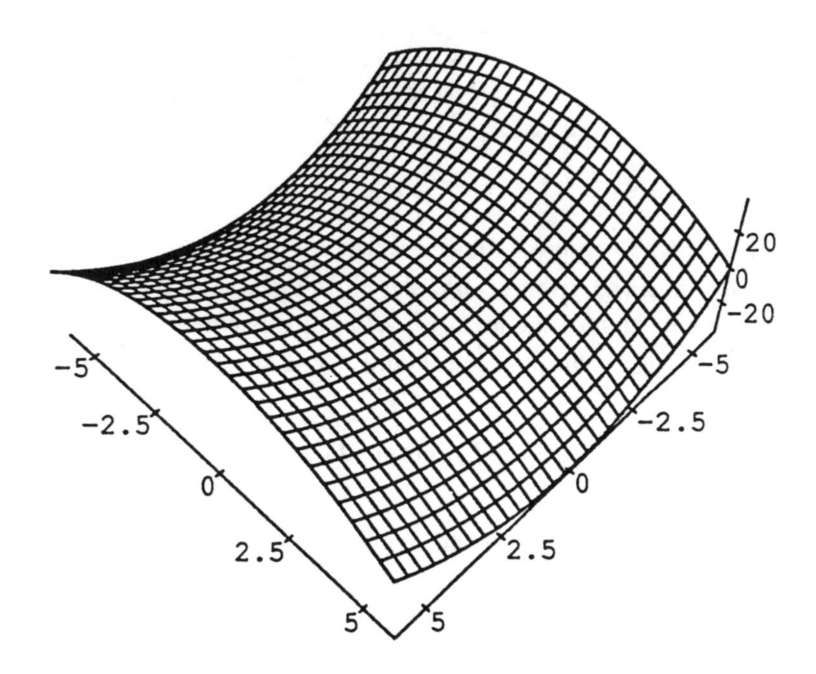

3. a. For the first surface in the preceding question, sketch the intersection of the surface with the plane $x = 0$. Repeat for the planes $x = 0.5$, $x = 1$, $x = 1.5$, $x = 2$, $x = 2.5$. Draw the graphs of all six intersections on the same set of axes.

   b. Repeat part a for the planes $y = 0$, $y = 0.5$, $y = 1$, $y = 1.5$, $y = 2$, $y = 2.5$.

   c. Can level curves be viewed as intersections like those in parts a and *b*? Explain your answer.

4. The following surface S is the graph of $z = 60xye^{-x^2-y^2}$.

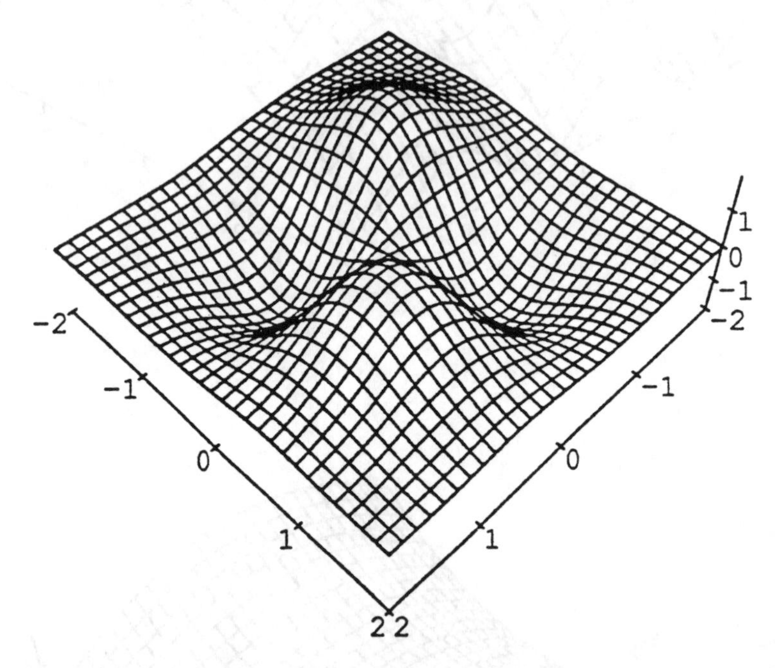

   a. Let PL_1 be the plane perpendicular to the xy-plane and containing the points $(0,0)$ and $(1,1)$. Using the given graph, sketch the graph of the intersection of PL_1 and S. Find the equation of the form $z = f(x)$ of this curve. Then, using this equation, obtain the graph of the function f and use it to check how well you were able to sketch $PL_1 \cap S$.

   b. Repeat part a for the plane PL_2, perpendicular to the xy-plane and containing the points $(0,0)$ and $(1,0)$.

   c. Once more, for PL_3, a plane perpendicular to the xy-plane and containing the points $(0,0)$ and $(-1,1)$.

5. The function $w = f(u)$ is given by its graph, shown below. Note that $f(u)$ is different from 0 only for $0 < u < 4$. We use $f(u)$ to define a function $w = g(x,t)$ of two variables x and t as follows: $w = g(x,t) = f(x-t)$. Regard t as a time variable.

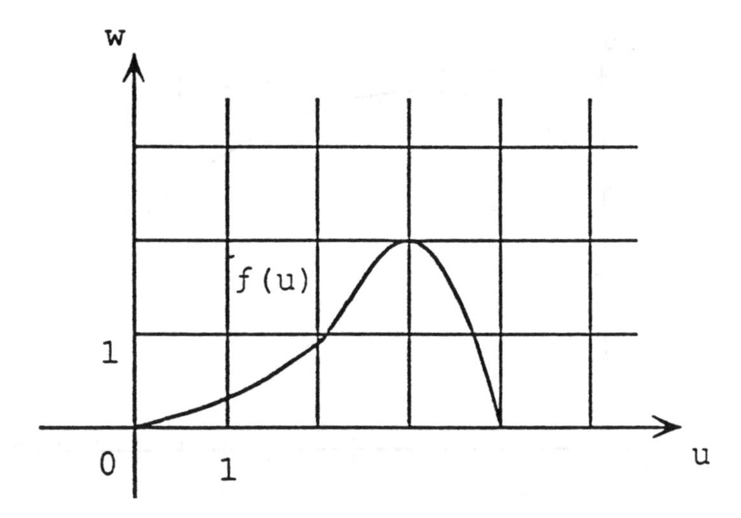

a. Into an xw-system sketch $w = g(x,1)$, $w = g(x,6)$, $w = g(x,8)$.

b. Into an xt-system sketch $w = g(6,t)$.

c. Into an xt-system sketch the level curves $w = 1$ and $w = 2$.

d. Find the *last* time t at which the point $(x,t) = (73,t)$ is vertically displaced 1 unit.

e. Find all values x for which the point $(x, 212)$ is vertically displaced by less than 0.5 unit.

6. The graph of the function $u = g(t)$ is shown below. We define a new function $f(x, y)$ of two variables by $f(x, y) = g(x^2 + y^2 - 4)$.

   a. Describe in words the shape of the level curve $f(x, y) = 3$.

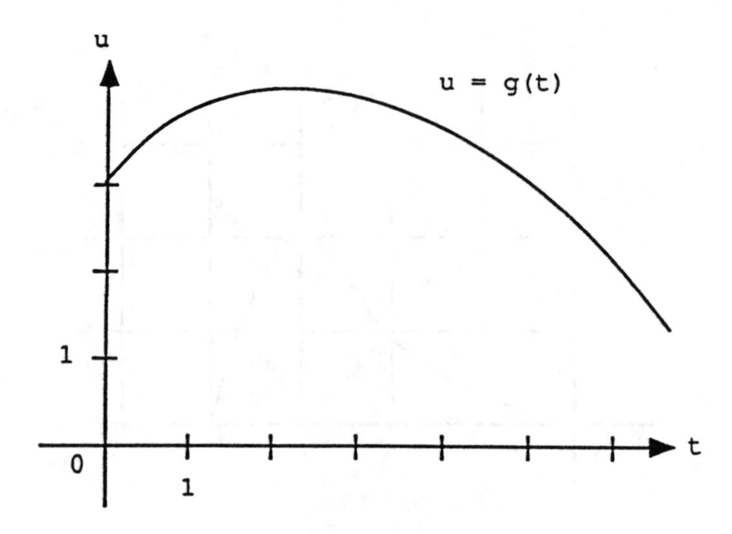

   b. Where does f appear to achieve its maximum value?

7. Suppose $f(x, y) = x^3y^3 - 9x^2y + 16y$. Find points P and Q on the line $x = 2$ so that the following is true: If the point (x, y) moves from P to Q, then $f(x, y)$ first decreases, then increases.

8. Suppose $F(x, y, z) = xyz$.

   a. Describe in words the level surface of $F(x, y, z)$ which passes through the origin.

   b. Repeat part a for the point $(1, 0, 7)$ instead of the origin.

   c. Let S be the level surface of $F(x, y, z)$ which passes through $(1, 1, 1)$. Write the equation of S in the form $z = g(x, y)$ and sketch the level curves $z = 1$ and $z = 433$ of S.

   d. A friend says: "I am confused. Weren't we told that functions of three variables have level *surfaces*. How come we draw level *curves* in part c?" What do you answer?

Section 2: Definition of Double and Triple Integrals

1. Let $f(x, y) = 100 - x^2 - y^2$, and let the region R in the xy-plane be the rectangle defined by the inequalities $1 \leq x \leq 5$, $2 \leq y \leq 4$. Let P be the partition of R defined by the lines bounding R together with the lines $x = 3$ and $y = 3$.

   a. Find M_0, the value of the Riemann sum $\sum f(x_i, y_j)\Delta A_{ij}$, if (x_i, y_j) is chosen as the lower left corner of R_{ij}, $i, j = 1, 2$. ΔA_{ij} is the area of R_{ij}.

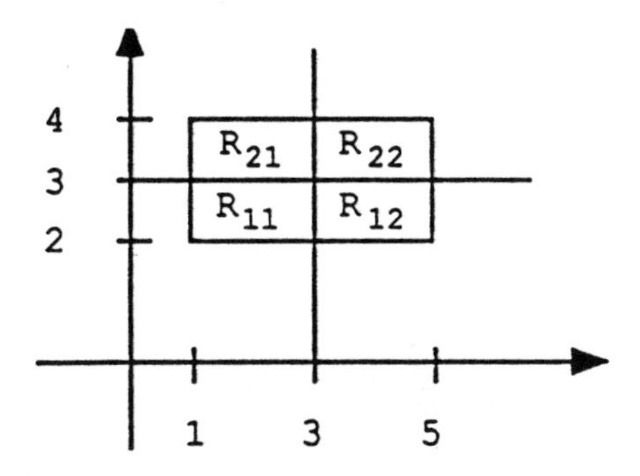

   b. Repeat part a to find m_0, the value of the Riemann sum if (x_i, y_j) is chosen to be the upper right corner of R_{ij}.

   c. Let $V = \iint_R f(x, y)\, dA$. (Do not calculate V.) Which of V, m_0, and M_0 is the largest? Which is the smallest? Justify your answers.

2. [C] This is a continuation of the preceding question for which f and R remain as before.

   a. Subdivide the partition P by introducing the lines $x = 2$, $x = 4$, $y = 2.5$, $y = 3.5$. This gives 16 subregions R_{ij}, $i, j = 1, 2, 3, 4$. For this new partition, calculate two new Riemann sums: M_1, for which (x_i, y_j) is chosen as the lower left corner of R_{ij}, and m_1, for which (x_i, y_j) is chosen to be the upper left corner of R_{ij}.

   b. Express the relation among m_0, m_1, M_0, M_1, V as a string of inequalities.

   c. Subdivide once again, in the spirit of part a, i.e. $x = 1.5, \ldots, y = 2.25, \ldots$ to get a partition of 64 subregions, R_{ij}, $i, j = 1, \ldots, 5$ and calculate the corresponding Riemann sums m_2, M_2.

   d. Write the m's, the M's, and V in a string of inequalities as you did in part *b*. Does this string help you to understand the sense in which V is the limit of a Riemann sum? (Write a sentence, rather than a "yes" or a "no.")

3. The function $u = g(t)$ is given by its graph below left. We define a function by $p(x,y) = g\left(\frac{3}{x^2+y^2}\right)$.

   a. Let D be the shaded domain shown below right. Find numbers m and M such that $m \leq \iint_D p\,dA \leq M$.

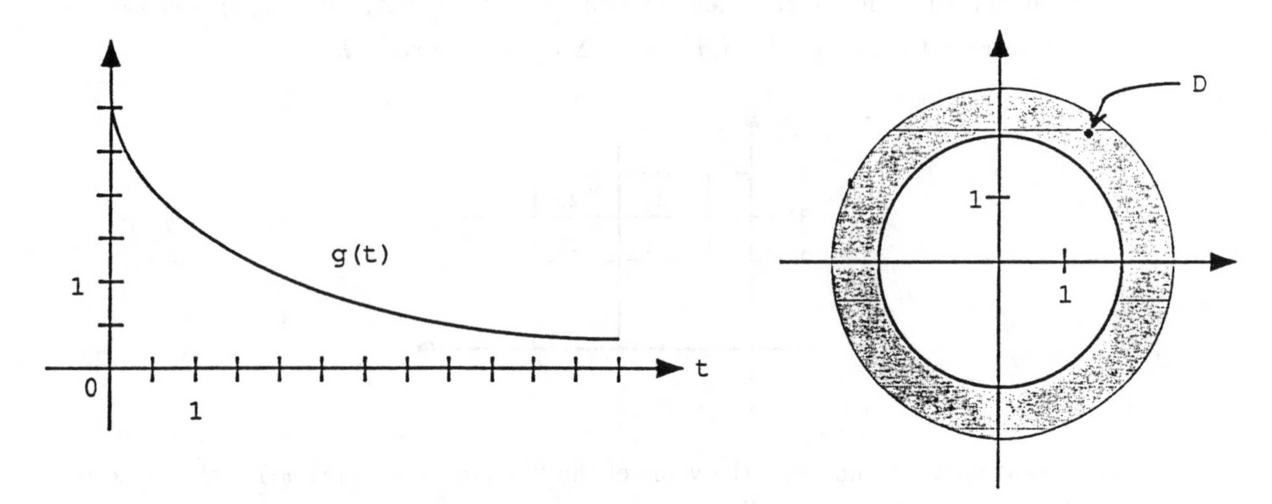

   b. Let D be shifted as shown below and repeat part a.

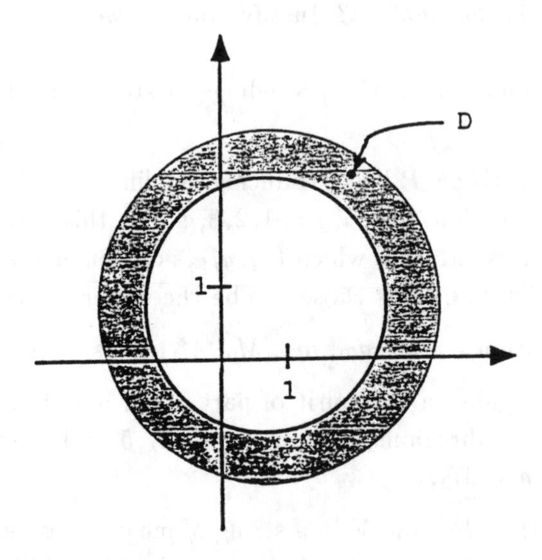

Section 3: Evaluation of Double Integrals

1. Fill in the required limit of integration for each of the following double integrals $\iint_D f(x,y)\,dA$ where D is the region indicated in the accompanying sketch.

a. $\int_0^1 dy \int_{\square}^{4y} f(x,y)\,dx + \int_1^2 dy \int_{\square}^{\square} f(x,y)\,dx$

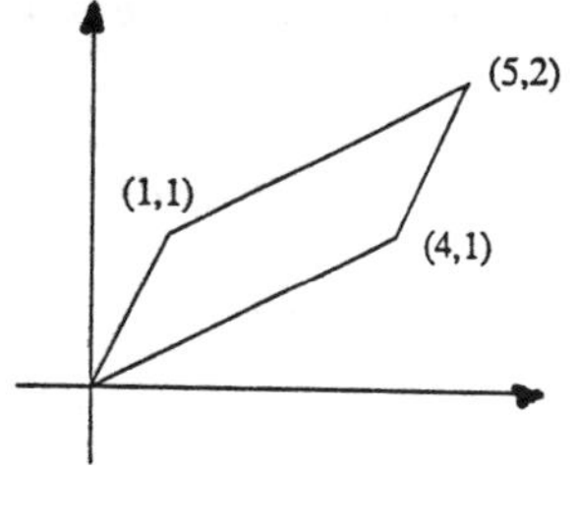

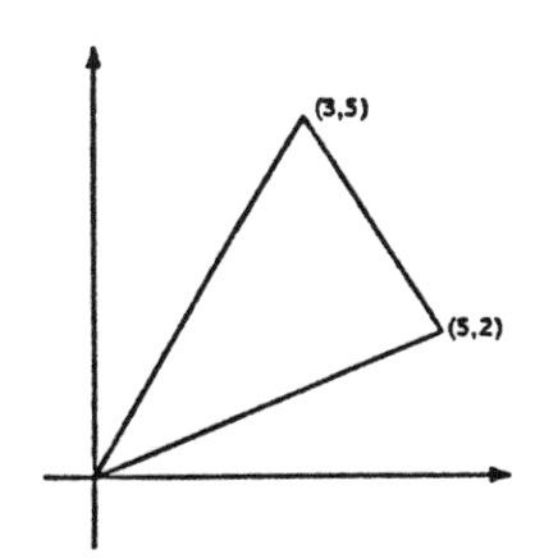

b. $\int_0^3 dx \int_{\square}^{\square} f(x,y)\,dy + \int_3^5 dx \int_{\square}^{\square} f(x,y)\,dy$

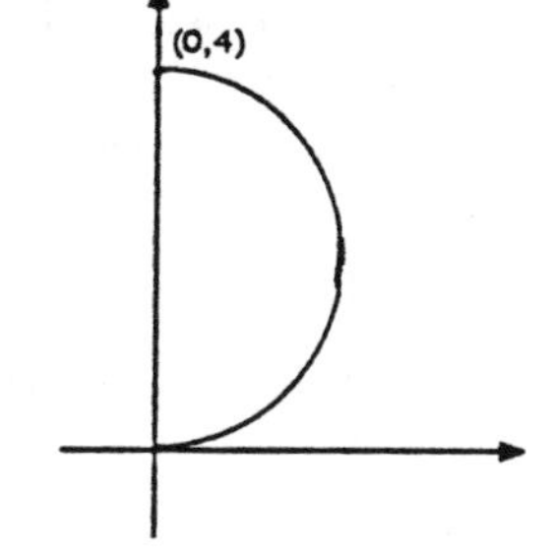

c. $\int_0^4 dy \int_0^{\square} f(x,y)\,dx$

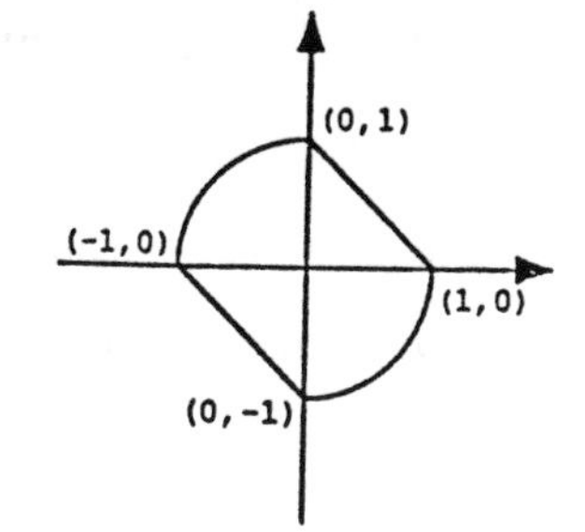

d. $\int_{-1}^0 dx \int_{\square}^{\square} f(x,y)\,dy + \int_{\square}^{\square} dx \int_{\square}^{\square} f(x,y)\,dy$

e. $\int_{-2}^{-1} dx \int_{\square}^{\square} f(x,y)\,dy + \int_{-1}^{1} dx \int_{\square}^{\square} f(x,y)\,dy + \int_{1}^{2} dx \int_{\square}^{\square} f(x,y)\,dy$

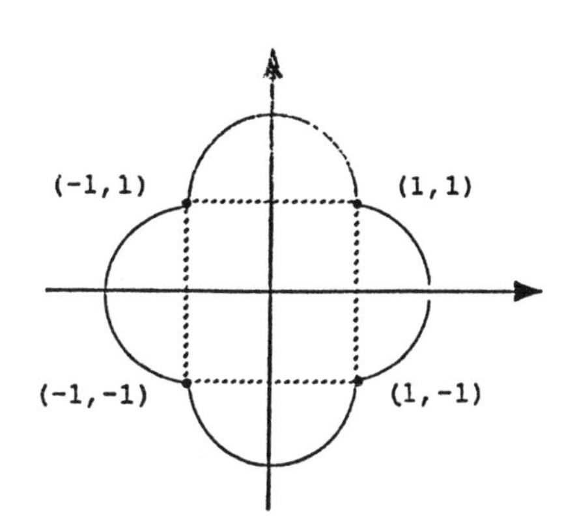

2. Each of the following integrals pertains to some region in 3-space. Write a question about each of these regions for which the corresponding integral is the answer. Note that, in your phrasing of the question, it will be necessary to give a clear geometric definition of the region involved. Do *not* evaluate the integrals.

   a. $\displaystyle\int_0^3 \int_0^{-\frac{2}{3}x+2} (6 - 2x - 3y)\, dy\, dx$

   b. $\displaystyle\int_0^{\pi} \int_0^{2\pi} \int_0^1 \rho^3 \sin\theta \, d\rho\, d\theta\, d\phi$

   c. $\displaystyle\int_0^{2\pi} \int_0^{1/\sqrt{2}} \int_r^{\sqrt{1-r^2}} r\, dz\, dr\, d\theta$

3. Show that $\displaystyle\int_0^1 \int_0^1 \int_0^1 \frac{dz\, dy\, dx}{1 - xyz} = \sum_{n=1}^{\infty} \frac{1}{n^3}$. Hint: The integral $\displaystyle\int_0^1 \frac{dz}{1 - xyz}$ is easily "do-able" and equals an expression of the type $\dfrac{\ln(1-u)}{u}$. The integral of this function is not "do-able" in the sense that there is no elementary antiderivative of $\dfrac{\ln(1-u)}{u}$. However, there is a Maclaurin expansion of $\dfrac{\ln(1-u)}{u}$ valid for $u \in (-1, 1)$, and use of this expansion yields the desired result.

4. Let $f(x, y) = \frac{1}{4}e^{-(|x|+|y|)}$ be the joint probability density function for the population distribution of a city called Mathematopolis. The probability that a person selected at random from Mathematopolis lives in the region R is given by $\displaystyle\iint_R f(x, y)\, dA$, where x and y are measured in kms.

   a. Find the probability that a person selected at random lives no more than 4 kms north of Main Street ($y = 0$) and no more than 4 kms east of Broadway ($x = 0$).

   b. Find the value of c such that the square bounded by the lines $x = y = -c$, $x = y = c$ contains 50% of Mathematopolis' population.

   c. Explain, but do not calculate, how you would determine the probability that a randomly selected person lives on Main Street.

5. Suppose that the following "average temperature" readings, in degrees Fahrenheit, are recorded for towns in the state of Wisconsin one year on August 1.

1.	Ashland	75°	10.	Madison	81°
2.	Eau Claire	81°	11.	N. Milwaukee	78°
3.	Fish Creek	72°	12.	Platteville	85°
4.	Fond Du Lac	78°	13.	Racine	79°
5.	Green Bay	75°	14.	Rhinelander	75°
6.	Hudson	83°	15.	Sheboygan	76°
7.	Janesville	82°	16.	Stevens Point	78°
8.	La Crosse	83°	17.	Superior	77°
9.	Lady Smith	78°	18.	Wausau	77°

Use the map below to answer the following questions:

a. *Guess* the average temperature for the state of Wisconsin for this day.

b. Calculate an estimate of the average temperature for the state. Explain carefully how you arrive at your answer.

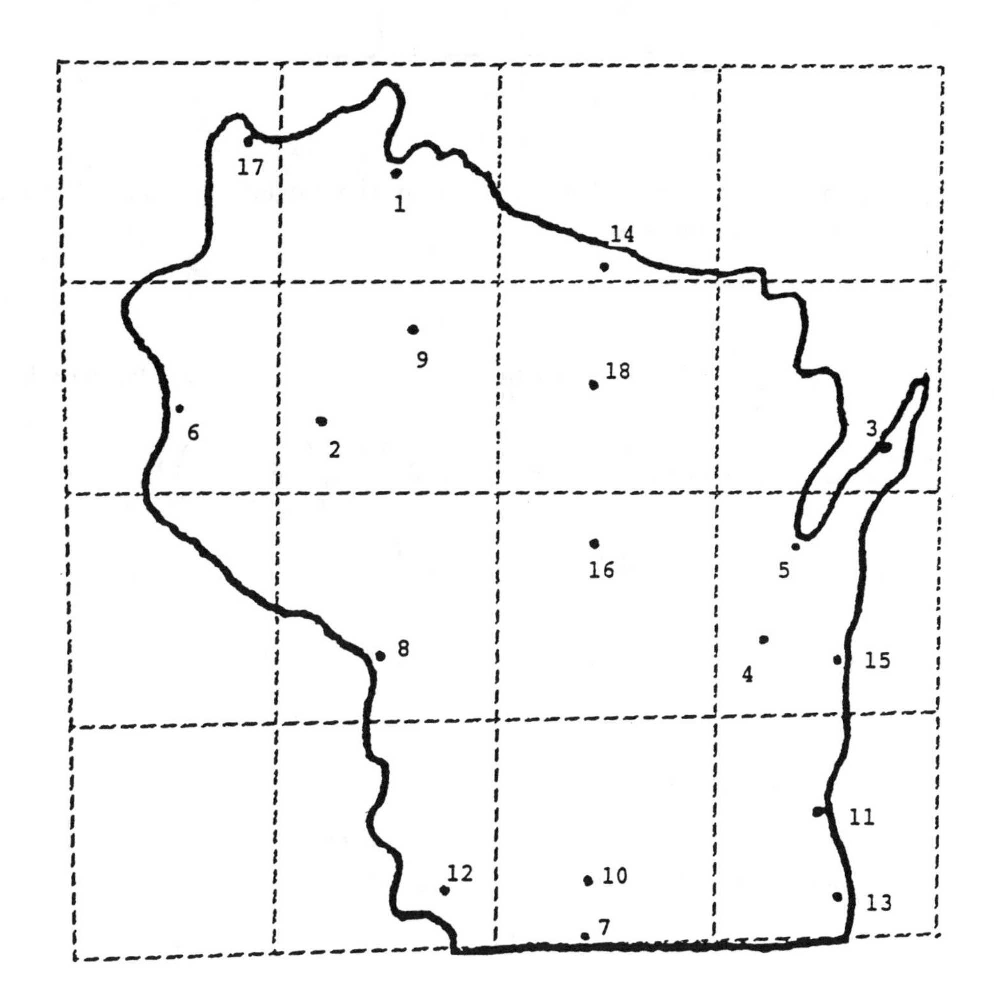

6. Suppose that on August 1, the average temperature T in degrees Fahrenheit is approximated at each point in the state of Wisconsin by the equation $T(x, y) = 86e^{-.0005(x+y)}$ where x and y are measured in miles and the origin $(0, 0)$ of the xy-plane is located at the southwest corner of the state. Calculate the average temperature for the state if the shape of the state is approximated as indicated below.

   a. A rectangle bounded by the lines $y = 0, y = 300, x = -60, x = 150$.

   b. [C] As indicated in the sketch below, a union of three trapezoidal subregions.

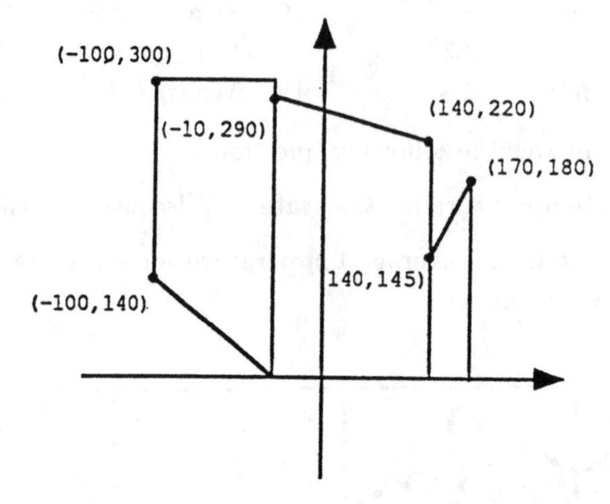

How do your answers compare with the answer in the previous question? Explain how to account for any discrepancies.

7. [CP]

   a. Calculate $\int_0^\infty \int_0^\infty e^{-(x^2+y^2)}\, dy\, dx$ by changing the integral to polar coordinates.

   b. Show that the given double integral is equal to $\left(\int_0^\infty e^{-x^2}\, dx\right)^2$.

   c. Using parts a and b, calculate the value of $\int_0^\infty e^{-x^2}\, dx$.

8. Suppose the region R has sides given by the following equations

$$y = 2x, \quad x = 1, \quad y = 1 + 2x, \quad x = 0.$$

   a. Find the corners of R and sketch R.

   b. Choose a, b, c, d so that the transformation T, where $T^{-1} : x = au + bv,\ y = cu + dv$, takes R into the square S given by $0 \le u \le 1, 0 \le v \le 1$.

   c. Find the area of R by calculating $\iint_S |J|\, dA$ where J is the Jacobian of T.

9. Let R be the region with sides given by $x = 0$, $x = 1$, $y = 0$, $y = 2x + 3$.

   a. Find the corners of R and sketch R.

   b. Choose a, b, c so that $x = u$, $y = au + bv + cuv$ changes R into the square S, $0 \le u \le 1, 0 \le v \le 1$.

   c. A nonlinear term uv is used in part b to change R into S. What regions in the xy-plane change to S if only linear transformations, $x = au + bv$, $y = cu + dv$, are used?

10. Let S be the square in uv-space with corners $(0,0), (1,0), (1,1), (0,1)$. Draw the region R in xy-space which the given transformation T maps onto S. Locate the corners of R and then the equations of its sides. (To make things simpler, the equations below define the inverse of T.)

    a. $x = 2u + v, \quad y = u + 2v$

    b. $x = 3u + 2v, \quad y = u + v$

    c. $x = e^{2u+v}, \quad y = e^{u+2v}$

    d. $x = 2uv, \quad y = v^2 - u^2$

    e. $x = u, \quad y = v(1 + u^2)$

    f. $x = u \cos v, \quad y = u \sin v$ (Three corners only.)

11. Define a disk of radius a in the $r\theta$-plane by $r \in [0, 2a\cos\theta]$, $\theta \in [0, \pi]$. $P = (0,0)$ is a point on the circle which is the boundary of that disk.

    a. Sketch a graph of the circle.

    b. Write down an equation for the average distance $\bar{r}$ from P to the points inside the circle: $\bar{r} = \dfrac{\iint \square}{\iint \square}$

    c. Evaluate $\bar{r}$.

12. Suppose that by a change of variables $x = \alpha(u,v)$, $y = \beta(u,v)$, we replace one double integral with another:

$$\iint_R f(x,y)\,dy\,dx = \iint_S g(u,v)\,dv\,du.$$

Which of the following are ALWAYS true, SOMETIMES true, or NEVER true? Justify your answers.

a. $\text{Area}(R) = \text{Area}(S)$.

b. $f\big(\alpha(u,v), \beta(u,v)\big) = g(u,v)$.

c. If $\text{Area}(R) = \text{Area}(S)$, then $f\big(\alpha(u,v), \beta(u,v)\big) = g(u,v)$.

13. Which of the following are ALWAYS true, SOMETIMES true, or NEVER true? Justify your answers.

a. $\displaystyle\iint_R dA \leq \iint_R x^2\,dA$

b. The average value of a function f over a region R is $\displaystyle\iint_R f(x,y)\,dA$.

c. The polar "rectangle" given by $0 \leq r_1 \leq r \leq r_2$, $\theta_1 \leq \theta \leq \theta_2$ has the same area as the quadrilateral with the same corners.

Chapter X: Vectors and Vector Geometry

Section 1: Vectors

In this section, the symbol $\mathbf{X} \cdot \mathbf{Y}$ will denote the scalar, or dot product, and the symbol $\mathbf{X} \times \mathbf{Y}$ will denote the vector, or cross product. $\|\mathbf{X}\|$ denotes the norm of the vector $\mathbf{X}$.

1. Use vectors to show that an angle inscribed in a semicircle is a right angle.

2. In each of the following situations, state whether the independent variable is a scalar or a vector, and whether the dependent variable describes a scalar or a vector field. If a vector quantity, describe the number of dimensions.

	Independent Variables	*Dependent Variables*
a.	Position on a thin wire	Temperature
b.	Position on a globe	Temperature
c.	Position on ocean surface	Pressure
d.	Position below ocean surface	Pressure
e.	Position on ocean surface	Wind velocity
f.	Position below ocean surface	Velocity of current

3. State and prove the Law of Cosines using the scalar product.

4. Consider the vector $a\,\mathbf{i} + b\,\mathbf{j}$ in $\mathbb{R}^3$. What are its direction cosines? How can this be related to the Pythagorean Theorem in two dimensions?

5. Describe two methods for finding the equation of the plane which contains three non-collinear points, A, B, and C.

6. a. Describe a method using vectors to determine whether three points lie on the same line.

   b. Describe a method for determining whether or not four points lie in the same plane.

7. Use the fact that the scalar product of three vectors is given by a determinant to establish Cramer's Rule for solving three equations in three unknowns:

$$a_{11}x + a_{12}y + a_{13}z = c_1$$

$$a_{21}x + a_{22}y + a_{23}z = c_2$$

$$a_{31}x + a_{32}y + a_{33}z = c_3$$

The remaining problems in this section all refer to the map below. The "tail" of the current vector is denoted by a (very) small circle, and the magnitude of the vector is proportional to its length.

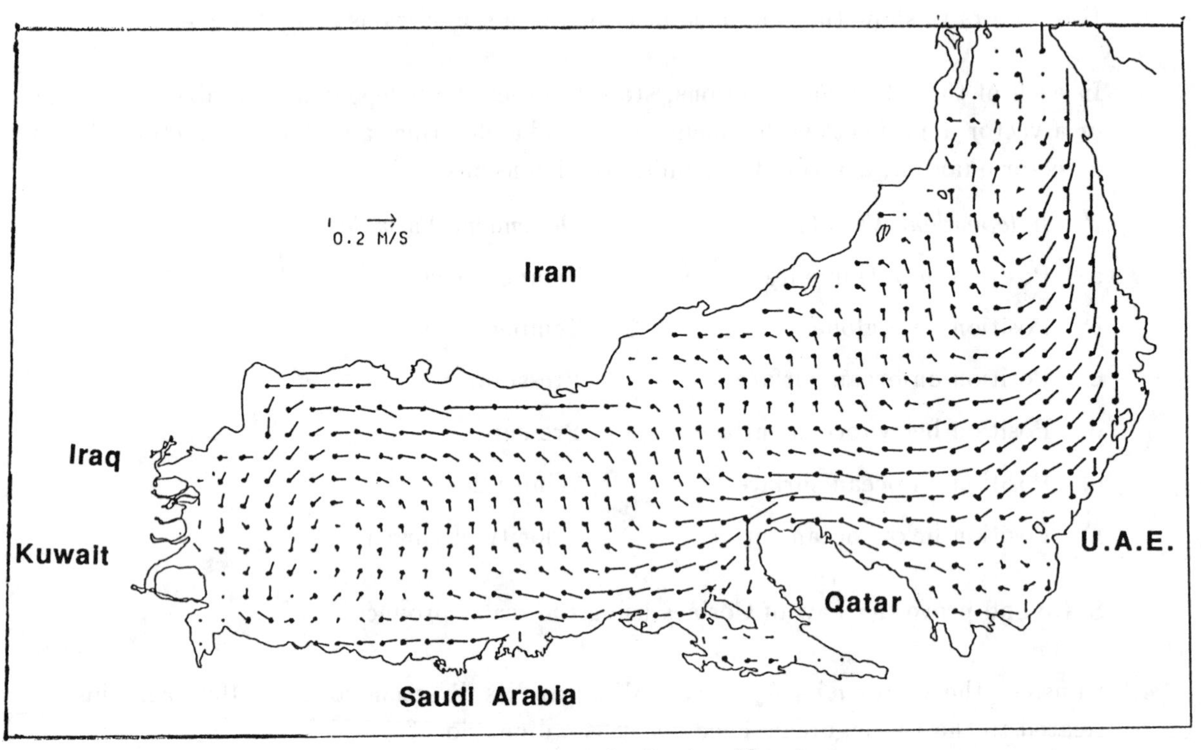

Annual average current in the top 10 meters.

8. In late January 1991, Saddam Hussein released a large oil slick from the general vicinity of Kuwait. Other than environmentalists, who was most upset, and why?

9. Locate the center of the major Persian Gulf gyre (whirlpool).

10. Off the coast of which country can one find a very small gyre?

11. Which Persian Gulf country would the Beach Boys most enjoy visiting? Which country would they least enjoy visiting? Why?

12. Which nation is most likely to be harmed by an oil spill taking place a short distance off their own shores? Which nations are least likely to be harmed by such catastrophes?

13. Political conditions notwithstanding, if your ship springs a leak, where are your chances of survival greatest?

14. Describe verbally the annual average current along the southern shore of the Gulf.

Section 2: Velocity and Acceleration

1. A space station is located at $(0,0,0)$. A rocket travels along the path

$$\mathbf{x}(t) = (\ln t - 2)\,\mathbf{i} + (\frac{2}{t} + 2)\,\mathbf{j} + (3t - 9)\,\mathbf{k},$$

where t is measured in appropriate units of time. If it shuts off its engines at $t = 1$, will the rocket coast into the space station?

2. (When Worlds Collide) Earth is moving on the path $\mathbf{x}(t) = t^2\,\mathbf{i} + 8\,\mathbf{j} + (5t - 6)\,\mathbf{k}$. Simultaneously the planet Xyra is moving along $\mathbf{y}(t) = (10-3t)\,\mathbf{i} + 2t^2\,\mathbf{j} + t^2\,\mathbf{k}$. In both cases, t is measured in unspecified units of time. When will the two worlds collide, and where? Which planet will be moving faster when they collide? At what angle will they collide?

3. [CP] Let $\mathbf{X}(t) = x(t)\,\mathbf{i} + y(t)\,\mathbf{j}$ be the equation of a curve in the plane. Assume that the velocity vector of the moving point is always perpendicular to the vector from the origin to the moving point. Show that the curve must lie on a circle centered at the origin. Hint: Note that $\mathbf{X}(t) \cdot \mathbf{X}'(t) = 0$. Write this as a scalar equation and go on to obtain an equation of the curve.

4. Suppose that the position of a particle is given by

$$\mathbf{x}(t) = a\cos rt\mathbf{i} + b\sin rt\mathbf{j}$$

where a, b, and r are constants and t is time measured in years.

   a. Explain why the path of the particle is an ellipse and show that the particle travels with constant angular velocity.

   b. Show that the particle is always accelerating toward the center of the ellipse.

5. Show that if a particle moves with constant speed, its acceleration vector is perpendicular to its velocity vector. Hint: Let $\mathbf{R}(t)$ denote the radius vector of the moving point. Then, if $\mathbf{v}(t)$ is the velocity vector and $\mathbf{a}(t)$ is the acceleration vector, $\|\mathbf{v}(t)\| = c$, a constant. So $\mathbf{v}(t) \cdot \mathbf{v}(t) = \|\mathbf{v}(t)\|^2 = c^2$.

6. Assume that a wheel of radius 1m is rotating at a constant angular velocity of b radians per minute, and that a bug is walking out from the center of the wheel (which is located at the origin) along a spoke of the wheel at a constant linear velocity of s m/minute relative to the center of the wheel.

   a. Show that $\mathbf{r}(t)$, the position vector from the center of the wheel to the bug is given by

   $$\mathbf{r}(t) = st(\cos bt)\,\mathbf{i} + st(\sin bt)\,\mathbf{j}$$

   where t is measured in minutes.

   b. Compute the velocity vector $\mathbf{v}_w$ and the acceleration vector $\mathbf{a}_w$ of the point on the end of the spoke along which the bug is walking.

   c. Show that the velocity vector $\mathbf{v}$ of the bug can be written as

   $$\mathbf{v} = s(\cos bt\,\mathbf{i} + \sin bt)\,\mathbf{j} + st\mathbf{v}_w.$$

   d. Show that the acceleration vector $\mathbf{a}$ of the bug can be written as $\mathbf{a} = 2s\mathbf{v}_w + st\mathbf{a}_w$.

Section 3: Arc Length

1. Which spring requires more material, one of radius 5 cm and height 4 m that makes three complete turns or one of height 3 cm and radius 4 cm that makes five complete turns?

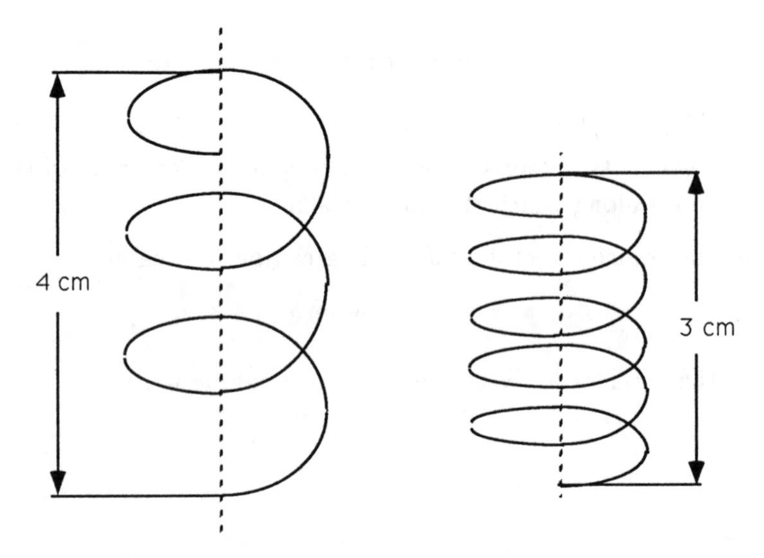

2. Can you see an important difference between the "upper and lower rectangles" approach to showing that the area under a curve can be evaluated as an integral, and the integral used for the computation of arc length? Can you think of some other situations in which this difference arises?

3. Let $\mathbf{x}(t) = (\cos 3t)\,\mathbf{i} + (\sin 3t)\,\mathbf{j} + 4t\,\mathbf{k}$.

   a. What type of curve is $\mathbf{x}(t)$? Reparametrize $\mathbf{x}(t)$ in terms of the arc length s.

   b. Use this formulation to compute the unit tangent and normal vectors $\mathbf{T}(s)$ and $\mathbf{N}(s)$, and show that $\mathbf{x}(s) + \mathbf{N}(s)$ lies on the z-axis.

   c. If you fit the best possible circle to the curve at any point, does the center of that circle lie on the z-axis? (This is called the "osculating circle".)

Chapter XI: The Derivative in Two and Three Variables

Section 1: Partial Derivatives

1. From the study of frost penetration in Iowa, the temperature T measured t days after the spring equinox and at a depth of x feet can be modeled by:

$$T(x,t) = T_0 + T_1 e^{-\lambda x} \sin\left(\tfrac{2\pi}{365}t - \lambda x\right)$$

with constant $\lambda \geq 0$.

a. What is the physical significance of $\dfrac{\partial T}{\partial x}$?

b. What is the physical significance of $\dfrac{\partial T}{\partial t}$?

c. What is the period of this function?

d. What is the physical significance of the term $-\lambda x$ within the sine function?

e. What is the physical significance of the factor $e^{-\lambda x}$ in this expression?

2. The intensity of light I at time t measured from sunrise and at an ocean depth of x feet is given by

$$I(x,t) = I_0 e^{-kx} \sin^3\left(\tfrac{\pi t}{D}\right),$$

where D is the length of daylight in hours, and I_0 and k are positive constants.

a. What is the physical significance of $\dfrac{\partial I}{\partial x}$?

b. What is the physical significance of $\dfrac{\partial I}{\partial t}$?

c. What is the physical significance of I_0 in the expression?

d. What is the physical significance of the factor e^{-kx} in the expression?

3. Each of the three functions listed below describes a surface. For each of them, find the slope of the curve formed by the intersection of the given surface with a plane perpendicular to the y-axis and passing through the given point.

   a. $f(x,y) = x^3y^2$ at $(1,2)$

   b. $f(x,y) = \cos(x+2y)$ at $\left(\frac{\pi}{4}, \frac{\pi}{2}\right)$

   c. $f(x,y) = ye^{xy}$ at $(2,3)$

4. The surface below is the graph of $z = f(x,y)$, over the region $-\pi \leq x \leq \pi$ and $0 \leq y \leq 2\pi$.

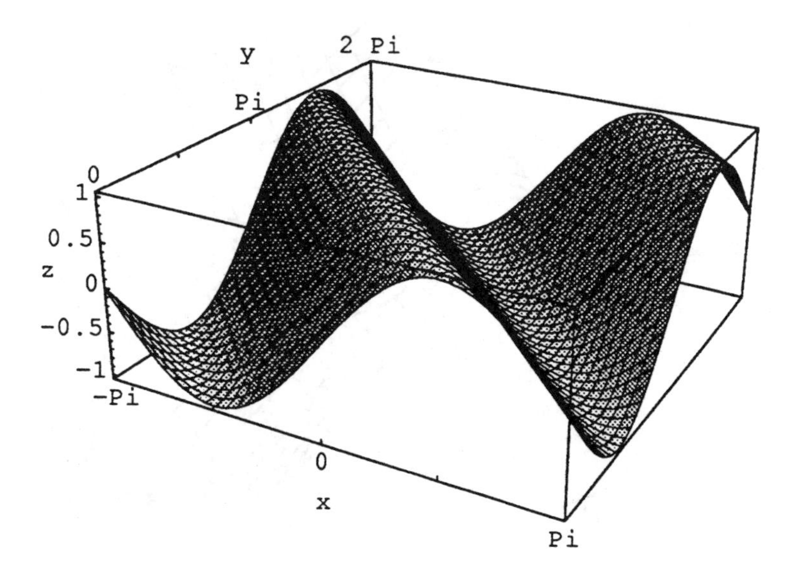

a. If a point P moves along the x-axis from $-\pi$ to π, then the partial derivative f_x at P is: (choose one)

A. always positive

B. first positive, then negative, and finally positive again

C. always negative

D. first negative, then positive, and finally negative again

E. none of the preceding four choices

b. If a point Q moves along the line $x = \pi$ (in the xy-plane) from $y = 0$ to $y = 2\pi$, then the partial derivative f_y at Q is: (Choose one of the possibilities from part a.)

c. Finally, if a point R moves along the y-axis from 0 to 2π, then describe the sign of the partial derivative f_y at R in a manner similar to the five choices of part a.

5. The graph of the surface $f(x, y) = \sin(\pi x) + y^2$ is given below.

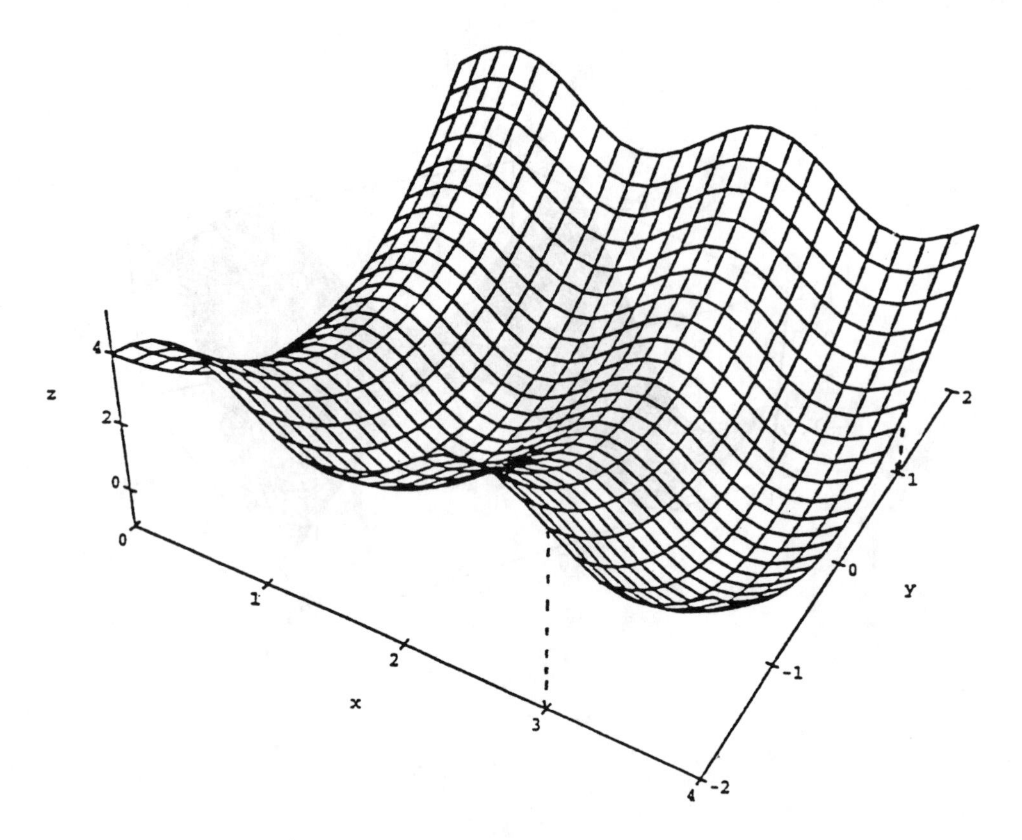

a. One of the curves drawn on this surface corresponds to the cross-section for which $x = 3$. Label that curve. Do the same for the curve for which $y = 1$.

b. Draw the tangent lines to these two cross-sections at the point $(3, 1, 1)$. Label the tangent line to the cross-section $x = 3$ by L_1 and to the cross-section $y = 1$ by L_2.

c. Find an equation for each of these tangent lines.

d. How would you describe the tangent plane to the surface at the point $(3, 1, 1)$?

6. The level curves of a surface $z = f(x,y)$ are given in the figure below.

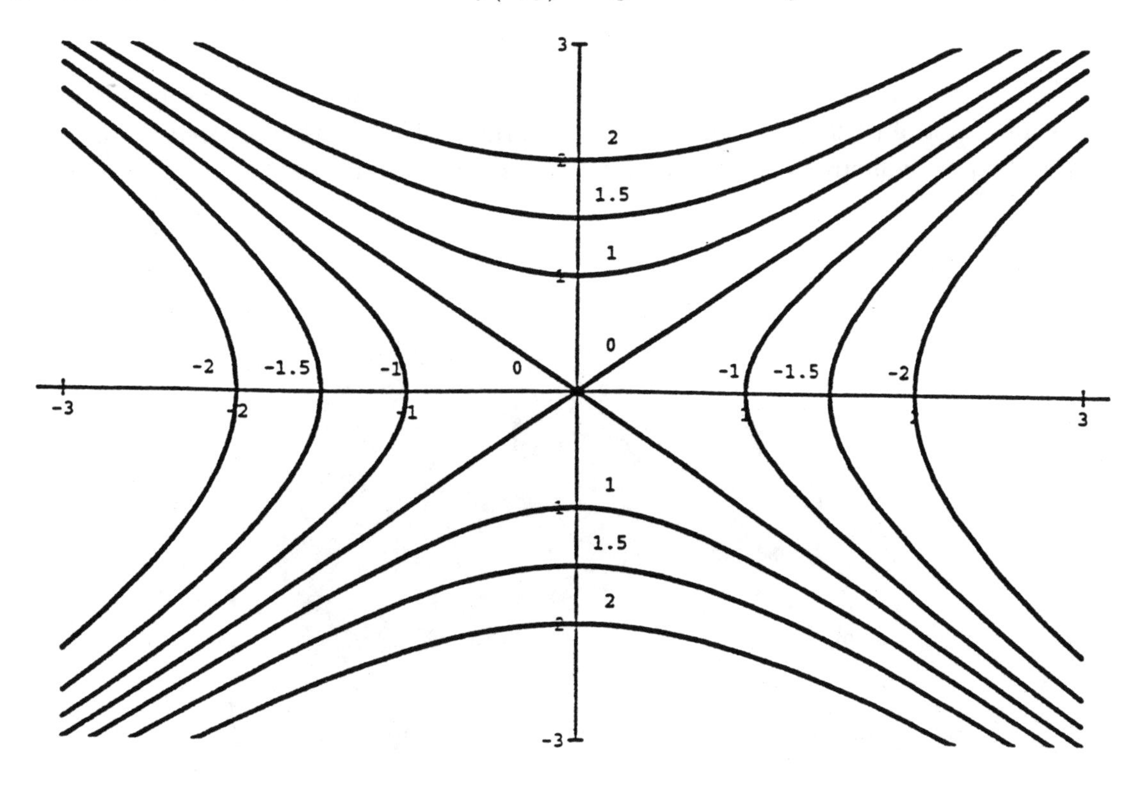

a. Estimate, as well as you can, f_x and f_y at the point $(1,1)$.

b. What are the values of f_x and f_y at the point $(0,0)$?

7. Suppose a region of vigorously rolling terrain can be modeled by

$$f(x,y) = 5 + \sin(\pi x + 2\pi y),$$

where $f(x,y)$ is the elevation in hundreds of feet at the point (x,y), where x is the distance east of $(0,0)$ and y is the distance north of $(0,0)$, both measured in hundreds of feet. See the figure below. (The vertical axis is marked in feet.)

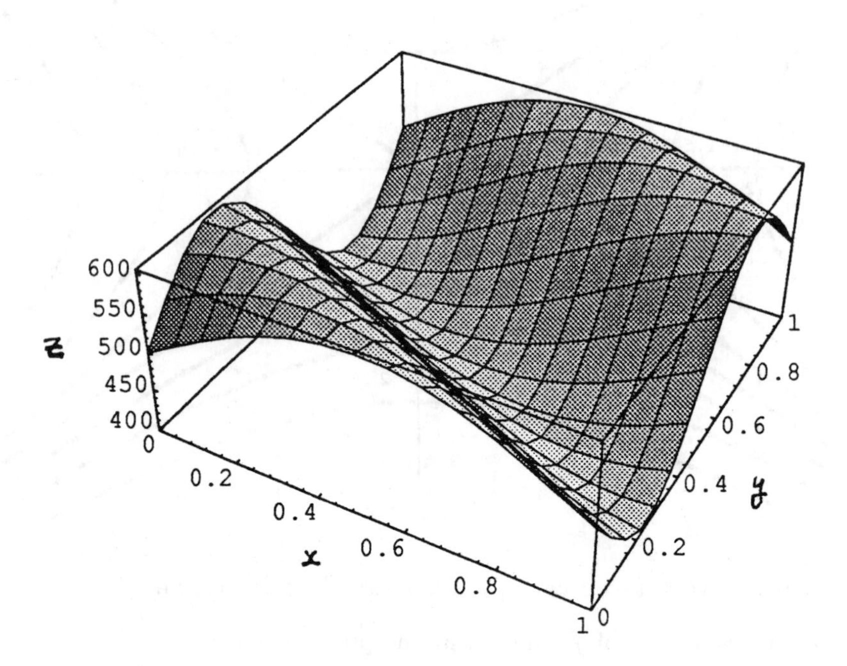

Suppose you are located 60 feet east and 80 feet north of $(0,0)$.

a. What is your elevation?

b. If you travel east from your location, how steep is the incline? Show this on the figure given.

c. If you travel north from your location, how steep is the incline? Show this on the figure given.

d. Which direction is steeper and why?

e. What can you say about the traveling west or south from your location?

f. In what general direction should you go to stay at the same elevation?

8. Suppose the barometric pressure (in mm of mercury) at a location (x, y) (where $(0, 0)$ is Des Moines, Iowa, and x and y are measured in hundreds of miles with x east and y north from Des Moines) at noon on July 4, 1991, can be modeled by:

$$P(x, y) = 30.20 - .1(x - 1)^2 - .2y^2$$

.

a. Draw the level curves, $P = 30.1$, 30.0, and 29.9 on the map of the state of Iowa below.. What is their significance in weather forecasting?

b. Des Moines (1), Iowa City (2), Ames(3), and Ottumwa (4) are indicated on the map. What is the average rate of change of barometric pressure from Des Moines to Iowa City, to Ames, and to Ottumwa?

c. Where is the barometric pressure a maximum? Where is it a minimum?

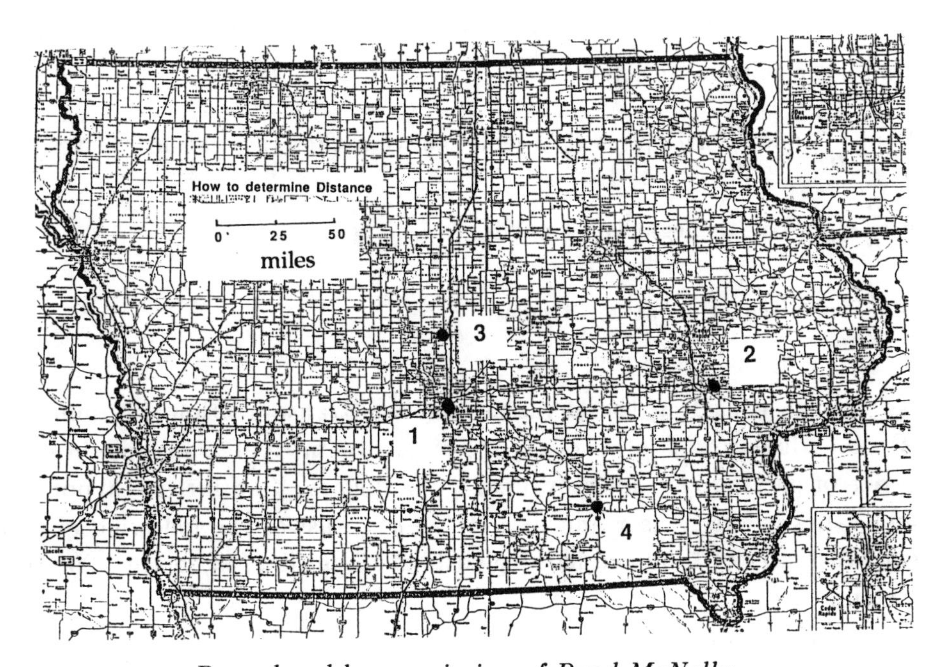

Reproduced by permission of Rand McNally

9. a. If $P(x,y) = x^2e^{3/y} - 4\sqrt{y}$, find the partial derivatives of P.

   b. Suppose you are manufacturing blackboards and whiteboards and that P is your monthly profit (in thousands of dollars) when your selling price for a blackboard is $\$x$ and for a whiteboard is $\$y$. In this setting, interpret $\frac{\partial P}{\partial y}$.

   c. Can $\frac{\partial P}{\partial y}$ be negative? Explain what this would mean.

   d. Can $\frac{\partial P}{\partial x}$ be negative? Explain what this would mean.

10. Suppose the number of crimes z committed in a city depends only upon the number x of unemployed people and the size y of the police force. This oversimplifies the problem, but let's make it even *simpler* by assuming that $z = f(x,y) = \frac{x^4y}{4e^y}$ where the domains of x and y are suitably selected.

    a. Find $\frac{\partial f}{\partial x}$ and $\frac{\partial f}{\partial y}$. Explain in a sentence what these partial derivatives measure.

    b. Is $\frac{\partial f}{\partial x}$ positive or negative? Explain.

    c. Is $\frac{\partial f}{\partial y}$ positive or negative? Explain.

11. A city official considers the demand for bus transportation (in number of bus riders) as a function f of two variables: x = price of a bus ride in dollars, and y = price of parking in a downtown garage in dollars. She further hypothesizes that demand varies inversely as x and directly as y^2. Answer the following questions if at present, with fares of \$ .50 and parking costs of \$6 per day, there are 7,200 riders using the system.

    a. Find $f(\frac{1}{2}, 5)$. What does this number mean?

    b. Find $\frac{\partial f}{\partial x}$ and $\frac{\partial f}{\partial y}$.

    c. Explain what $\frac{\partial f}{\partial x}$ measures. Why is it negative?

    d. Explain what $\frac{\partial f}{\partial y}$ measures. Why is it positive?

Section 2: Gradient and Directional Derivatives

1. You need to explore the vicinity of your campus to find a region where the terrain varies in elevation, and yet you can easily walk over it. (This will be easier to do in some states than in others). After locating this region, find a path of constant elevation and walk along it a couple of times to get a feel for your path. Then answer the following questions.

   a. How would you describe the path in mathematical terms?

   b. If you stop at some point along your path, in which direction would you have to turn to find the direction of steepest incline up the hill?

   c. At the location where you stopped for part *b*, in which direction would you have to turn to find the direction of steepest decline down the hill?

   d. What theorem have you just demonstrated by this exploration?

2. The level curve of $f(x, y) = 3xe^{x^2 y}$ through the point $(-1, 0)$ is shown below.

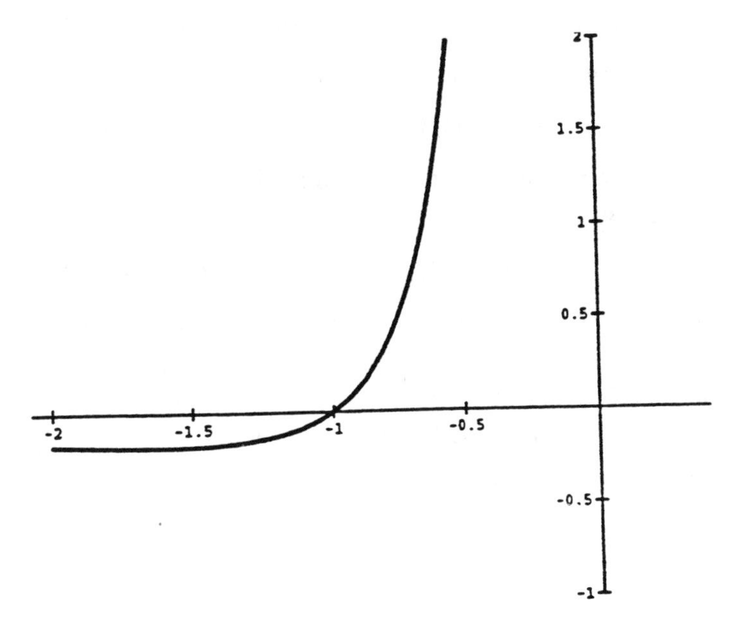

   a. Draw the unit vectors $\left(\frac{1}{\sqrt{2}}, \frac{1}{\sqrt{2}}\right)$ and $\left(\frac{1}{\sqrt{2}}, -\frac{1}{\sqrt{2}}\right)$ with their tails at $(-1, 0)$.

   b. Which one of these vectors points in the direction in which f is increasing most rapidly at $(-1, 0)$? Why?

   c. At what rate is f increasing in the direction of the other vector? Explain why from the figure.

3. Suppose you are at the point with coordinates $(.6, .8)$ in a region where the altitude is given by $f(x, y) = \sin(\pi x + 2\pi y)$. In what direction(s) should you go in order to stay at the same elevation? Justify your answer with a brief description of how you solve this problem.

4. a. How are the directional derivative and the gradient of a function related?

   b. When is the directional derivative of a function: a maximum, a minimum, 0, and $\frac{1}{2}$ its maximum value?

5. Suppose that you are given only the following information about a function f: $f(8, 5) = 33.1$, $f(8.01, 5) = 33.3$, and $f(8, 5.02) = 33.0$. Estimate $f_x(8, 5)$, $f_y(8, 5)$, $\nabla f(8, 5)$, $D_{\mathbf{u}} f(8, 5)$ where $\mathbf{u} = \frac{1}{5}(3\mathbf{i} + 4\mathbf{j})$.

6. The depth in feet of a crater lake is given by $f(x, y) = 100 - .015x^2 - .001y^2$ where x and y are measured in feet from the center of the lake. Jimmy, who is not a very good swimmer, is in the water at the point $(25, 90)$.

   a. Sketch a contour map of the lake showing its depths.

   b. How deep is the water at Jimmy's location?

   c. Find the rate at which the depth changes if he swims towards the point $(26, 90)$.

   d. Find the rate at which the depth changes if he swims toward a diving platform located at $(26, 89)$.

   e. In which direction would the depth of water increase most rapidly for Jimmy?

   f. Since Jimmy is not a very good swimmer, in which direction should he swim?

7. The equations $z = f(x, y)$ and $F(x, y, z) = f(x, y) - z = 0$ describe the same surface in 3-space. Yet ∇f and ∇F are not the same vectors.

   a. What is the difference? In what way are the two gradients related?

   b. If $\nabla f = 2\mathbf{i} - 3\mathbf{j}$, find a vector normal to the surface $z = f(x, y)$?

8. At a given point, the directional derivatives of $f(x, y)$ are known in two nonparallel directions, given by unit vectors $\mathbf{u}$ and $\mathbf{v}$. Can you determine ∇f at this point? If so, then how do you do it?

Section 3: Equation of the Tangent Plane

1. To find the equation of the tangent plane to a surface, you need some information about the plane.

   a. What is the most convenient form of this information?

   b. What other forms might this information take and how can you change it into the form you need?

2. You are given only the following information about a function f:

$$f(8,5) = 33.1, \qquad f(8.01,5) = 33.3, \qquad f(8,5.02) = 33.0.$$

   a. Approximate the equation of the tangent plane to the surface at $(8,5)$.

   b. Given only the three pieces of information about f in this problem, it is impossible to find the value of f at the point $(8.01, 5.02)$; however, you can *approximate* the value of f there. Do so.

3. Suppose that you are employed on an earth removal project in western Illinois. From data obtained at the work site, you know the following: The elevation at a point $(0,0)$ is 720 ft. The rate of change of elevation is 10 ft per 100 feet towards the east and 8 ft per 100 feet to the north. Your job is to approximate the amount of overburden (soil and rock) that must be moved to level an area 250 feet east and 500 feet north from $(0,0)$ to an elevation of 720 ft. The following parts will lead you through a procedure for solving this problem.

   a. Assume that the current surface in this area is a plane. Find the equation of this plane.

   b. Draw a rectangle 250 by 500 and then draw level curves in this rectangle for elevations of 725, 730, 735, etc.

   c. What is the maximum elevation over this region and what is the average elevation?

   d. Find the volume of overburden that needs to be removed.

   e. Suppose that our goal is not to *remove* soil and rock but only to *redistribute* it over the work site so that the site is level. How much overburden must be moved?

4. What is the equation of the tangent plane to the surface S: $z = |x^2 - y^2|$ at $(0,0)$?

Section 4: Optimization

1. Use what you know about simple polynomials, exponential and trigonometric functions to determine the absolute maxima and/or minima of the following functions (for all values of x and y).

   a. $f(x,y) = x^2 + y^2$

   b. $f(x,y) = e^{x^2+y^2}$

   c. $f(x,y) = e^{-(x^2+y^2)}$

   d. $f(x,y) = \cos x \sin y$

   e. $f(x,y) = x^2 - y^2$

   f. $f(x,y) = (\cos x)(\cos y)e^{-\sqrt{x^2+y^2}}$

2. Consider the problem of finding the maximum and minimum value of $f(x,y) = 4x^2 - 3y^2 + 2xy$ on the square $0 \le x \le 1$, and $0 \le y \le 1$.

   a. Does f have a critical point in the interior of the square?

   b. Find critical values for f on the boundary by examining how f behaves on each of the line segments which make up the boundary. First set $x = 0$, then $x = 1$, next $y = 0$, and finally $y = 1$.

   c. What are the absolute maxima and minima of f over the square?

3. Let $T(x,y) = x^2 - 2xy$ be the temperature at the point (x,y) in the region bounded by the curves $y = x$ and $y = x^2$. Suppose that a bug is crawling around the region.

   a. At $\left(\frac{1}{2}, \frac{1}{3}\right)$, in what direction should the bug go to cool down as quickly as possible?

   b. At $\left(\frac{1}{2}, \frac{1}{3}\right)$, in what directions should the bug go to maintain its current temperature?

   c. Where is the hottest point in the region? Explain your answer.

   d. If, at $\left(\frac{1}{2}, \frac{1}{3}\right)$, the bug moves in such a way that for each change in its x-direction of 2 units and the change in its y-direction is -1 unit, find $\dfrac{dT}{dt}$, the change in the temperature from the bug's point of view.

4. Find the shortest distance between the origin $(0,0,0)$ and each of the following surfaces S. As you work through these questions, keep in mind what S looks like and what techniques are the most appropriate to find the distance required.

   a. $S:\ x + 2y - z = 6$

   b. $S:\ x^2 + y^2 + z^2 = 9$

5. Continue, as with the previous problem, to find the shortest distance between the origin and each of the surfaces given below.

   a. $S: \ xyz = 4$

   b. $S: \ \cos x + \sin y + z = 2$

Chapter XII: Line Integrals

Section 1: Line Integrals

1. Some values of a continuous scalar-valued function $f(x,y)$ are given in the following table.

	$x=\frac{1}{4}$	$x=\frac{3}{4}$	$x=\frac{5}{4}$	$x=\frac{7}{4}$
$y=\frac{7}{4}$	-1	0	1	0
$y=\frac{5}{4}$	0	1	1	-1
$y=\frac{3}{4}$	2	1	-1	-2
$y=\frac{1}{4}$	4	3	1	-1

a. Estimate the value of $\int_{C_1} f(x,y)\,ds$, where C_1 is the vertical line segment running from $(\frac{1}{4},0)$ to $(\frac{1}{4},2)$. Justify your answer.

b. Estimate the value of $\int_{C_2} f(x,y)\,ds$, where C_2 is the horizontal line segment running from $(0,\frac{3}{4})$ to $(2,\frac{3}{4})$. Justify your answer.

c. Estimate the value of $\int_{C_3} f(x,y)\,ds$, where C_3 is the diagonal line segment running from $(0,0)$ to $(2,2)$. Justify your answer.

2. Suppose f is a continuous scalar function and let C be a level curve of f, i.e., $C=\{(x,y)\mid f(x,y)=k\}$ for some constant k. Express $\int_C f(x,y)\,ds$ in terms of k and the arc length of C.

3. Let f be a continuous real-valued function. Find a formula for the length of the graph of f between $(a,f(a))$ and $(b,f(b))$.

4. Let f be a continuous non-negative function. Find a formula for the length of the curve given in polar coordinates by $r=f(\theta)$ between $\theta=a$ and $\theta=b$.

5. Let $f(x,y) = \begin{cases} 2 & \text{if } x + y \leq 1, \\ 1 & \text{otherwise.} \end{cases}$

   a. Let C_1 be the line segment running from $(-2,0)$ to $(3,0)$. Find $\int_{C_1} f(x,y)\,ds$.

   b. Let C_2 be the upper half of the circle of radius 1 and center $(1,0)$. Find $\int_{C_2} f(x,y)\,ds$.

6. a. One end of a straight wire is attached to a pole 1 meter above the ground, and the other end is attached to the ground 1 meter from the base of the pole. Beads are attached to the wire at the top end and are dropped down the wire. The speed v of such a bead traveling down the wire depends on its distance to the ground according to the formula $v = \sqrt{2g(1-y)}$, where $g = 9.8\,\text{m/sec}^2$ is the acceleration due to gravity and y is the height of the object above the ground. How long does it take a bead to travel from the top of the wire to the bottom?

   b. Suppose that instead of the straight wire of part a, the wire is curved to form one fourth of the circle $(x-1)^2 + (y-1)^2 = 1$, where x is the horizontal distance from the base of the pole and y the height above the ground. The two ends of the wire remain in the same positions they had in part a. Now how long will it take a bead to travel down the wire? (The formula for speed given in part a is valid regardless of the shape of the wire.) Set up the appropriate integral and estimate it using a numerical method. Do not attempt to work this integral by hand.

   c. Galileo thought that the circular path of part *b* is the shape of the wire for which the bead would reach the ground faster than any other path with the same endpoints. Try some other paths to test out Galileo's conjecture. The problem to find out the fastest possible path is a famous one known as the *Brachistochrone Problem.*

7. You are traveling in an all-terrain vehicle in a swamp. The speed at which you can travel depends on how wet the ground is, which gets worse the closer you are to the center of the swamp. More precisely, if you are r miles from the center of the swamp, you can go $v(r) = 5 + 10r$ miles per hour. You are presently located 2 miles due west of the center of the swamp and wish to get to the point 2 miles due east of the center of the swamp.

   a. How long would it take you to travel on a straight line to the point 2 miles due east of the center of the swamp? Would it be faster to get to this point by avoiding the worse parts of the swamp by traveling on a semicircle of radius 2 centered at the center of the swamp? Assume that you always travel at the maximum possible speed.

   b. Another possible path would be to travel on a straight line to the point 1 mile due south of the center and then on another straight line to the point two miles due east of the swamp. How long would it take you to travel along this path at the maximum possible speed? Set up the appropriate integral and estimate it using a numerical method. Do not attempt to work this integral by hand.

8. Determine whether the following statements are always true or at least sometimes false. Justify your answers.

   a. If γ_1 and γ_2 are two parametrizations of the same curve, then for any continuous scalar field f, $\int_{\gamma_1} f\, ds = \int_{\gamma_2} f\, ds$.

   b. If γ_1 and γ_2 are two parametrizations of the same curve, then for any continuous vector field $\mathbf{f}$, $\int_{\gamma_1} \mathbf{f}(\mathbf{r}) \cdot d\mathbf{r} = \int_{\gamma_2} \mathbf{f}(\mathbf{r}) \cdot d\mathbf{r}$.

   c. If C is a differentiable curve whose tangent vectors are orthogonal to a continuous vector field $\mathbf{f}$ at each point on C, then $\int_C \mathbf{f}(\mathbf{r}) \cdot d\mathbf{r} = 0$.

   d. If C is a differentiable curve whose tangent vectors are parallel to a continuous vector field $\mathbf{f}$ at each point on C (i.e., $\mathbf{f}$ is a non-negative scalar multiple of the tangent to C at each point of C), then $\int_C \mathbf{f}(\mathbf{r}) \cdot d\mathbf{r} = \int_C \|\mathbf{f}\|\, ds$.

   e. If $\gamma(t) = (\cos t, \sin t)$ and $\alpha(t) = (\cos 2t, \sin 2t)$, then for any continuous vector field $\mathbf{f}$, $2\int_0^{2\pi} \mathbf{f}(\gamma(t)) \cdot \gamma'(t)\, dt = \int_0^{2\pi} \mathbf{f}(\alpha(t)) \cdot \alpha'(t)\, dt$.

   f. If f is a continuous scalar field on $\mathbb{R}^2$ and $\int_C f(x,y)\, ds = 0$ for any curve C, then $f(x,y) = 0$ for all (x,y) in $\mathbb{R}^2$.

9. Let $\mathbf{f}$ be the vector field given in the diagram below.

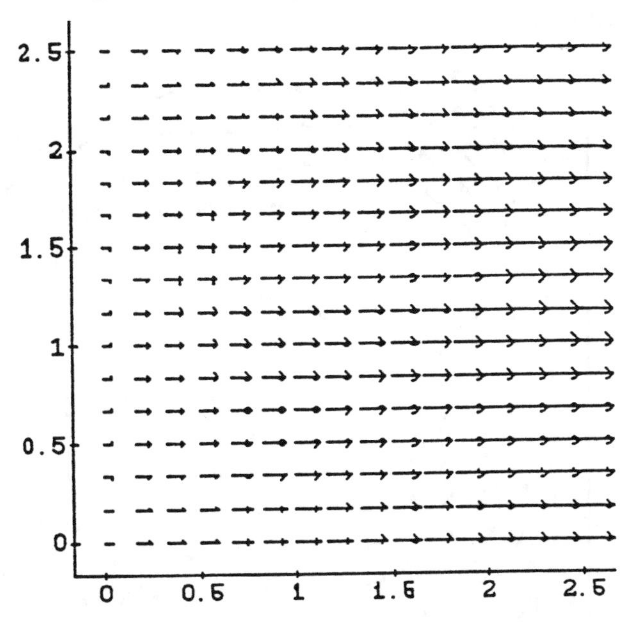

a. Let C be any vertical line segment. Is $\int_C \mathbf{f}(\mathbf{r}) \cdot d\mathbf{r}$ positive, negative, or zero? Justify your answer.

b. Let C_1 be the line segment running from $(0,0)$ to $(2,1)$. Is $\int_{C_1} \mathbf{f}(\mathbf{r}) \cdot d\mathbf{r}$ positive, negative or zero? Justify your answer.

c. Let C_2 be the horizontal line segment between $(0,1)$ and $(2,1)$ and let C_1 be the line segment given in part *b*. Is $\int_{C_1} \mathbf{f}(\mathbf{r}) \cdot d\mathbf{r}$ greater than, less than, or equal to $\int_{C_2} \mathbf{f}(\mathbf{r}) \cdot d\mathbf{r}$? Justify your answer.

d. Is $\mathbf{f}$ a conservative vector field? Justify your answer.

10. Let $\mathbf{g}$ be the vector field given in the diagram below. Note that the origin is in the center of the diagram.

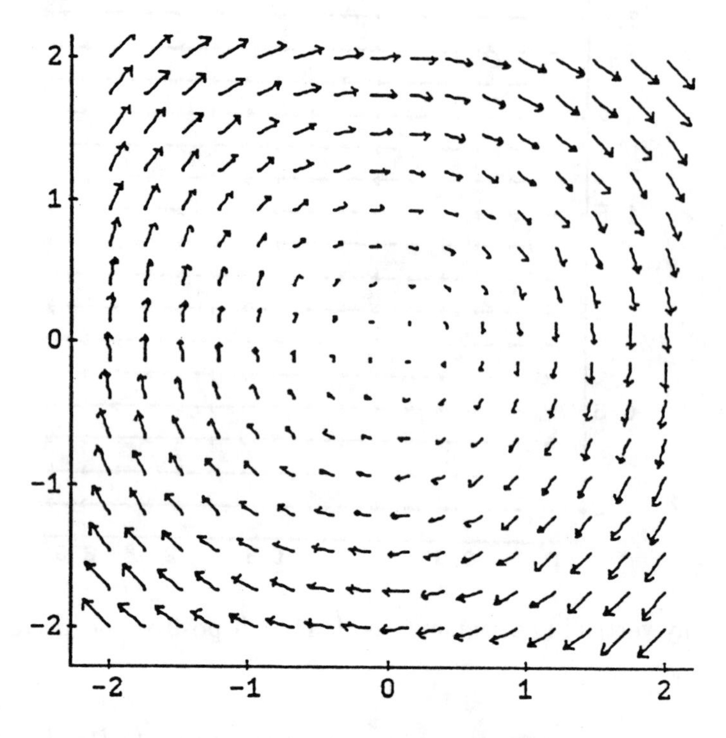

a. Let C_1 be the horizontal line segment running from $(0, \frac{3}{2})$ to $(2, \frac{3}{2})$. Is $\int_{C_1} \mathbf{g}(r) \cdot d\mathbf{r}$ positive, negative, or zero? Justify your answer.

b. Let C_2 be the circle with radius 1 and center $(0, 0)$, oriented counterclockwise. Is $\int_{C_2} \mathbf{g}(\mathbf{r}) \cdot d\mathbf{r}$ positive, negative, or zero? Justify your answer.

c. Is $\mathbf{g}$ a conservative vector field? Justify your answer.

11. Let $\mathbf{h}$ be the vector field given in the diagram below. Note that the origin is in the center of the diagram.

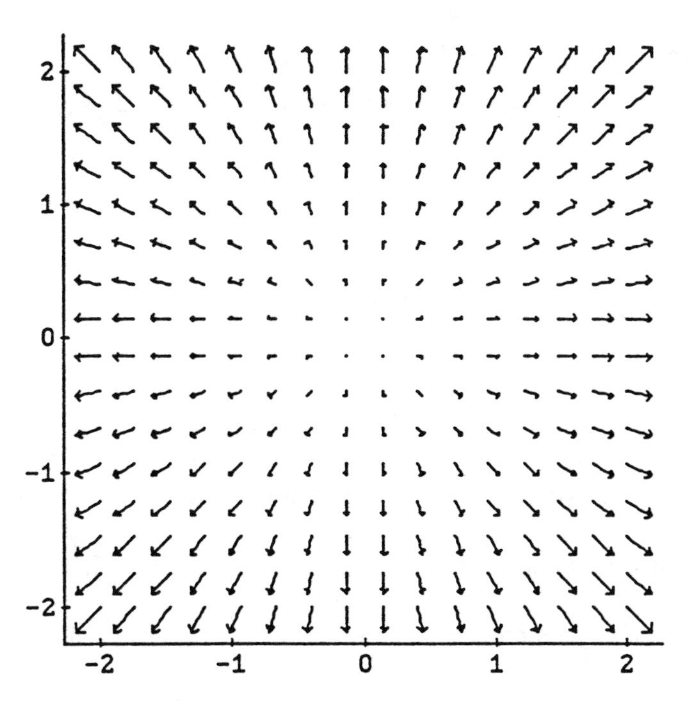

a. Let C_1 be the horizontal line segment running from $(0,1)$ to $(2,1)$. Is $\int_{C_1} \mathbf{h}(\mathbf{r}) \cdot d\mathbf{r}$ positive, negative, or zero? Justify your answer.

b. Let C_2 be the horizontal line segment running from $(-2,1)$ to $(2,1)$. Is $\int_{C_2} \mathbf{h}(\mathbf{r}) \cdot d\mathbf{r}$ positive, negative, or zero? Justify your answer.

c. Let C_3 be the circle with radius 1 and center $(0,0)$, oriented counterclockwise. Is $\int_{C_3} \mathbf{h}(\mathbf{r}) \cdot d\mathbf{r}$ positive, negative, or zero? Justify your answer.

d. Is $\mathbf{h}$ a conservative vector field? Justify your answer.

Section 2: Conservative Vector Fields and Green's Theorem

1. Let $\mathbf{f}(x, y) = (2x + y)\cos(x^2 + xy)\,\mathbf{i} + (x\cos(x^2 + xy) + 1)\,\mathbf{j}$.

   a. Is $\mathbf{f}$ a conservative vector field? Justify your answer.

   b. Let C be the curve parametrized by $\gamma(t) = (\sin t, 1 - \cos t)$, $0 \le t \le \pi$. Find $\int_C \mathbf{f}(\mathbf{r}) \cdot d\mathbf{r}$.

   c. Let $\mathbf{g}(x, y) = (2x + y)\cos(x^2 + xy)\,\mathbf{i} + (x\cos(x^2 + xy) + 1 + x)\,\mathbf{j}$ and let C be the curve given in part *b*. Find $\int_C \mathbf{g}(\mathbf{r}) \cdot d\mathbf{r}$.

2. Let $P(x, y) = \dfrac{-x}{(x^2 + y^2)^{3/2}}$ and $Q(x, y) = \dfrac{-y}{(x^2 + y^2)^{3/2}}$.

   a. Show that, except at the origin (where the functions are not defined), $\dfrac{\partial P}{\partial y} = \dfrac{\partial Q}{\partial x}$.

   b. Find a potential function for $\mathbf{f}(x, y) = P(x, y)\,\mathbf{i} + Q(x, y)\,\mathbf{j}$.

   c. Suppose C is a continuous curve running from $(1, 0)$ to $(3, 4)$. Find $\int_C \mathbf{f}(\mathbf{r}) \cdot d\mathbf{r}$.

3. Let $\mathbf{f}(x, y, z) = \dfrac{-1}{(x^2 + y^2 + z^2)^{3/2}}(x\,\mathbf{i} + y\,\mathbf{j} + z\,\mathbf{k})$.

   a. Show that except at the origin (where $\mathbf{f}$ is not defined), $\nabla \times \mathbf{f} = \mathbf{0}$.

   b. Find a potential function for $\mathbf{f}$.

   c. Give a physical situation that will give rise to this vector field.

4. Let $P(x,y) = \dfrac{-y}{x^2+y^2}$ and $Q(x,y) = \dfrac{x}{x^2+y^2}$.

   a. Show that except at the origin (where the functions are not defined), $\dfrac{\partial P}{\partial y} = \dfrac{\partial Q}{\partial x}$.

   b. Let C be any circle centered at the origin, oriented counterclockwise. Find $\displaystyle\int_C P\,dx + Q\,dy$.

   c. Explain why part a doesn't imply that the integral in part *b* is zero.

   d. Let C_1 be any simple closed curve that does *not* contain the origin. Find $\displaystyle\int_{C_1} P\,dx + Q\,dy$.

   e. Let C_2 be any simple closed curve that contains the origin and is oriented counterclockwise. Find $\displaystyle\int_{C_2} P\,dx + Q\,dy$.

5. Suppose $\mathbf{F}$ is a continuously differentiable *radial* vector field on $\mathbb{R}^2$. This means that there is a continuously differentiable function of one variable f such that $\mathbf{F}(x,y) = f(x^2+y^2)(x\,\mathbf{i} + y\,\mathbf{j})$. Show that $\mathbf{F}$ is conservative.

6. Suppose f and g are continuously differentiable scalar fields and C is a closed curve. Prove that $\displaystyle\int_C \big(f(\mathbf{r})\nabla g(\mathbf{r})\big)\cdot d\mathbf{r} = -\int_C \big(g(\mathbf{r})\nabla f(\mathbf{r})\big)\cdot d\mathbf{r}$.

7. Which of the following statements are always true and which are at least sometimes false? Justify your answers. Assume that all functions are continuously differentiable.

   a. If P and Q are scalar fields defined for all points in $\mathbb{R}^2$ except the origin, and $\dfrac{\partial P}{\partial y} = \dfrac{\partial Q}{\partial x}$ for all points where these functions are defined, then for any closed curve C in $\mathbb{R}^2 - \{0\}$, $\displaystyle\int_C P\,dx + Q\,dy = 0$.

   b. If C is a closed curve and $\mathbf{f}$ is a vector field in $\mathbb{R}^2$ such that $\displaystyle\int_C \mathbf{f}(\mathbf{r})\cdot d\mathbf{r} = 0$, then $\mathbf{f}$ is the gradient of a scalar field in the interior of C.

   c. If $\displaystyle\int_C \mathbf{f}(\mathbf{r})\cdot d\mathbf{r} = 0$ for every closed curve C in $\mathbb{R}^2$, then $\mathbf{f}$ is the gradient of a scalar field.

   d. If $\mathbf{f}$ is a vector field on $\mathbb{R}^3$ whose curl is never zero and C is any closed curve in $\mathbb{R}^3$, then $\displaystyle\int_C \mathbf{f}(\mathbf{r})\cdot d\mathbf{r} \neq \mathbf{0}$.

8. a. Show that for any closed simple curve C oriented counterclockwise, the area inside C is equal to $\int_C x\,dy$.

   b. Use part a to calculate the area inside the ellipse C which has a parametric representation $\psi(t) = (a\cos t, b\sin t)$, $0 \le t \le 2\pi$, where a and b are positive constants.

   c. Show that $\int_C x\,dy = -\int_C y\,dx$.

9. Let C_r be the circle of radius r in $\mathbb{R}^2$ centered at the origin, oriented counterclockwise, and suppose P and Q are continuously differentiable scalar fields on $\mathbb{R}^2$. Find

$$\lim_{r\to 0} \frac{1}{\pi r^2} \int_{C_r} P\,dx + Q\,dy$$

in terms of the partial derivatives of P and Q.

Chapter I: Functions and Graphs

Section 1: Domain and Range. Elementary Functions

1. This problem is primarily designed to help students realize that functions can be used to represent correlations between measurements and to get them to develop the habit of translating English phrases into mathematical relationships.

   a. $A = (H\cos X)(H\sin X)/2 = (H^2 \sin 2X)/4$
   Domain $= \{x \mid 0 < x < \pi/2\}$, Range $= \{x \mid 0 < x < H^2/4\}$.

   b. $H = \sqrt{L^2+4}$, Domain $= \{x \mid x > 0\}$, Range $= \{x \mid x > 2\}$.

   c. $L = \sqrt{H^2-25}$, Domain $= \{x \mid x > 5\}$, Range $= \{x \mid x > 0\}$.

   d. $F = 1.8C + 32$, Domain $= \{x \mid x > -273\}$, Range $= \{x \mid x > -459.4\}$.

   Some quibbling can be allowed on this one.

   e. $y = \text{max}(x, 1-x) = \begin{cases} 1-x & \text{if } x < \frac{1}{2} \\ x & \text{if } x \geq \frac{1}{2} \end{cases}$
   Domain = all real numbers, Range $= \{x \mid x \geq \frac{1}{2}\}$

   This is a good example of a piecewise-defined function.

   f. $A = \frac{1}{2}XR^2$ (Note that X must be measured in radians.)
   Domain $= \{x \mid 0 < x < 2\pi\}$, Range $= \{x \mid 0 < x < \pi R^2\}$.

2. $f(x) = 1$ if $x = -1.5 + 3n$, where n is an integer. $f(x) = 0.5$ if $x = -1.75 + 3n$ or $x = -0.25 + 3n$, where n is an integer. Range $= \{x \mid 0 \leq x \leq 1\}$.

   [MN]

3. A quadratic and a linear function have at most two points in common. Tangency occurs when they have precisely one point in common. This means that the equation $x^2 + kx = 5$ must have precisely one solution, which will occur when the discriminant of $x^2 + kx - 5$ is 0. This can never happen. To see this geometrically, observe that $x^2 + kx$ opens upward and must pass through the origin.

   [MN]

4. In slope-intercept form, $y = -\frac{2}{k}x - 1$.

   a. $-\frac{2}{k} = 3 \Longrightarrow k = -\frac{2}{3}$.

   b. The y-intercept is *always* -1.

   c. The slope $-\frac{2}{k}$ can never be 0.

   d. If $k = 0$, the line $x = 0$ is vertical.

   e. The slope of the line perpendicular to the given line is $\frac{k}{2}$. If this perpendicular line passes through the origin, the equation of the line is $y = \frac{k}{2}x$. The intersection of this line with the original line occurs when $-\frac{2}{k}x - 1 = \frac{k}{2}x$. Therefore $x = -1/(\frac{k}{2} + \frac{2}{k}) = -2k/(k^2 + 4)$. The y-coordinate of this point is $-k^2/(k^2 + 4)$. This point lies within 1 unit of the origin if

$$x^2 + y^2 = (4k^2 + k^4)/(k^2 + 4)^2 \leq 1,$$

   which simplifies to $4k^2 \geq -16$, so the line always passes within 1 unit of the origin! This result can easily be seen from the original equation since the line passes through $(0, -1)$, which is exactly 1 unit from the origin.

[MN]

5. The primary purpose of this exercise is to see if the student is comfortable with elementary transformations applied to functions.

a.

```
Plot[f[x],{x,-2,4}]
```

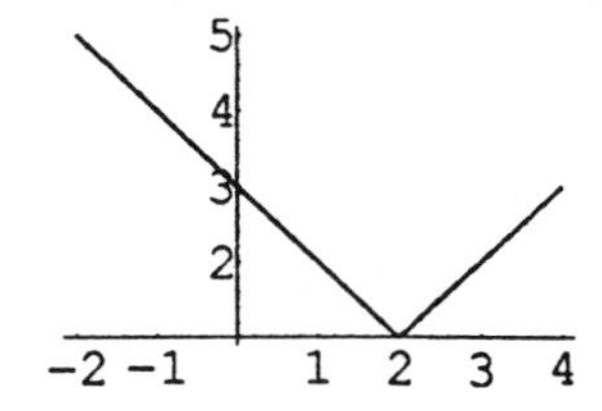

b.

```
Plot[-f[-x],{x,-4,2}]
```

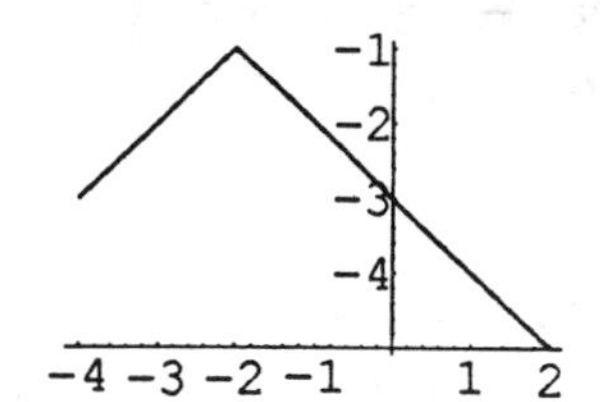

c.

```
Plot[f[x+2],{x,-3,3}]
```

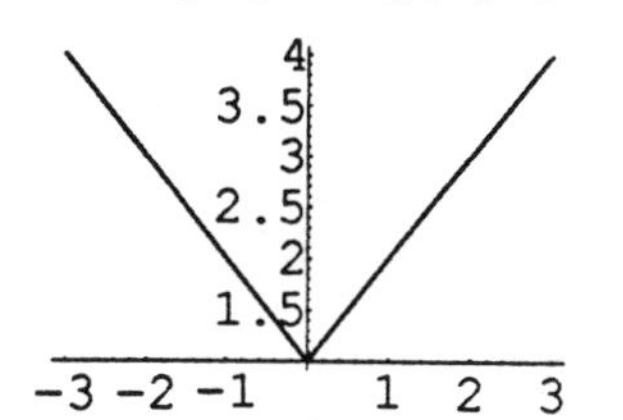

d.

```
Plot[f[2x],{x,-1,3}]
```

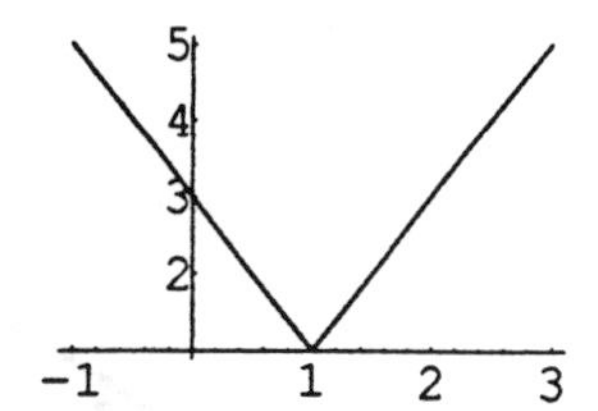

e.

```
Plot[f[3x-6],{x,0,5}]
```

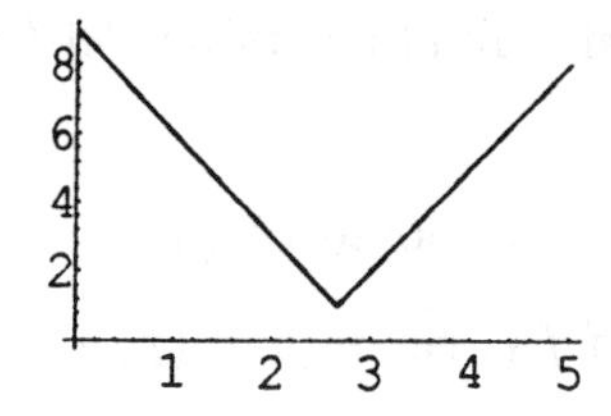

Section 2: Trigonometric Functions

1. Two reasons are that polynomials have only a finite number of zeros, whereas the cosine has an infinite number of zeros and that any non-constant polynomial is unbounded, whereas $-1 \leq \cos x \leq 1$. This shows that the cosine function is outside the class of polynomials.

[MN]

2. This reacquaints the student with right-triangle trigonometry. Unroll the stripe to form a right triangle with height 100 cm and base the circumference of a circle with radius 30. The base of the triangle is 60π. Thus $\tan A = \frac{100}{60\pi} = 0.5305$. The angle A is therefore approximately $27.947°$, and the stripe's length is $\sqrt{100^2 + (60\pi)^2}$, which is approximately 213.379 cm.

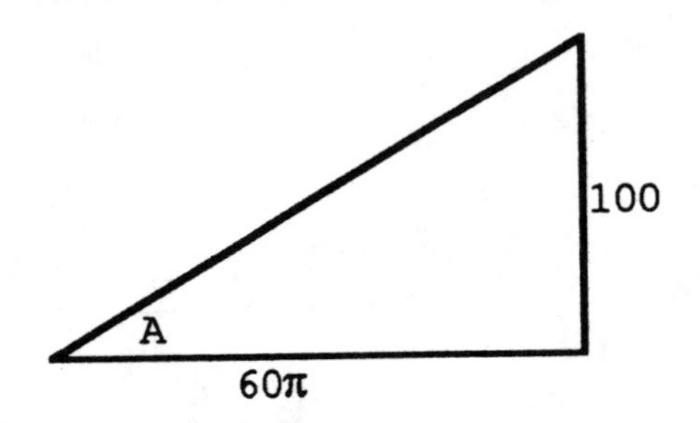

[MN]

3. Write $y = f(x) = a \sin[b(x - (-\frac{c}{b})]$. Then the amplitude of the graph is a, the period is $2\pi/b$, and the phase angle is $-\frac{c}{b}$. The graph G of f is obtained from S, the graph of the sine function ($y = \sin x$), by shifting S $\frac{c}{b}$ to the left, dividing its period (2π) by b, and multiplying its amplitude (1) by a.

   a. The graph G is shifted twice as far to the left ($\frac{2c}{b}$ as opposed to $\frac{c}{b}$).

   b. G "oscillates" twice as much because the period of f is cut in half (π/b as opposed to $2\pi/b$).

   c. The amplitude doubles, so G is "stretched" twice as far.

[MN]

4. a. One can show that $\sin(\frac{\pi}{2} - x) = \cos x$ by using the addition formula for the sine, but it is also important that students see that the defining relation for sine in a right triangle (side opposite the angle divided by the hypotenuse) is the same as the defining relation for the cosine (side adjacent the angle divided by the hypotenuse) of the complementary angle in the same triangle.

   b. The same comment as made in part a applies here. One can either show that $\cos(\frac{\pi}{2} - x) = \sin x$ by using the addition formula for cosine or by looking at the defining relationship for the cosine of an angle and the sine of the complementary angle in a right triangle.

   c. Work backwards from $\cos 2x = \cos(x+x) = \cos^2 x - \sin^2 x = \cos^2 x - (1 - \cos^2 x) \Longrightarrow \cos^2 x = (1 + \cos 2x)/2$.

   d. As before, $\cos 2x = \cos^2 x - \sin^2 x = 1 - 2\sin^2 x$, so $\sin^2 x = (1 - \cos 2x)/2$.

   e. $\cos x = \cos(2\frac{x}{2}) = \cos^2(\frac{x}{2}) - \sin^2(\frac{x}{2}) = 2\cos^2(\frac{x}{2}) - 1 = 2/\sec^2(\frac{x}{2}) - 1 = 2/(\tan^2(\frac{x}{2}) + 1) - 1 = \dfrac{1 - \tan^2(\frac{x}{2})}{1 + \tan^2(\frac{x}{2})}$. This identity, like the one in part f, is attributed to Weierstrass, who showed as a consequence that all integrals of rational functions of trigonometric functions could be transformed to integrals of rational functions of $\tan x/2$.

   f. $\sin x = 2\sin\frac{x}{2}\cos\frac{x}{2} = 2\tan\frac{x}{2}\cos^2(\frac{x}{2}) = \dfrac{2\tan\frac{x}{2}}{\sec^2(\frac{x}{2})} = \dfrac{2\tan\frac{x}{2}}{\tan^2(\frac{x}{2}) + 1}$. The identities of parts e andf are useful, and they haul out a lot of the artillery in the trigonometric identities arsenal, thus supplying students with good practice.

5. In the diagram, we see $OC = 1$ on the unit circle, so $\sin x = AC$ and $\cos x = OA$ when defined using the unit circle. $BD/OD = AC/OC = AC$ using similar triangles, so $AC = \dfrac{\text{side opposite } x}{\text{hypotenuse}}$ in triangle OBD. We likewise see $OA = \dfrac{\text{side adjacent } x}{\text{hypotenuse}}$ in triangle OBD. The Pythagorean Theorem shows that $OA^2 + AC^2 = 1$. This problem is not profound, but it is used incessantly.

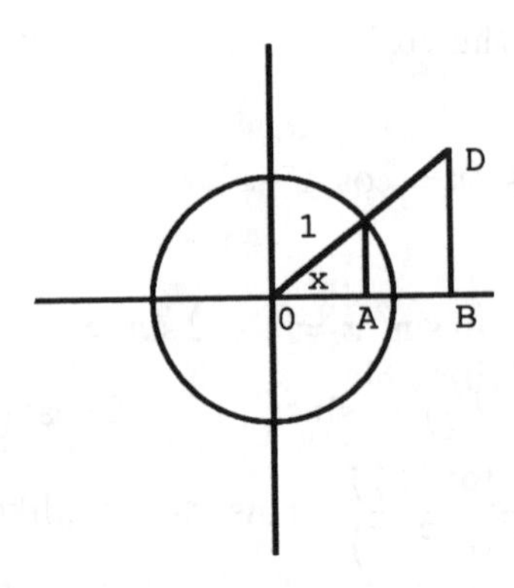

6. This is a useful problem for familiarizing students with properties of the sine function. It helps to write $f(x) = -4\sin\left[4\left(x + \frac{\pi}{24}\right)\right]$. Having a graphics package available makes problems such as this easy to visualize, but there is nonetheless merit in solving it without relying on visual aids, as it is important that students become familiar with the properties of functions, as well as their graphs.

   a. The zeros of f can be found by knowing that the zeros of the sine are integral multiples of π. We would thus need $4(x + \frac{\pi}{24}) = n\pi$, where n is an integer, in order that x be a zero of the function. Therefore $x = n\pi/4 - \pi/24 = (6n-1)\pi/24$. When $n = 1$, we obtain the smallest positive value for x, namely $5\pi/24$.

   b. The factor -4 indicates that the maximum value of f occurs when $\sin\left[4\left(x + \frac{\pi}{24}\right)\right]$ achieves its minimum value of -1, which occurs at angles $n\pi/2$, where n is congruent to 3 mod 4. (For students, use the set $\{\ldots, -5, -1, 3, 7, \ldots\}$.) Solving, we get $x = n\pi/8 - \pi/24 = (3n-1)\pi/24$, where n is restricted to integers congruent to 3 mod 4. The smallest positive value of x occurs when $n = 3$, for which $x = \frac{\pi}{3}$.

   c. The factor -4 indicates that the minimum value of f occurs when $\sin\left[4\left(x + \frac{\pi}{24}\right)\right]$ achieves its maximum value of 1, which occurs at angles $n\pi/2$, where n is congruent to 1 mod 4. (For students, use the set $\{\ldots, -7, -3, 1, 5, \ldots\}$.) Solving, we get $x = n\pi/8 - \pi/24 = (3n-1)\pi/24$, where n is restricted to integers congruent to 1 mod 4. The smallest positive value of x occurs when $n = 1$, for which $x = \frac{\pi}{12}$.

[MN]

7. In the accompanying diagram, the height of the tree is $TG = TX + XG$. But $TX/180 = \tan 22°$, and $XG/180 = \tan 7°$. Therefore, $TG = 180(\tan 22° + \tan 7°) \approx 94.83$ feet. Make the students draw the picture. A very common error is to construct a right triangle in which the angle is $22° + 7° = 29°$. This gives a tree height of $180(\tan 29°) = 99.78$ feet, about 5 feet more than the correct answer.

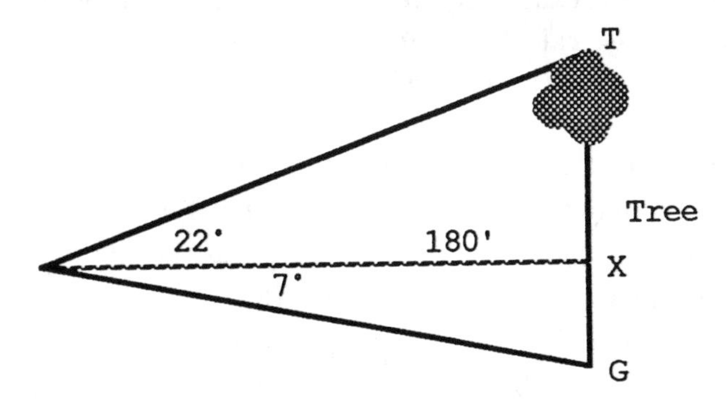

8. Although it is important that students become radian-conscious, some problems are easily solved using degree measure. Three rpm is the same as $3 \cdot 360 = 1080$ degrees per minute, so in four seconds the beam moves $1080/15 = 72$ degrees. If D = distance of lighthouse from shore, $300/D = \tan 72°$, so $D = 300/\tan 72° \approx 97.48$ feet.

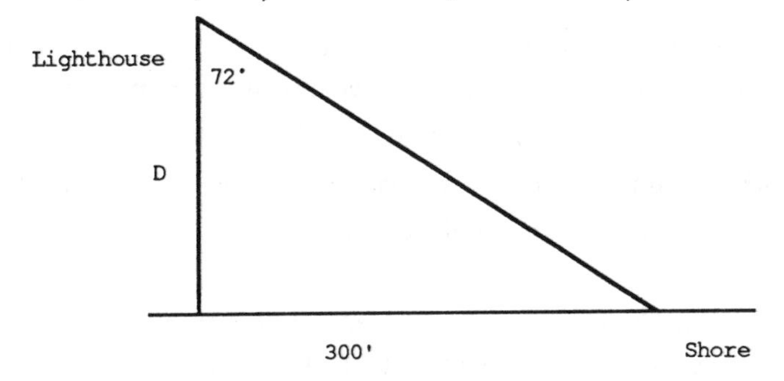

It can easily be seen that, if S is the distance down the shore that the light beam moves, and A is the angle through which the beam moves in the given period of time, then $D = S/\tan A$. Thus the remainder of the problem consists of analyzing whether S or A increases or decreases, and what effect that has on D.

a. $S > 300$, so the lighthouse is further from shore.

b. $S < 300$, so the lighthouse is nearer to shore.

c. The beam sweeps through less than 1080 degrees per minute, so A is less than 72 degrees, $\tan A < \tan 72°$, and the lighthouse is further from shore.

d. The beam sweeps through a lesser angle in 3 seconds than it does in 4, so $\tan A < \tan 72°$, and the lighthouse is further from shore.

Section 3: Exponential and Logarithmic Functions

1. There are several different ways to see that $\log_b x < x^b < b^x$. This offers a good opportunity to haul out a graphics package to get students to formulate a conjecture, and then see if they can verify it. This is also a good problem to revisit when students have learned calculus, since the correspondence between positive derivative and increasing functions can be used to supply an easy proof.

2. Like the one above it, this question deals with what might be called the hierarchy of size although students may treat it as a question of shape. $e^{1000} = e^{990+10} = (e^{990})(e^{10}) \approx (e^{990})(2.203)(10^3)$. $(1000)^{144} = (990 \bullet \frac{1000}{990})^{144} = (990)^{144}(\frac{1000}{990})^{144} \approx (990)^{144}(4.251)$ Thus the graph which "looks" exponential is that of $y = e^x$ whereas the one which looks linear is that of $y = x^{144}$. It is important that students get some sense of the relative growth of functions, even before they encounter calculus, and there are variations on this problem which may help. For example, you may want to ask your students to estimate $\lim_{x \to \infty} x^p \dfrac{e^{-x}}{p}$ where p is a positive number. Start with values of p like 1, then work up to values like 100.

3. $N = I_0 2^{kt}$ is a good model for radioactive decay, where I_0 is the initial quantity of pure material, N the amount of radioactive material remaining, and t the time in hours. If $t = 15$, then $N = I_0/2$, and so $I_0/2 = I_0 2^{15k}$, so $15k = -1$. The danger period expires when $N = I_0/10$, so we must solve $I_0/10 = I_0 2^{-t/15}$. Therefore $-\frac{t}{15}\ln 2 = \ln\frac{1}{10}$; $t = 15\ln 10/\ln 2 = 49.83$ hours.

4. Here are two concepts for the price of one problem. The domain of logarithmic functions is the set of all positive numbers, so we must have $(x-5)(x-3) > 0$, which occurs if $x > 5$ or $x < 3$.

[AP]

5. This problem is a good refresher for basic properties of the exponential function and graphical symmetries.

   a. $y = .2^x = \left(\frac{1}{5}\right)^x = 5^{-x}$, so reflect the graph of $y = 5^x$ about the y-axis.

   b. $y = 3 - 5^x$. First reflect the graph of $y = 5^x$ about the x-axis, then translate it 3 units upward.

   c. $y = 7(5^x)$. Since $7 = 5^{\log_5 7}$, $y = 5^{x+\log_5 7} = 5^{x-(-\log_5 7)}$, so translate the graph of $y = 5^x$ a total of $\log_5 7$ units to the left.

   d. $y = 5(7 + 5^{x-4}) = 35 + 5^{x-3}$, so translate the graph of $y = 5^x$ three units to the right and then 35 units upward.

[MN]

6. This problem is related to the previous one.

   a. $y = -4^x$ b. $y = 4^{-x}$ c. $y = 10 - 4^x$

   d. $y = 4^{-(4+x)}$ e. $y = -4^{-x}$ f. $y = -4^{-x}$

   Note: Students may observe that horizontal and vertical reflections commute, but if the axes of reflection are skewed, this is not necessarily the case.

[AP]

7. The answers for parts a and b are easily obtained from the way the bacteria grow.

   a. 11:50:00 A.M.

   b. 11:40:00 A.M.

   Parts c and d should probably be handled simultaneously. The equation for growth is given by $N = I_0 2^{kt}$ (see Problem 3 from this section). If t is measured in hours, and $t = 0$ corresponds to noon, then $N = I_0/2$ when $t = -\frac{1}{6}$, so $I_0/2 = I_0 2^{-k/6}$. Therefore $k = 6$, and $N = I_0 2^{6t}$. Note that I_0 is the quantity of bacteria present at noon. The time t at which rI_0 bacteria are present is given by solving $rI_0 = I_0 2^{6t}$. The solution is $t = \dfrac{\ln r}{6 \ln 2}$.

   c. If $r = .05$, then $t = -.72032$ hours $= 2593$ seconds before noon, or 11:16:47 A.M.

   d. If $r = .01$, then $t = -1.1073$ hours $= 386$ seconds before 11 A.M., or 10:53:34 A.M.

8. a. If the equation $N = I_0 2^{kt}$ holds, then $I_0/2 = I_0 2^{kH}$, and $I_0/3 = I_0 2^{kT}$. Since $\frac{1}{2} = 2^{kH}$, $kH = -1$, and so $k = -1/H$. Therefore $N = I_0 2^{-t/H}$. If $t = T$ (the third-life), then $I_0/3 = I_0 2^{-T/H}$, so $\ln \frac{1}{3} = -\frac{T}{H} \ln 2$. Therefore $T = H\dfrac{\ln 3}{\ln 2}$.

   b. The identical calculation used above applies, except that the numbers $\frac{1}{2}$ and $\frac{1}{3}$ are replaced by 2 and 3 respectively. We then find that $T = D\dfrac{\ln 3}{\ln 2}$. Students should be aware that dividing by a factor of N during exponential decay corresponds to expanding by a factor of N during exponential growth.

Section 4: Composite Functions

1. This manipulational problem is designed to familiarize the students with composition. The domain of the function is the set of all real numbers unless otherwise stated.

   a. $(f \circ f)(x) = 3(3x + 2) + 2 = 9x + 8$

   b. $(f \circ g)(x) = 3\sin(4x) + 2$

   c. $(f \circ h)(x) = 3\ln x + 2$ Domain is the set of all positive numbers.

   d. $(g \circ f)(x) = \sin(4(3x + 2)) = \sin(12x + 8)$

   e. $(g \circ g)(x) = \sin(4\sin 4x)$

   f. $(g \circ h)(x) = \sin(4\ln x)$ Domain is the set of all positive numbers.

   g. $(h \circ f)(x) = \ln(3x + 2)$ Domain $= \{x \mid x > -\frac{2}{3}\}$.

   h. $(h \circ g)(x) = \ln(\sin(4x))$ The domain consists of the set of those numbers for which $\sin 4x > 0$. Since $\sin 4x > 0$ if $n\pi < 4x < (n + 1)\pi$, where n belongs to the set $A = \{\ldots, -4, -2, 0, 2, 4, \ldots\}$, we must have $n\pi/4 < x < (n + 1)\pi/4$, where n belongs to A.

   i. $(h \circ h)(x) = \ln(\ln x)$ Domain $= \{x \mid x > 1\}$.

2. Part a of this problem helps prepare students for the Chain Rule. The remaining parts basically improve their "feel" for functions and composition.

   a. $f(x) = \ln x + 4$ Domain is all positive numbers.

   b. Yes. If $f(x) = ax + b$ and $g(x) = cx + d$, then $(g \circ f)(x) = g(ax + b) = c(ax + b) + d = acx + bc + d$.

   c. No. If $f(x) = g(x) = x^2$, then $(g \circ f)(x) = g(x^2) = x^4$.

   d. Yes. The proof is similar to b, but messier.

3. a. Use right triangles. If $A = \arctan x$, then $\tan A = x$. The hypotenuse is $\sqrt{x^2+1}$, and so $f(x) = x/\sqrt{x^2+1}$.

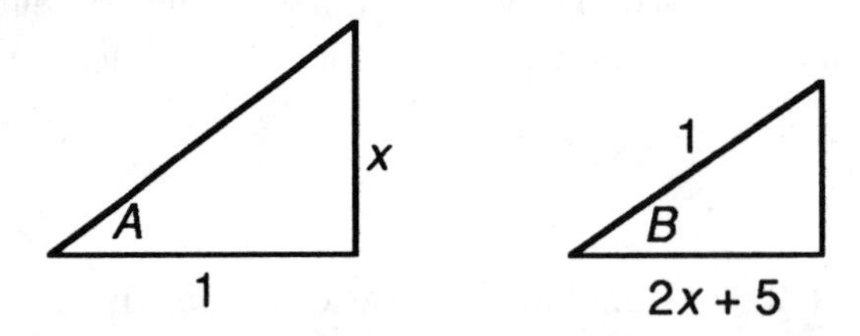

b. Let $B = \arccos(2x+5)$. Then $\cos B = 2x+5$. Complete the triangle, letting $z = 2x+5$. Then $g(x) = z/\sqrt{1-z^2}$.

c. The arcsin is defined only on $\{x \mid -1 \le x \le 1\}$. In order that $-1 \le \tan x \le 1$, the domain of f must be $-\frac{\pi}{4} + n\pi \le x \le \frac{\pi}{4} + n\pi$ for n an integer.

d. The arcsec is defined only on $\{x \mid |x| \ge 1\}$. However, $|\sin x| \ge 1$ only if x is an odd multiple of $\frac{\pi}{2}$, so the domain of g consists of odd multiples of $\frac{\pi}{2}$.

4. Since $f(g(x)) = \ln[g(x)^2] = \ln(x^2+4)$, $(g(x))^2 = x^2+4$. Since $g(x) > 0$, $g(x) = \sqrt{x^2+4}$.

[AP]

5. If f and g are even, $(f \circ g)(-x) = f(g(-x)) = f(g(x))$, so $f \circ g$ is even. If f and g are odd, $(f \circ g)(-x) = f(g(-x)) = f(-g(x)) = -f(g(x))$, so $f \circ g$ is odd. Similar calculations show that if f is odd and g is even, or vice-versa, the composite is even.

6. This property is frequently used, but it often goes unproved, and students are sometimes under the impression that it must be a triviality.

If g is the inverse function of f and (a,b) lies on the graph of f, then $b = f(a)$, so $g(b) = g(f(a)) = a$. Therefore, (b,a) lies on the graph of g. The line perpendicular to $y = x$ must have slope -1, and so the equation of such a line which goes through (a,b) must be $y - b = -(x-a)$. Note that both (a,b) and (b,a) lie on this line. The intersection of this line with the line $y = x$ is $x = y = (a+b)/2$, and (a,b) and (b,a) are equidistant from this intersection.

If the graph of g is the reflection of f about the line $y = x$, the argument above shows that if (a,b) lies on the graph of f, then (b,a) lies on the graph of g. Then $b = f(a) = f(g(b))$, and $a = g(b) = g(f(a))$, so f and g are inverses.

7. a. $f \circ g(x) = f(cx+d) = a(cx+d)+b = acx+ad+b$.
   $g \circ f(x) = g(ax+b) = c(ax+b)+d = acx+bc+d$.
   For $f \circ g = g \circ f$, we must have $ad+b = bc+d$. One way to express this is that $a(d-1) = c(b-1)$.

   b. If $f \circ g = f$, then $acx+ad+b = ax+b$, so $c = 1$ and $ad = 0$. Since there is no *a priori* guarantee that $a = 0$, we must have $d = 0$. Therefore $g(x) = x$.

   c. A similar argument shows that if $f \circ g = g$, then $f(x) = x$.

8. $(f \circ (g+h))(x) = f(g(x)+h(x))$; $((f \circ g)+(f \circ h))(x) = f(g(x))+f(h(x))$. As long as f is not additive, this identity has virtually no chance. For instance, let $f(x) = \ln x$, $g(x) = x$, $h(x) = x$. Then $(f \circ (g+h))(x) = \ln 2x$, but $((f \circ g)+(f \circ h))(x) = 2\ln x$. These two functions are not equal.

Problems 7 and 8 help to make students aware of the fact that "universal linearity" (all properties are linear) is false.

9. Since $\sin x = 0$ if x is an integer multiple of π, we see that $\sin(\frac{1}{x}) = 0$ if $\frac{1}{x}$ is an integer multiple of π. Therefore $\sin(\frac{1}{x}) = 0$ if $x = \frac{1}{n\pi}$ for any integer n. Since any interval $(0, c)$ contains infinitely many of these points, the function $f(x) = \sin(\frac{1}{x})$ cannot be one-to-one on this interval, and so cannot have an inverse.

   Whenever one introduces a pathological example in an elementary calculus course, one takes a risk. It is important to stress what *can* be done in calculus, rather than where one runs into difficulties. Functions using $\sin(\frac{1}{x})$ possibly represent the limits of pathology that students can tolerate.

10. Suppose $x < y$. Either $f^{-1}(x) < f^{-1}(y)$, $f^{-1}(x) = f^{-1}(y)$, or $f^{-1}(x) > f^{-1}(y)$. Equality is impossible, because both f and f^{-1} are one-to-one. Suppose $f^{-1}(x) > f^{-1}(y)$. Since f is increasing, $f(f^{-1}(y)) < f(f^{-1}(x))$. But this says that $y < x$, a contradiction. Therefore, $f^{-1}(x) < f^{-1}(y)$.

11. The standard technique for finding inverse functions for $y = f(x)$ is to "solve" the equation for x in terms of y, and then interchange the roles of x and y. If one sketches the graph of the parabola $y = x(x-4)$, one sees that it is one-to-one both to the left and the right of $x = 2$. One can solve $y = x^2 - 4x$ for x in terms of y either by the quadratic equation or by completing the square. The roots are $x = 2+\sqrt{y+4}$ and $x = 2-\sqrt{y+4}$. Interchanging the roles of x and y gives two possible functions: $y = 2 + \sqrt{x+4}$ and $y = 2 - \sqrt{x+4}$. When the plus sign is used, the function is increasing, and when the minus sign is used, the function is decreasing. Since the parabola decreases to the left of $x = 2$ and increases to the right, we have $g(x) = 2 - \sqrt{x+4}$ and $h(x) = 2 + \sqrt{x+4}$. Both have the domain $\{x \mid x > -4\}$, $g(x)$ is the inverse of the left branch of $f(x)$, and $h(x)$ is the inverse of the right branch.

Section 5: Functions Described by Tables or Graphs

1. a. All real numbers except -2.

   b. $\{x \mid x \geq 0 \text{ or } x < -5\}$

   c. $\{x \mid x < -2\}$, $\{x \mid x > 0\}$

   d. $\{x \mid -2 < x \leq 0\}$

   e. Yes. Domain $= \{x \mid x > 2\}$, range $= \{x \mid x < -2\}$.

   f. No. The function is not one-to-one on this interval.

   g. Yes. Domain $= \{x \mid x < -5 \text{ or } x \geq 0\}$, range $= \{x \mid x > -2\}$.

   h. Yes. domain $= \{x \mid x < -5\}$, range $= \{x \mid x > 0\}$.

   Problems with domain, range, and the existence of inverses on subsets become especially important when studying logarithms, inverse trigonometric functions, and inverse hyperbolic trigonometric functions.

   Students are sometimes unaware of the "flip and trace" method of determining whether the inverse of a function exists, and what the domain and range of the inverse functions will be. Graph the function using dark ink on thin paper—draw the line $y = x$ lightly. Flip the paper over leaving the line $y = x$ invariant. Trace the original curve on the flip side of the paper. This is the graph of the inverse relation. If it satisfies the vertical line test for functions, then the original function has an inverse. The same technique can be used for subsets of the original domain simply by failing to trace that portion of the curve whose domain does not belong to the desired subset.

   This problem can serve as a template for other problems of this type. Good examples can be constructed using simple functions such as cubic polynomials, or the sine and cosine functions.

2. The graph on the left represents the number of doctors who have received information via the mass media, and the graph on the right represents the number of doctors who have received information via word of mouth. Initially, word of mouth spreads slowly, whereas mass media promulgates information initially to a large number of people.

   [GL&S]

3. The graph on the left is the learning curve for unskilled labor, where much of the required knowledge is learned quickly. The graph on the right is the learning curve for semi-skilled or skilled jobs, where abilities are initially acquired slowly, but after a certain amount of time the skills increase rapidly.

4. This problem can be revisited when the student has learned the interpretations of the slope of the tangent as an instantaneous rate of change.

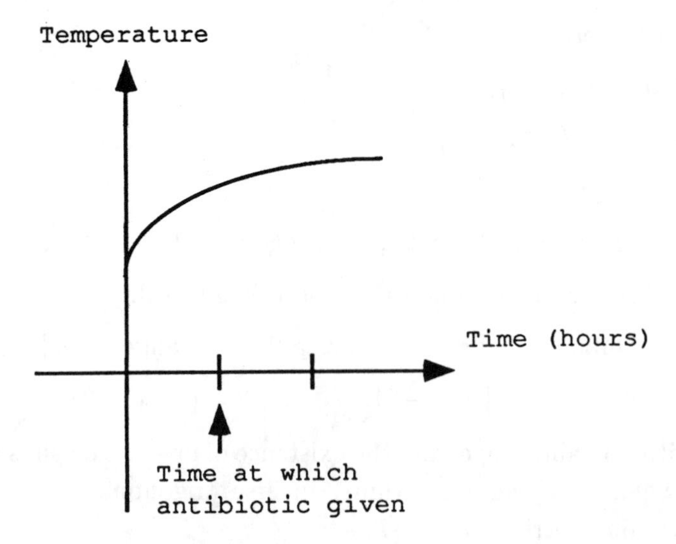

5. The container is wider at the base than it is at the top, as the volume differences between equal increments in depth decrease at greater depths.

6. a. $V(D)$ = the volume of water in the bowl if the depth of the water is D inches. $V(0)$ should always be 0, and this means that Graph A is not a viable candidate to represent V as a function of depth D for any bowl.

   b. The bowl used for Graph C might be hemispherical in shape. The bowl for Graph B might be cylindrical or "box-like"; in any event, it must have a constant cross-sectional area since the volume depends linearly on depth. The bowl for Graph D might look like an erlenmeyer flask since its cross-sectional area increases as you push your finger down into the bowl.

7. All five data lie on the line given by $N = 27 + 4(Y - 1983)$, where N is the number of unindicted co-conspirators, and Y is the year. The students should conclude that if the first differences are constant (or the second differences are zero), then a linear function fits the data. The more perceptive might try for the obvious generalization, which is that if the nth differences are zero, then a polynomial of degree $n - 1$ fits the data.

Year	1983	1984	1985	1986	1987
Number of UCC	27	31	35	39	43
First Differences	—	4	4	4	4

8. a. The deer population is decreasing at an average rate of 50 deer per year during intervals (ii) and (v).

   b. The deer population is increasing most rapidly between times $t = 3$ and $t = 4$ years since the slope of the tangent line appears visually to be greatest in that interval. This problem can be revisited after students have learned about inflection points to obtain the greatest instantaneous rate at approximately $t = 3.5$ years, as the curve appears to have an inflection point at approximately this time.

   c. At $t = 1.5$, the deer population is decreasing at a rate of approximately 200 deer per year. This estimate can be obtained by computing the difference quotient $\frac{P(2) - P(1)}{2 - 1} = \frac{400 - 600}{1} = -200$ where $P(t)$ is the deer population at time t. Notice that the rate of change at $t = 1.5$ is approximated by computing the difference quotient using data from $t = 1$ and $t = 2$. This may raise some questions when the derivative is first introduced since computing the derivative at t involves the difference quotients using data from t and $t + h$.

9. a. The average annual income in terms of 1982 dollars. If Y is the year, T the total income, N the number of lawyers, and A the average annual income, then $A = T/[N(1.05^{Y-1982})]$

   b.

Year	1982	1983	1984	1985	1986
Avg. Ann. Inc.	$62,500.00	$57,142.86	$61,678.00	$64,787.82	$59,234.58

   c. 1985

   d. 1984

   e. A matter of interpretation. 1986 sees the worst *absolute* decline from the preceding year, but 1983 sees the worst *percentage* drop from the preceding year, although one must visit the 8th decimal place to prove it!

Section 6: Parametric Equations

1. It requires only simple algebra to see that each of the parametrizations in this problem describes some portion of the graph of the parabola $y = x^2$.

   a. The curve described is the complete parabola. The point P moves downward along the left branch until it reaches the origin, and then upward along the right branch.

   b. The curve described is the complete parabola, but the motion of P is opposite in direction to that in part a.

   c. The curve described is the positive branch of the parabola (only those points with non-negative x-coordinates). The point P moves in from the right to the origin, then back out again.

   d. The curve is that portion of the parabola for which the x-coordinate lies between -1 and 1. The point P oscillates from $(-1, 1)$ to $(1, 1)$ and back again.

   e. The curve is that portion of the parabola for which the x-coordinate lies between 0 and 1. The point P oscillates between $(0, 0)$ and $(1, 1)$.

2. Students should be encouraged to draw their own pictures. Although the description of this one is lengthy, the picture itself is quite easy to draw. It is clear that $P(t) = (a\cos t, a\sin t)$, $Q(t) = (b\cos t, b\sin t)$. Thus the desired equations are $R(t) = (b\cos t, a\sin t)$, $S(t) = (a\cos t, b\sin t)$. Therefore, if $R = (x, y)$, then $x = b\cos t$, $y = a\sin t$. Since $\frac{x}{b} = \cos t$ and $\frac{y}{a} = \sin t$, $\left(\frac{x}{b}\right)^2 + \left(\frac{y}{a}\right)^2 = \cos^2 t + \sin^2 t = 1$. This is the equation of an ellipse with intercepts $(b, 0)$, $(0, a)$, $(-b, 0)$, and $(0, -a)$. Similarly, the equation for S describes an ellipse with intercepts $(a, 0)$, $(0, b)$, $(-a, 0)$, and $(0, -b)$, which is the ellipse above, rotated 90° about the origin.

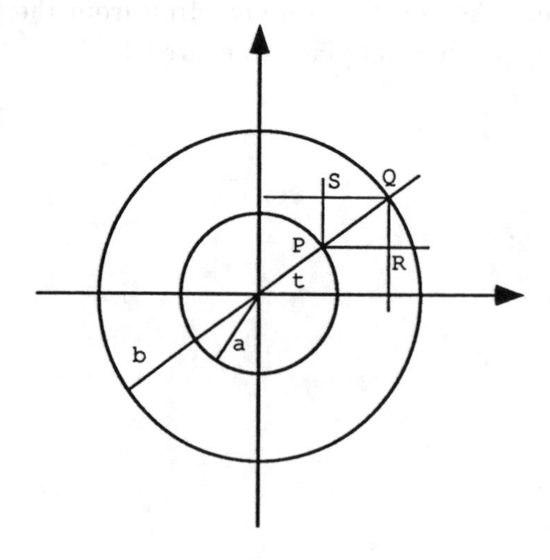

[MN]

3. There are two key issues to understanding this problem. The first is that the taut string is always tangent to the circle, and the tangent is perpendicular to the radius at the intersection. The second is that the length of the taut string BP is equal to the length unwound BX. This is a recurrent theme in many such problems. If $P = (x(t), y(t))$, then $x(t) = OQ + RP = r\cos t + BP\sin t$. Since $BP = BX = rt$, $x(t) = r\cos t + rt\sin t$. $y(t) = RQ = BQ - BR = r\sin t - BR = r\sin t - BP\cos t = r\sin t - rt\cos t$. The graph of the involute is given in the figure below.

   This problem is a good "warm-up" for many of the problems from the cycloid family because it hits all the basic ideas at a relatively elementary level. The involute concept can be revisited when the student has learned how to compute arc length.

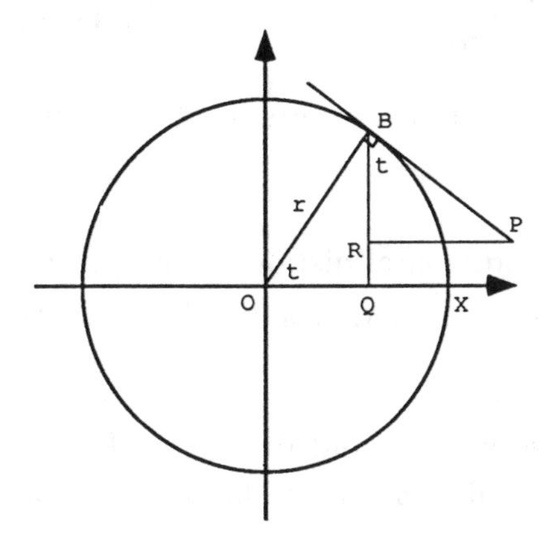

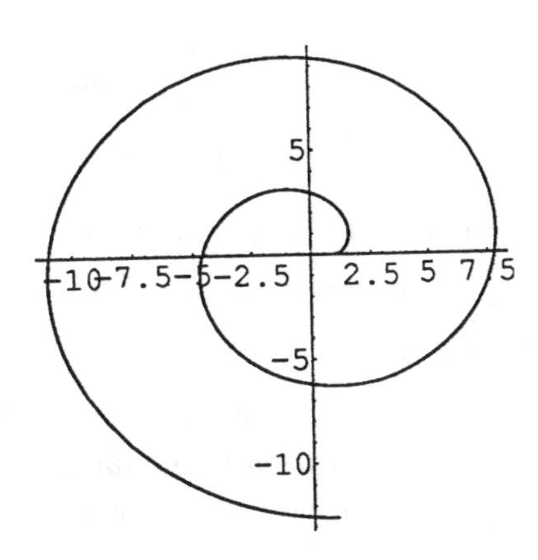

4. a. The required sketch is given below.

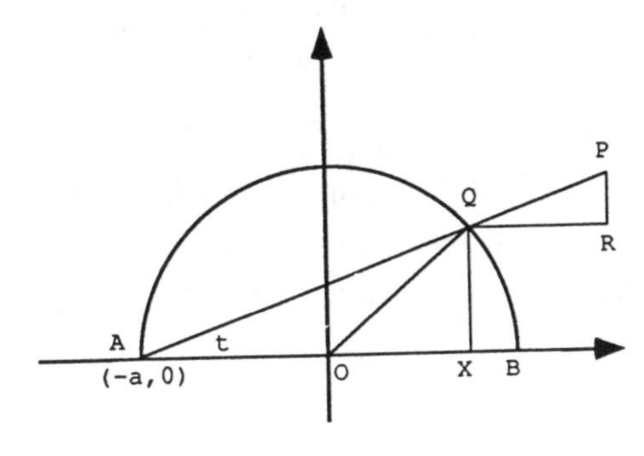

   b. Since $OQ = OA = a$, the triangle AOQ is isosceles, and so $\angle AQO = t$. Since AQX is a right triangle, $\angle AQX = \frac{\pi}{2} - t$, so $\angle OQX = \angle AQX - \angle AQO = \frac{\pi}{2} - t - t = \frac{\pi}{2} - 2t$, so $\angle QOX = \frac{\pi}{2} - \angle OQX = 2t$. So $x(t) = OX + QR = a\cos 2t + h(t)\cos t$, and $y(t) = QX + PR = a\sin 2t + h(t)\sin t$, since $\angle PQR = t$.

[MN]

5. a. Note that $t = x/(v \cos A)$. Substituting, we obtain

$$y = \frac{(v \sin A)x}{v \cos A} - \tfrac{1}{2}g\left(\frac{x}{v \cos A}\right)^2 = (\tan A)x - \left(\frac{g \sec^2 A}{2v^2}\right)x^2.$$

This is the graph of a parabola.

b. Factoring the above equation yields $y = x\left[\tan A - \left(\frac{g \sec^2 A}{2v^2}\right)x\right]$. The projectile will hit the ground when $y = 0$, so $x = \dfrac{2v^2 \tan A}{g \sec^2 A}$ is the x-coordinate of the point of impact.

c. Simplifying the above equation, we get $x = \dfrac{2v^2}{g}\dfrac{\sin A}{\cos A}\cos^2 A = \dfrac{2v^2}{g}\sin A \cos A = \dfrac{v^2}{g}\sin 2A$. The maximum value of $\sin 2A$ is 1, and this occurs at $A = \frac{\pi}{4}$, with a maximum horizontal distance of v^2/g.

d. From part c, the angle of launch is $\frac{\pi}{4}$, and the horizontal distance is v^2/g. Solving the parametric equation of x for t, we have $x = (v \cos A)t$, $v^2/g = vt/\sqrt{2}$, $t = \sqrt{2}\,v/g$.

e. From part c, $300 = v^2/32$, so $v = 97.98$ feet per second, and from part d, $t = 4.33$ seconds. Note that a strong-armed outfielder should *not* throw the ball at an angle of $\frac{\pi}{4}$ radians. (See part f below.)

f. First do a unit conversion. 90 mph equals 132 feet per second. From part c, $300 = (\frac{v^2}{g})\sin 2A$, so $A = \frac{1}{2}\arcsin(\frac{(300)(32)}{(132)^2}) \approx 16.7°$, and from part d, we have that $x = (v \cos A)t$. Thus $t = \dfrac{300}{v \cos A} \approx 2.37$ seconds. There is a second solution in which $A = 73.3°$ and $t = 7.90$ seconds, which is long enough for the runner to tag up and score from second base and for the general manager to sign the papers giving the centerfielder his unconditional release.

Section 7: Polar Coordinates

1. This problem, like all the other problems in this section, is designed to acquaint students with the similarities and differences between Cartesian and polar coordinates. The key idea here is that there are infinitely many different polar representations of the same point (in Cartesian coordinates, each point has only a single representation). If $r > 0$, all points (r, θ) such that $r\cos\theta = x$ and $r\sin\theta = y$ represent the point (x, y). If $r < 0$, all points $(-r, \theta + \pi)$ such that $-r\cos(\theta + \pi) = x$ and $-r\sin(\theta + \pi) = y$ represent the point (x, y). Notice that $-r\cos(\theta + \pi) = r\cos\theta$ and $-r\sin(\theta + \pi) = r\sin\theta$.

2. Students should think about graphs as they generate them.

   a. Translating the segment of the graph between $x = a$ and $x = a + 2\pi$ by 2π units (either to the right or the left) will place it directly on top of another portion of the graph.

   b. The entire graph can be obtained simply by looking at the set $\{(r, \theta) \mid 0 \leq \theta < 2\pi\}$. For a periodic function with period 2π, any ray extending from the origin intersects the curve in at most one point, and this property characterizes such periodic functions.

3. This is yet another problem requiring students to think about how algebraic properties of functions are reflected in the geometric properties of their graphs.

   a. An even function is symmetric about the y-axis.

   b. Since (r, θ) on the graph implies that $(r, -\theta)$ also lies on the graph, then the graph is symmetric about $\theta = 0$.

4. a. An odd function is symmetric with respect to the origin.

   b. Since (r, θ) on the graph implies that $(-r, -\theta)$ also lies on the graph, then the graph is symmetric about $\theta = \frac{\pi}{2}$.

5. Transforming to rectangular coordinates, we see that the graph of $r = \cos\theta$ is

$$\sqrt{x^2+y^2} = \frac{x}{\sqrt{x^2+y^2}}$$
$$x^2+y^2 = x$$
$$x^2 - x + \tfrac{1}{4} + y^2 = \tfrac{1}{4}$$
$$(x-\tfrac{1}{2})^2 + y^2 = \left(\tfrac{1}{2}\right)^2$$

This is the graph of a circle. The graph of $r = \cos\theta + 3$ transforms to

$$\sqrt{x^2+y^2} = \frac{x}{x^2+y^2} + 3$$
$$x^2+y^2 = x + 3\sqrt{x^2+y^2}$$

This is not the graph of a circle, so the graph of $r = \cos\theta + 3$ is not a translate of the graph of $r = \cos\theta$.

6. Cross-multiplying and transforming to rectangular coordinates yields

$$r\sin\theta + r\cos\theta = 5$$
$$y + x = 5$$

This is the equation of a line. In rectangular coordinates, it has slope -1 and passes through the points $(0,5)$ and $(5,0)$. It has singularities when $\sin\theta + \cos\theta = 0$, or $\tan\theta = -1$, which occurs at $\theta = \frac{3\pi}{4} + n\pi$ for integer values of n. If n is an integer such that $\frac{3\pi}{4} + 2n\pi < \theta < \frac{7\pi}{4} + 2n\pi$, the direction in which a point on the line moves is from the fourth quadrant toward the second quadrant; in other words, the point moves in a northwest direction along the line. If n is an integer such that $-\frac{\pi}{4} + 2n\pi < \theta < \frac{3\pi}{4} + 2n\pi$, the point comes in from the 2nd quadrant and exits in the 4th quadrant, moving southeast along the line.

7. To insure that (r,θ) on the graph implies that $(r,\theta+\pi)$ also lies on the graph, we need to have $r = f(\theta) = f(\theta+\pi)$. In other words, the function f must be periodic with period π.

8. a. The graph can be obtained by reflecting the original graph in the x-axis.

   b. Since $(-r, \theta) = (r, \theta - (-\pi))$, this part of the problem is solved once part *d* is solved. This order of presentation may seem a bit unfair: If part *d* were presented first, this part would then be easier. You may want to reverse the order in which the question appears.

   c. The graph is translated c units to the right. An easy way to see this is to look at the graph of $y = x$, the line through the origin, as compared with the graph of $y = x - c$, which is the line with the same slope through $(c, 0)$.

   d. The new graph is generated by a counterclockwise rotation around the origin through an angle of c radians. Applying this to part *b*, we realize that the graph of $r = -f(\theta)$ is generated by rotating the original graph counter-clockwise through an angle of $-\pi$, or a clockwise rotation through an angle of π.

   e. $y = \sin x = \cos\left(\frac{\pi}{2} - x\right) = \cos\left(x - \frac{\pi}{2}\right)$ since the cosine function is even. Thus the graph of $y = \sin x$ can be obtained from the graph of $y = \cos x$ by translating $\frac{\pi}{2}$ units to the right.

   f. Part *e* shows that $r = \sin\theta = \cos(x - \pi/2)$, and so the graph of $r = \sin\theta$ can be obtained by rotating the graph of $r = \cos\theta$ (which we know is a circle from Problem 5) counter-clockwise around the origin through an angle $\frac{\pi}{2}$.

9. One situation involves a ship in the middle of an ocean. Many students will be familiar with this from watching naval war films when torpedoes approach. It is worth remarking that the angular second coordinate one normally sees in polar coordinates has been replaced by clock designations "torpedo—range 2000 yards, bearing three o'clock"). This is a fairly well-known example but one that so strongly motivates polar coordinates that it bears repeating.

Chapter II: The Derivative

Section 1: Average Rates of Change

1. This problem is intended to get students used to the relationships between a function described in words, its graph, and the graph of its derivative.

   a. b.

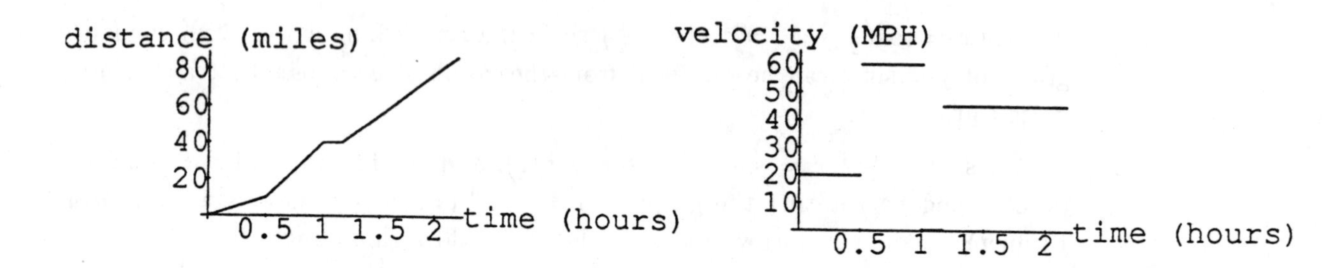

   c. She travels 85 miles in a total of 130 minutes $= 130/60$ hours, so her average velocity is $85/(130/60) \approx 39.23$ MPH.

2. a. It appears from the graph that $p(1) = -1$ and $p(4) = 6$, so the average velocity between $t = 1$ and $t = 4$ is $\frac{p(4) - p(1)}{4 - 1} = \frac{7}{3}$ m/sec.

   b. An equation of the secant line is $y - 1 = \frac{7}{3}(x + 1)$.

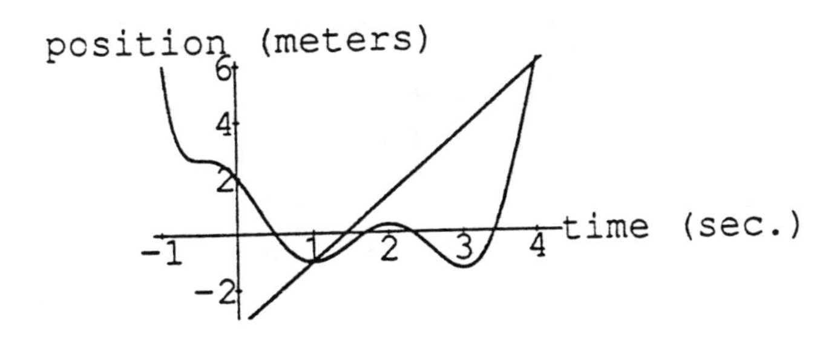

   c. It appears that the object's velocity is positive when $1 < t < 2$ and when $t > 3$; it is negative when $t < 1$ and when $2 < t < 3$.

3. $f(5) = f(1) + 3(5 - 1) = 14$

4. a. $p(3) - p(1) = 28 - 4 = 24$ meters.

   b. $\frac{p(3) - p(1)}{3 - 1} = \frac{28 - 4}{2} = 12$ m/sec.

   c. average velocity $= \frac{p(t + \Delta t) - p(t)}{\Delta t} = \frac{3(t + \Delta t)^2 + 1 - (3t^2 + 1)}{\Delta t}$

   $= \frac{3t^2 + 6t\Delta t + 3\Delta t^2 + 1 - 3t^2 - 1}{\Delta t} = 6t + 3\Delta t$ m/sec.

This question can be asked for other functions $p(t)$, e.g. $p(t) = t^3$.

5. a. $\dfrac{f(2) - f(-1)}{2 - (-1)} = \dfrac{4 - (-2)}{3} = 2.$

   b.

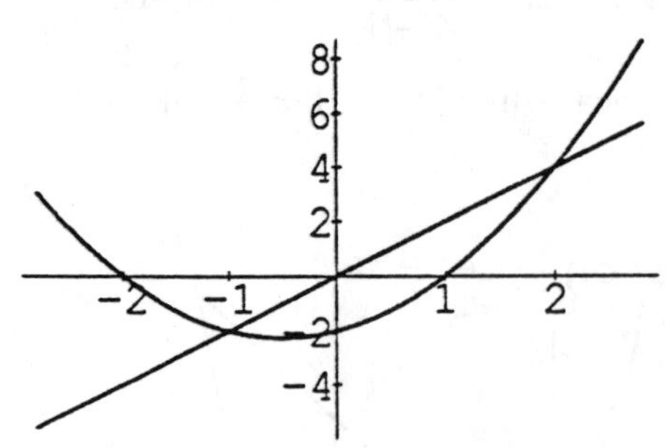

   c. By part a, the slope of the secant line is 2. Since the secant line goes through the point $(-1, -2)$, its equation is $y + 2 = 2(x + 1)$.

   d. $\dfrac{f(x + \Delta x) - f(x)}{\Delta x} = \dfrac{(x + \Delta x)^2 + (x + \Delta x) - 2 - (x^2 + x - 2)}{\Delta x} = 2x + 1 + \Delta x.$

   This question can be repeated for other functions f, e.g. $f(x) = x + 1/x$.

6. The car travels $(30\,\text{MPH})(\frac{1}{3}\,\text{hour}) = 10\,\text{miles}$ during the first 20 minutes and $(50\,\text{MPH})(\frac{1}{2}\,\text{hour}) = 25\,\text{miles}$ during the last 30 minutes. Thus the car travels a total of 35 miles with an average velocity of $(35\,\text{miles})/(\frac{5}{6}\,\text{hour}) = 42\,\text{MPH}$.

7. This is a classic problem that gets students to really think about what they are doing as the answer in part a is probably not what they expect. Part c caused a controversy when discussed in the "Ask Marilyn" column of *Parade* magazine.

   a. The total distance traveled is 60 miles. It takes $\frac{30}{40} = \frac{3}{4}$ hours to go the first 30 miles, but only $\frac{30}{60} = \frac{1}{2}$ hours to go the second 30 miles. Thus the trip takes $\frac{5}{4}$ hours, so the average velocity is $60/\frac{5}{4} = 48\,\text{MPH}$.

   b. This time the car travels 20 miles at 40 MPH and then 30 miles at 60 MPH. Thus the average velocity is 50 miles/1 hour = 50 MPH.

   c. Unless the car can travel a mile in literally no time, this is not possible. Since the car averages 30 MPH in the first mile, it takes the car two minutes to travel the first mile. Yet the car must traverse the full two miles in this time in order to average 60 MPH for the full trip.

8. At the beginning of the fifth week, the tumor has been growing for four weeks, so its weight is $w(4) = \frac{16}{15}$ grams. At the end of the week, it weighs $w(5) = \frac{25}{15}$ grams. Thus the average rate of change is $\dfrac{\frac{25}{15} - \frac{16}{15}}{5 - 4} = \frac{3}{5}$ grams/week.

9. Average cost $= \dfrac{c(64000) - c(27000)}{64000 - 27000} = \dfrac{4980 - 2840}{37000} = \dfrac{2140}{37000} \approx .0578$ dollars/pound.

10. a. True

b. False. For example, if $f(x) = x^2$, the average rate of change of the function between -1 and 1 is zero, but the function is not constant.

c. True. If the average rate of change of the function between any two points is a constant m, then for any number $x \neq 0$, $\dfrac{f(x) - f(0)}{x - 0} = m$; so $f(x) = mx + f(0)$, which is the slope-intercept form of a straight line.

Section 2: Introduction to the Derivative

1. a. $\frac{dV}{dt}$ is the rate at which water is flowing into the tank (i.e., the rate of change of the volume of water in the tank), while $\frac{dH}{dt}$ is the rate at which the water level is changing.

   b–c. Water is flowing *into* the tank, so both $\frac{dV}{dt}$ and $\frac{dH}{dt}$ are positive.

   d. Since water is flowing into the tank at a *constant* rate, $\frac{dV}{dt}$ is constant. On the other hand, $\frac{dH}{dt}$ is not constant; the water level will rise most slowly when the tank is half full, since that is where the tank is the widest. Similarly, $\frac{dH}{dt}$ will be greatest when the tank is (almost) full or empty.

2. a. Let q be the total quantity of the product that is produced, and let $p(q)$ be the price (per unit) of the product. Then $p'(q) < 0$.

   b. Let $T(t)$ be the child's temperature as a function of the time t. Then $T'(t)$ is positive, but is smaller during the last hour.

   c. Let $H(t)$ be the cost of a given health insurance policy at a given time t. Then $H'(t)$ is positive and increasing.

   d. Let $p(t)$ be the position of the car at time t. Then $p'(t)$ is gradually going to zero.

3. a. $\frac{2^1 - 2^0}{1} = 1$

   b. $\frac{\sqrt{2} - 1/\sqrt{2}}{1} = \frac{1}{\sqrt{2}} \approx .707$

   c. $f'(0) \approx .69$ (actually equal to $\ln 2$).

   d. Since the graph of f is concave up, the average rate of change of f between 0 and 1 must be larger than $f'(0)$. Using an interval symmetric about $x = 0$ such as $[-\frac{1}{2}, \frac{1}{2}]$ should (and does) give a value closer to $f'(0)$.

4. a. Using the results from Problem 4 of Section II.1, we get the following:

$$p'(-1) = \lim_{\Delta t \to 0} \frac{p(-1+\Delta t) - p(-1)}{\Delta t} = \lim_{\Delta t \to 0} (6(-1) + 3\Delta t) = -6.$$

b. $$p'(t) = \lim_{\Delta t \to 0} \frac{p(t+\Delta t) - p(t)}{\Delta t} = \lim_{\Delta t \to 0} (6t + 3\Delta t) = 6t$$

This problem can be repeated if the position $p(t)$ of an object is given by $p(t) = t^3$.

5. a. $$f'(x) = \lim_{\Delta x \to 0} \frac{f(x+\Delta x) - f(x)}{\Delta x} = \lim_{\Delta x \to 0} (2x + 1 + \Delta x) = 2x + 1$$

b. By part a, $f'(-1) = -1$, so an equation of the tangent line is:
$y + 2 = -(x+1)$ or $y = -x - 3$.

c.

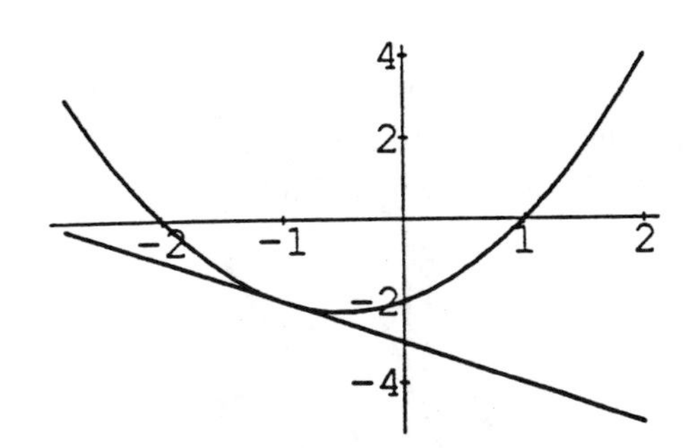

This problem can be repeated for other functions f and points, e.g. $f(x) = x + 1/x$ at the point $(1, 2)$.

6. a. The average rate of change of f between 1.97 and 2 is $\dfrac{7 - 6.905}{0.03} \approx 3.17$ and between 2 and 2.02, it is $\dfrac{7.059 - 7}{0.02} = 2.95$. So it looks as if $f'(2) \approx 3$.

b. The average rate of change of f between 3.99 and 4 is $\dfrac{9 - 8.98}{4 - 3.99} = 2$ while between 4 and 4.01 it is $\dfrac{9.2 - 9}{4.01 - 4} = 20$. So it does not look as if f is differentiable at $x = 4$.

7. a. $f(0+0) = f(0) + f(0) + 2 \cdot 0 \cdot 0$, so $f(0) = 0$. Note that this cannot be obtained from the assumption that $\lim_{h \to 0} \frac{f(h)}{h} = 7$, since f has not been assumed *a priori* to be continuous.

   b. By assumption, $f(x + \Delta x) = f(x) + f(\Delta x) + 2x\Delta x$, so

$$\lim_{\Delta x \to 0} \frac{f(x+\Delta x) - f(x)}{\Delta x} = \lim_{\Delta x \to 0} \frac{f(\Delta x) + 2x\Delta x}{\Delta x} = \lim_{\Delta x \to 0} \frac{f(\Delta x)}{\Delta x} + 2x = 7 + 2x.$$

[AP]

8. $$f'(1) = \lim_{\Delta x \to 0} \frac{f(1+\Delta x) - f(1)}{\Delta x} = \lim_{\Delta x \to 0} \frac{3\Delta x + 4(\Delta x)^2 - 5(\Delta x)^3}{\Delta x} = 3$$

9. This problem would best be done with the aid of a computer graphing package. Students can actually read off the value of $f'(1)$ from the graph of $g(\Delta x)$. Alternatively, students could be asked to use a calculator to estimate $f'(1)$ by calculating the value of $g(\Delta x)$ for values of Δx near zero. Parts a–*d* are graphed below:

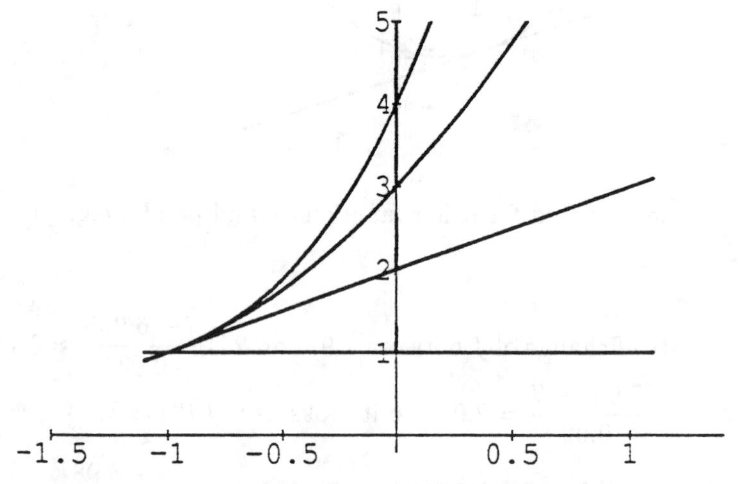

From these graphs, it is a reasonable assumption that the derivatives at $x = 1$ of the functions given in parts a, *b*, c and *d* are 1, 2, 3 and 4, respectively. The other parts are done similarly.

[AZ]

10. This formula is equal to the derivative and is useful in numerical computations. Note that the assumption of differentiability at x is necessary. The limit in $(*)$ may exist even if $f'(x)$ does not, e.g. $f(x) = |x|$ at $x = 0$.

[AZ]

11. It will become apparent after graphing just a few functions that

$$\lim_{\Delta x\to 0}\frac{f(x-\Delta x)-f(x)}{\Delta x} = -f'(x).$$

This can be seen more rigorously by making the substitution $h = -\Delta x$.

[AZ]

12. a. NEVER can be true; by the definition of the derivative, $f'(2) = 0$.

    b. MIGHT be true; $f(2)$ could be any number.

    c. MUST be true since f is differentiable and hence continuous at $x = 2$.

    d. MIGHT be true; we have been given no information about the values of f at or near 0.

    e. MUST be true; see part c.

[AP]

13. These problems would be good to give when studying one-sided limits, provided the derivative has already been introduced.

    a. When $\Delta x > 0$, $\dfrac{f(0+\Delta x)-f(0)}{\Delta x} = \dfrac{(\Delta x)^3-0}{\Delta x} = (\Delta x)^2$, which approaches 0 as Δx approaches 0. Similarly, when $\Delta x < 0$,

$$\frac{f(0+\Delta x)-f(0)}{\Delta x} = \frac{(\Delta x)^2-0}{\Delta x} = \Delta x,$$

which also approaches 0 as Δx approaches 0. Thus,

$$f'(0) = \lim_{\Delta x\to 0}\frac{f(0+\Delta x)-f(0)}{\Delta x} = 0.$$

    b. Proceeding as we did in part a, we find that $\displaystyle\lim_{\Delta x\to 0^+}\frac{f(0+\Delta x)-f(0)}{\Delta x} = 1$ while $\displaystyle\lim_{\Delta x\to 0^-}\frac{f(0+\Delta x)-f(0)}{\Delta x} = 2$ so the two-sided limit doesn't exist. Hence f is not differentiable at 0.

    c. Since $\displaystyle\lim_{\Delta x\to 0^-}\frac{f(0+\Delta x)-f(0)}{\Delta x} = \lim_{\Delta x\to 0^-}\frac{\Delta x+1}{\Delta x}$ does not exist, f cannot be differentiable at 0.

    d. $\displaystyle\lim_{\Delta x\to 0}\frac{f(0+\Delta x)-f(0)}{\Delta x} = \lim_{\Delta x\to 0}\frac{(\Delta x)^2\sin(1/\Delta x)-0}{\Delta x} = \lim_{\Delta x\to 0}\Delta x\sin(\frac{1}{\Delta x}) = 0.$ So f is differentiable at 0 and $f'(0) = 0$.

14. $f(1.1) \approx f(1) + f'(1)(.1) = 2.3$ and $f(.95) \approx f(1) + f'(1)(-.05) = 1.85$

Section 3: Graphical Differentiation Problems

These problems are intended to get students to think about derivatives in an intuitive and geometrical way, as opposed to thinking of them just in terms of algebraic formulas.

1. As with most problems of this type, any reasonable answer is acceptable. The important point is that students get a qualitative feel for the derivative.

   a. $f'(-2) = -4$ b. $f'(-1) = 0$ c. $f'(0) = -2$

   d. $f'(1.5) = -1.25$ e. $f'(2) = 0$ f. $f'(3) = 4$

2. It appears that $h(x) = f(x) - g(x)$ is constant, so its derivative is zero.

3. The graph of g is simply the graph of f translated up 3 units.

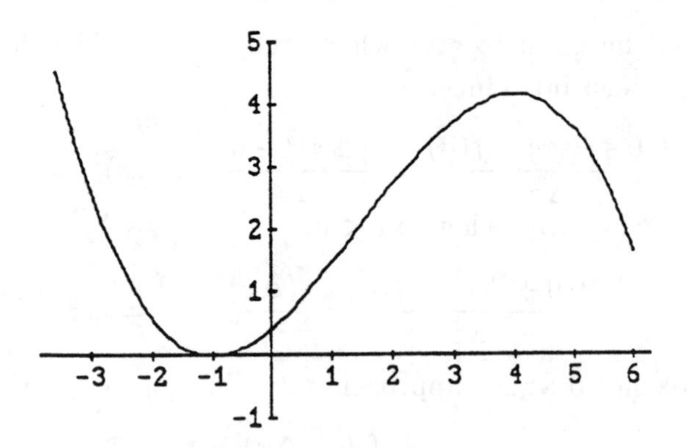

4. This problem is intended to get students used to the relationship between a function and its derivative. It would be a good exercise to do with a graphing package or on a graphing calculator.

5. a. There are two such intervals, namely (iii) $[1, 3]$ and (v) $[5, 6]$.

   b. Between years 3 and 4, probably about year 3.5.

   c. It is decreasing at a rate slightly more than 200 deer per year, perhaps about 230 deer per year.

6. a. (iv)
   b. (ii)
   c. (v)
   d. (i)
   e. (iii)

7. The graphs of the derivatives of the functions are given below.

a.

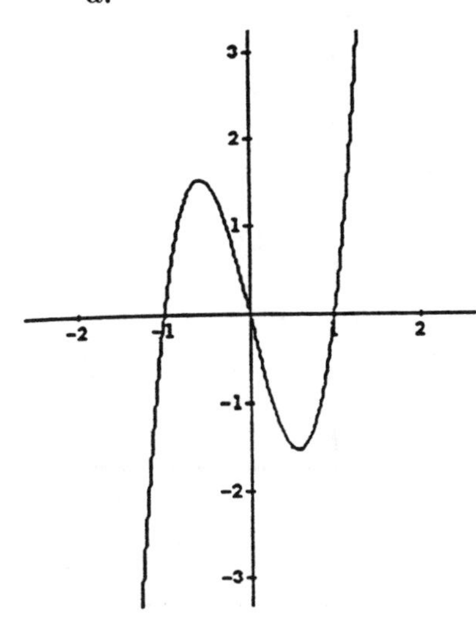

b.

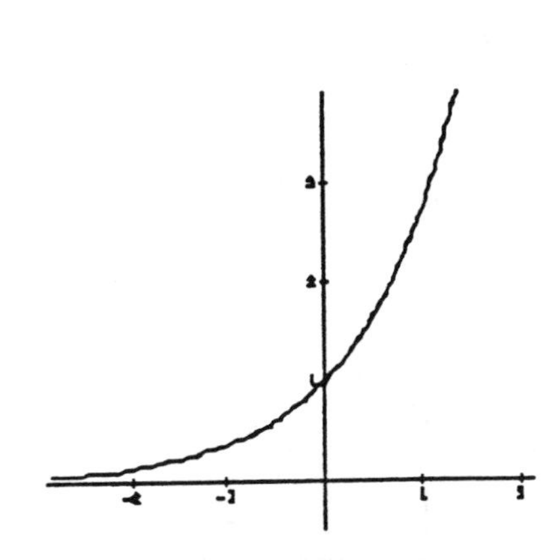

c.

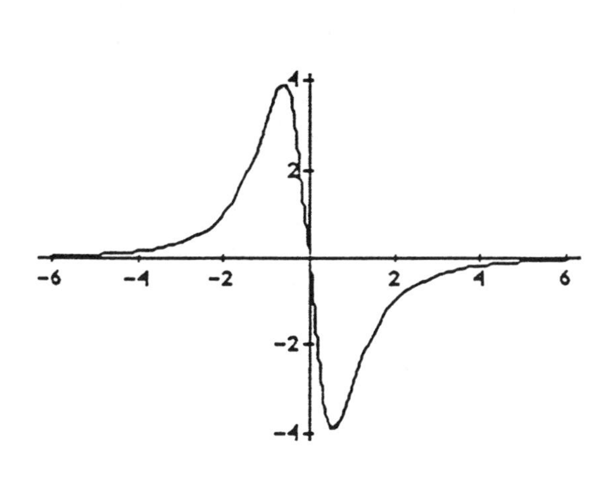

d.

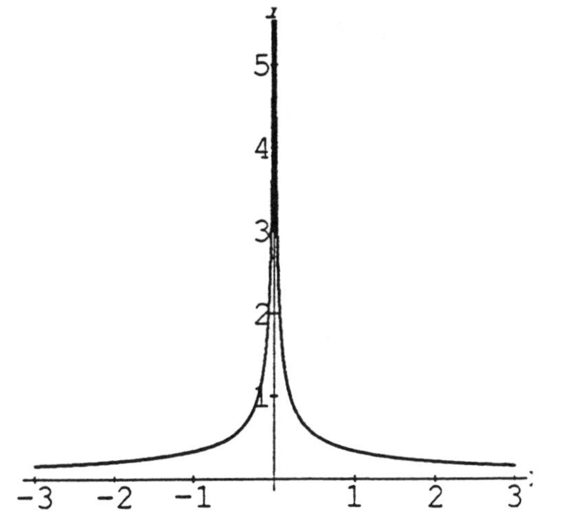

7. (continued) The graphs of the derivatives of the functions are given below.

e.

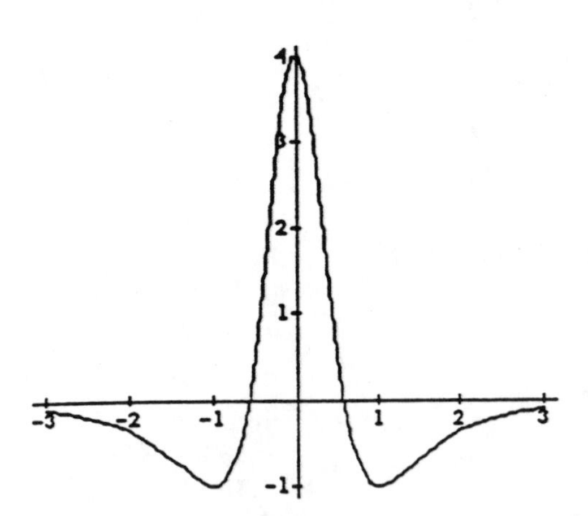

f.

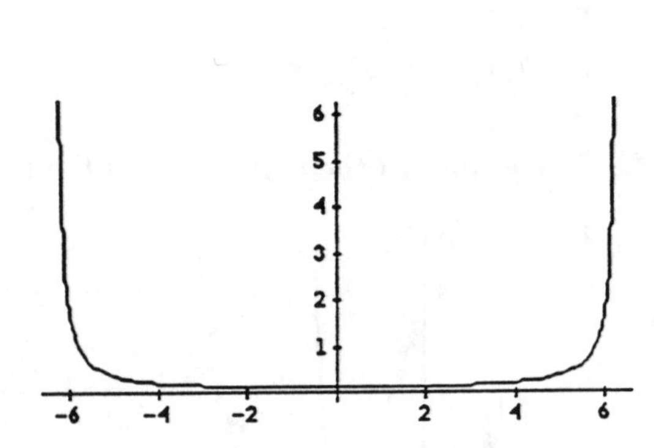

g.

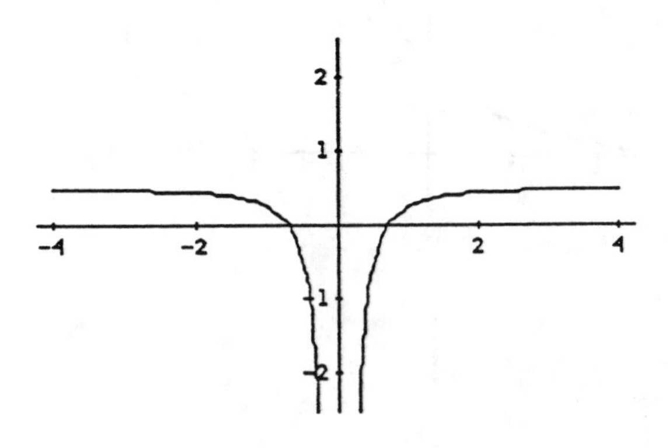

h.

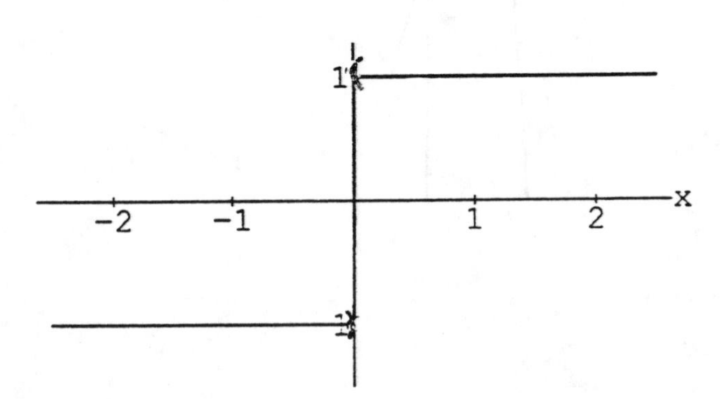

8. f is not differentiable at $x = -2$ (vertical tangent), $x = -1, 0, 2$ (sharp corner) and at $x = 1$ (discontinuity).

9. It would be very difficult to obtain the graph above just by graphically differentiating the C.P.I. graph in the problem; however, the qualitative features of this graph, with the peak inflation rate occurring in 1980–81 and the low inflation of 1986, can be easily seen by graphical differentiation.

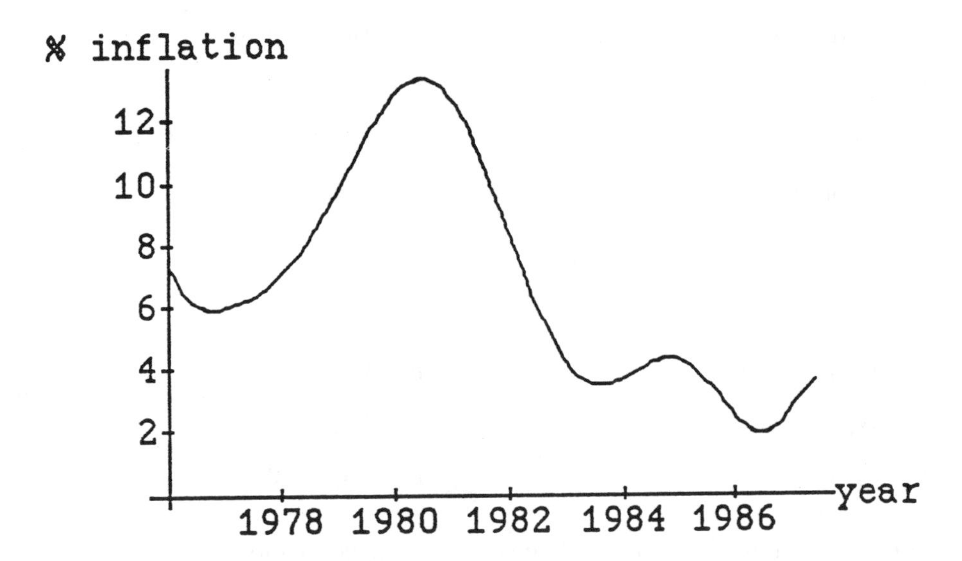

Section 4: Limits

1. This problem gets students to think about what a limit is since they have no formulas to work with. Parts e through *l* cover the various limit rules; parts *h* and *l* require some thought.

   a. Does not exist.
   b. Does not exist.
   c. Exists; 1
   d. Exists; −1
   e. Does not exist.
   f. Exists; 4
   g. Does not exist.
   h. Exists; 0
   i. Does not exist.
   j. Exists; 0
   k. Exists; 0
   l. Exists; 2

2. a. Exists; 1
   b. Exists; 2
   c. Exists; 1
   d. Exists; 1
   e. Exists; 1
   f. Exists; 0

3. This problem is one that every calculus student needs to understand.

   a. The left hand side is not defined at $x = 1$ while the right hand side is defined for all x. Nonetheless the discontinouity at $x = 1$ is removable, and this can be pointed out to students after they have grasped the point of this question.

   b. The expression $\frac{x^2 - 1}{x - 1}$ is equal to $x + 1$ for all x except $x = 1$. Since the limit measures what happens when x is near but is not equal to 1, this equation is correct.

4. Some of your students are bound to attempt to answer some of these questions like (a) with their calculators in degree mode. Take this opportunity to advise them that *all* calculus questions about trigonometric functions *must* be answered in radians.

   a. 1 b. 0

   c. 1 d. e

   e. 4 f. Does not exist.

   g. $\ln 2 \approx .6931$ h. 0

   i. This one is tricky in that one has to use very small x values before the function gets close to its limit of 0. Some students may think that the limit is 1 on the basis of the function values at $x = 0.1$ or even $x = 0.01$.

   j. Does not exist. Some students will want to make their conclusion on the basis of one small value for x. This problem should convince them that one needs to check several values to make sure that the function is really approaching a limit.

   k. Does not exist. Students who plug in only "round" numbers such as 10^{-n} for x will get zero for an answer. Intellegent use of a calculator is needed here.

   l. 0

5. Since the denominator is approaching zero as x approaches 3, the numerator must also be approaching zero as x approaches 3 (even for an infinite limit to exist this must still be true since the denominator changes sign at 3). As the numerator is continuous, this means that we need to solve the equation $2 \cdot 3^2 - 9a + 3 - a - 1 = 0$. The only solution to this equation is $a = 2$. It can be easily verified that when $a = 2$, the limit exists and is equal to $7/4$.

6. a. Since $\lim_{x \to 0+} \frac{\sin x}{|x|} = 1$ and $\lim_{x \to 0-} \frac{\sin x}{|x|} = \lim_{x \to 0-} \frac{\sin x}{-x} = -1$, the limit does not exist.

   [DA]

   b. $\lim_{x \to 0} \frac{1 - \cos^2(3x)}{x^2} = \lim_{x \to 0} \left(\frac{\sin 3x}{x}\right)^2 = 3^2 = 9.$

   [DA]

   c. Since $\cos x$ is continuous, $\lim_{x \to 0} \cos\left(\frac{1 - \cos x}{x}\right) = \cos\left(\lim_{x \to 0} \frac{1 - \cos x}{x}\right) = \cos 0 = 1.$

7. a. 0

   b. As in Problem 6, since $\cos x$ is continuous, $\lim_{x\to 0} \cos(f(x)) = \cos\left(\lim_{x\to 0} f(x)\right) = \cos(0) = 1$.

   c. Since $\sin x$ approaches but does not equal zero (except at zero itself) in a neighborhood of zero, $\lim_{x\to 0} f(\sin x) = \lim_{y\to 0} f(y) = 0$.

   d. This is the flip side of part c. For any $x \neq 0$, $f(f(x)) = f(0) = \pi$. Hence $\lim_{x\to 0} f(f(x)) = \pi$.

8. $\lim_{x\to 0} \frac{\sin(\sin x)}{x^k} = \lim_{x\to 0} \frac{\sin(\sin x)}{\sin x}\frac{\sin x}{x}x^{1-k} = \lim_{x\to 0} x^{1-k}$. Thus the only positive integer for which this limit exists is $k = 1$, in which case the limit is 1.

9. a.

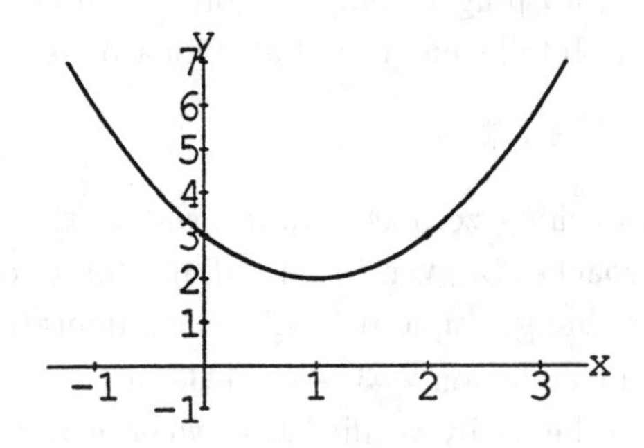

   b. 3

   c. (2.81,3.21)

   d. Any interval contained in (1.95, 2.04) will work.

[PF]

10. a. MIGHT be true (it is true if and only if f is continuous at a).

b. MIGHT be true. $f(0)$ may be assigned any value without changing the limit.

c. MUST be true. If $\lim_{x\to 0} f(x) \neq 0$, then for some $\epsilon > 0$ and any $\delta > 0$, there will be a number x such that $0 < |x| < \delta$ but $|f(x)| > \epsilon$, so $|\frac{f(x)}{x}| > \frac{\epsilon}{\delta}$. Since δ is arbitrary, this contradicts $\lim_{x\to 0} \frac{f(x)}{x} = 1$. More intuitively, since $\lim_{x\to 0} \frac{f(x)}{x} = 1$, $f(x)$ must be close to x when x is close to but not equal to 0; i.e., $\lim_{x\to 0} f(x) = 0$.

d. MUST be true by the definition of the derivative.

e. Can NEVER be true. The two-sided limit does not exist since the left- and right-hand limits are different.

Section 5: Continuity

1. a. $x = -2$, $x = -1$, $x = 2$

   b. $x = -2$, $x = -1$, $x = 2$, $x = 0$, $x = 1$

2. a. Yes, $f(a) = 0$.

   b. Yes, $\lim_{x \to a} f(x) = \lim_{x \to a} \frac{x^2 - a^2}{x - a} = \lim_{x \to a} x + a = 2a$.

   c. No, since $f(a) \neq \lim_{x \to a} f(x)$.

   d. No, f cannot possibly be differentiable at a since it is not continuous there.

[AP]

3. Yes. Define $f(0) = \lim_{x \to 0} \frac{\sin x}{x} = 1$.

4. None of the statements is necessarily true. Refer to Problem 2 for counterexamples.

5. By the definition of continuity, statements (i), (ii) and (iii) are true. Statement (iv) is not necessarily true.

6. $f(a) = L$. This problem is simply asking students whether or not they know and understand the definition of continuity.

7. a. Since $\lim_{x \to 0^-} f(x) = 2 \neq 3 = \lim_{x \to 0^+} f(x)$, $\lim_{x \to 0} f(x)$ does not exist and hence f cannot be continuous at 0.

   b. $\lim_{x \to 1} f(x) = 2$ since both one sided limits exist and are equal to 2. Moreover, since $f(2) = 2$, f is continuous at 1.

8. a. Sometimes false.

   b. Must be true.

9. a. f will be continuous at 1 if and only if $f(1) = \lim_{x \to 1^-} f(x) = \lim_{x \to 1^+} f(x)$. Hence $a = \lim_{x \to 1^-} ax = \lim_{x \to 1^+} (bx^2 + x + 1) = b + 2$, which is satisfied if and only if $a = b + 2$.

   b.

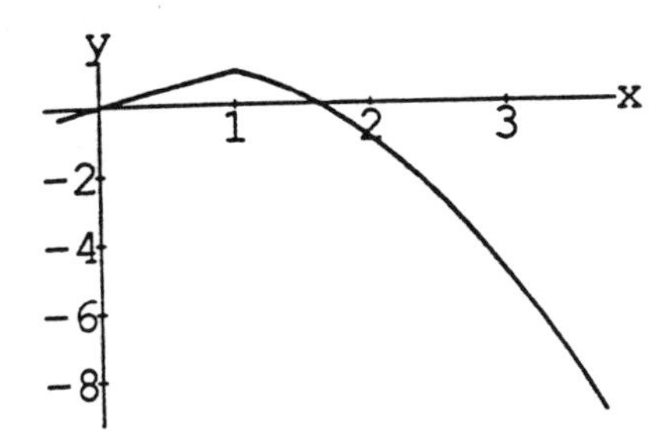

   c. If f is differentiable at 1 then its left and right hand derivatives at 1 must be equal, so $a = 2b + 1$. Since f must also be continuous at 1, $a = 2(a - 2) + 1 = 2a - 3$, so $a = 3$ and $b = 1$ are the only choices of a and b that make f differentiable.

   d.

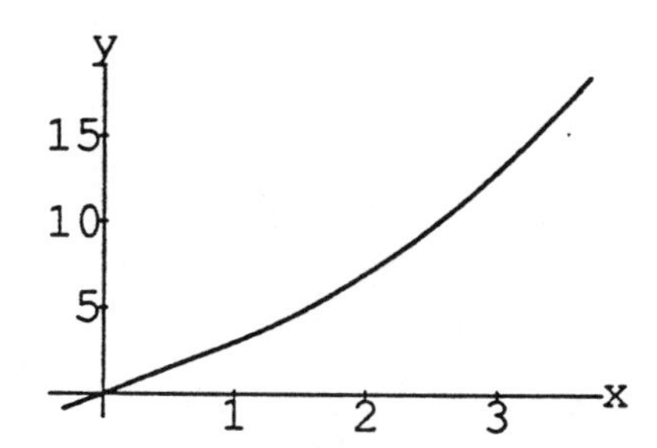

10. This question is a classic Intermediate Value Theorem problem. Let $p(t)$ denote George's distance along the road from his starting point in St. Louis at time t, where t denotes the time elapsed (in hours) since 9 AM Monday. Similarly, let $q(s)$ denote his distance along the road from his St. Louis starting point at time s, where s is the time elapsed (in hours) since 9 AM Tuesday. Intuitively, since $p(0) = q(5)$ and $q(0) = p(5)$ and p and q are continuous, the graphs of p and q must cross. More rigorously, let $f(x) = p(x) - q(x)$. Now $f(0) = p(0) - q(0) = 0 - q(0) < 0$, while $f(5) = p(5) - q(5) = p(5) - 0 > 0$. Therefore, by the Intermediate Value Theorem, there must be a time c for which $f(c) = 0$, which says that $p(c) = q(c)$, i.e. at time c he was at the same point on the road both days.

11. a. False. If f is not continuous, this need not be true since f may "jump" over 0.

b. True by the Intermediate Value Theorem.

c. False. For example take $f(x) = (x-3/2)^2$ or $f(x) = (x-3/2)^2 - 1/5$. Many students may give a graphical counterexample, which in fact might be more illuminating than one of the above formulas.

d. True by the Intermediate Value Theorem.

Section 6: Power, Sum and Product Rules

1. a. 1

   b. 7

   c. 1

2. 1

3. 33 knots by the addition rule.

4. 22 cm/sec. Normally such a problem would appear in a "related rates" section; but since this problem is an application of the product rule rather than the chain rule, it is inserted here.

5. a. $(-1)^n \dfrac{n!}{x^{n+1}}$

   b. $(-1)^n \dfrac{n!}{(x+1)^{n+1}}$

   c. $(-1)^{n+1} \dfrac{1 \cdot 3 \cdot 5 \cdots (2n-3)}{2^n} x^{-(2n-1)/2} \quad (n = 2, 3, \ldots)$

6. Differentiating both sides of $f(x)g(x) = x$ and evaluating at $x = 1$ gives $f'(1)g(1) + f(1)g'(1) = 1$, so $g(1)$ and $f(1)$ cannot both be zero.

7. a. $f(0) = f(0+0) = f(0)f(0)$. Since $f(0) \neq 0$, this implies that $f(0) = 1$.

   b. $f(0) = f(x+(-x)) = f(x)f(-x)$. Since $f(0) \neq 0$, this shows that $f(x) \neq 0$ as well.

   c. $$f'(x) = \lim_{\Delta x \to 0} \frac{f(x+\Delta x) - f(x)}{\Delta x} = \lim_{\Delta x \to 0} \frac{f(x)f(\Delta x) - f(x)}{\Delta x}$$
   $$= f(x) \lim_{\Delta x \to 0} \frac{f(\Delta x) - 1}{\Delta x} = f(x)f'(0) = f(x).$$

   d. $k(x)$ is defined for all x since $g(x) \neq 0$ by part (b). Using the quotient rule and part c yields $k'(x) = \dfrac{f'(x)g(x) - f(x)g'(x)}{g^2(x)} = \dfrac{f(x)g(x) - f(x)g(x)}{g^2(x)} = 0$. Since $k(0) = 1$, $k(x) = 1$ for all x, and this shows that $f(x) = g(x)$.

   e. The exponential function, $y = e^x$, satisfies conditions (i)–(iii), and in view of part d, it is the only function to do so.

[TA]

8. a. $h''(x) = f''(x)g(x) + 2f'(x)g'(x) + f(x)g''(x)$.

b. $h'''(x) = f'''(x)g(x) + 3f''(x)g'(x) + 3f'(x)g''(x) + f(x)g'''(x)$.

c. $h^{(n)}(x) = f^{(n)}(x)g(x) + nf^{(n-1)}(x)g'(x) + \binom{n}{2} f^{(n-2)}(x)g''(x)$

$+ \binom{n}{3} f^{(n-3)}(x)g'''(x) + \cdots + nf'(x)g^{(n-1)}(x) + f(x)g^{(n)}(x)$

Note the similarity to the Binomial Theorem.

[CB]

9. a. $h(-2) = f(-2)g(-2) = (1)(3) = 3$. $h(3) = f(3)g(3) = (\frac{1}{2})(-1) = -\frac{1}{2}$.

b. $f'(-2) \approx -1$, $f'(3) \approx -1$, $g'(-2) \approx 3/5$, $g'(3) \approx \frac{3}{10}$

c. $h'(-2) = f(-2)g'(-2) + f'(-2)g(-2) \approx (1)(3/5) + (-1)(3) = -12/5$; $h'(3) = f(3)g'(3) + f'(3)g(3) \approx (1/2)(3/10) + (-1)(-1) = 23/20$

Section 7: The Chain Rule

1. a. $h'(2) = f'(g(2))g'(2) = f'(1)g'(2) = 2 \cdot -3 = -6$

   b. $k'(2) = g'(f(2))f'(2) = g'(-1)f'(2) = 5 \cdot 4 = 20$

2. $\dfrac{d}{dx} \sin |x| = \begin{cases} \cos x & \text{if } x > 0, \\ -\cos x & \text{if } x < 0. \end{cases}$ The derivative does not exist at 0.

   [DA]

3. $12\dfrac{dy}{dx} = \dfrac{d}{dx}y^3 = 3y^2\dfrac{dy}{dx}$, so $y = 2$.

   [AP]

4. $\dfrac{d}{dx}f(h(x)) = f'(h(x))h'(x) = 2xg(x^2)$

   [AP]

5. a. $\dfrac{d}{dx}f_2(x) = f'(f(x))f'(x)$.

   b. $\dfrac{d}{dx}f_3(x) = f'(f_2(x))f'(f(x))f'(x)$.

   c. $\dfrac{d}{dx}f_n(x) = f'(f_{n-1}(x))f'(f_{n-2}(x)) \cdots f'(x)$.

6. a. That f is an even function means $f(-x) = f(x)$. Differentiating both sides of this expression gives $-f'(-x) = f'(x)$, which shows that f' is an odd function.

   b. Similarly, if f is an odd function, then $f(-x) = -f(x)$ and $-f'(-x) = -f'(x)$, so f' is an even function.

7. a. $h(-2) = f(g(-2)) = f(3) \approx 1/2$; similarly $h(3) \approx 1/4$.

   b. $h'(-3) = f'(g(-3))g'(-3) = f'(2)g'(-3) = 0g'(-3) = 0$

   c. $h'(-1) = f'(g(-1))g'(-1) = f'(3)g'(-1)$. Since both $f'(3)$ and $g'(-1)$ are clearly negative, $h'(-1) > 0$.

   Refer to the problem in II.6 in which f and g are used to define a product function h. The idea in both these problems can be extended to other situations, e.g. a function f of two variables, x and y, each of which is a function of a variable t. Given graphs of f, x, and y, students can be asked to approximate $F'(t_o)$, where $F(t) = f(x(t), y(t))$ and t_o is some value of t.

8. a. $h'(x) = 2f(x)f'(x) + 2g(x)g'(x) = 2f(x)g(x) - 2g(x)f(x) = 0$. Since $h(0) = 1$, $f^2(x) + g^2(x) = 1$ for all x.

   b. $k'(x) = 2[F(x) - f(x)][F'(x) - f'(x)] + 2[G(x) - g(x)][G'(x) - g'(x)]$
   $= 2[F(x) - f(x)][G(x) - g(x)] + 2[G(x) - g(x)][f(x) - F(x)] = 0.$
   Since $k(0) = 0$, k is identically zero and hence $F(x) = f(x)$ and $G(x) = g(x)$ for all x.

   c. $f(x) = \sin x$ and $g(x) = \cos x$ satisfy properties (i) and (ii). In light of the answer to part *b*, these are the only functions to satisfy both these properties.

[TA]

Section 8: Implicit Differentiation and Derivatives of Inverses

1. Differentiating both sides of $x = t^3 - t$ with respect to x gives $\frac{dt}{dx} = \frac{1}{3t^2 - 1}$. Hence $\frac{dy}{dx} = \frac{dy}{dt}\frac{dt}{dx} = \frac{3}{2\sqrt{3t+1}}\frac{1}{3t^2-1}$. Evaluating at $t = 1$ gives $\left.\frac{dy}{dx}\right|_{t=1} = \frac{3}{8}$.

[AP]

2. a. It is clear from the graph (a circle with center at the origin) that the derivative should be positive in the second and fourth quadrants, and negative in the first and third quadrants.

   b. $\frac{dy}{dx} = \frac{-x}{y}$

[AZ]

3. a. The graph is a hyperbola that intersects the x axis, and from its graph the derivative should be positive in the first and third quadrant, and negative in the second and fourth quadrant.

   b. $\frac{dy}{dx} = \frac{x}{y}$

[AZ]

4. a. $\frac{dy}{dx} = \frac{x}{y} \cdot \frac{6 - (x^2 + y^2)}{6 + (x^2 + y^2)}$

   b.

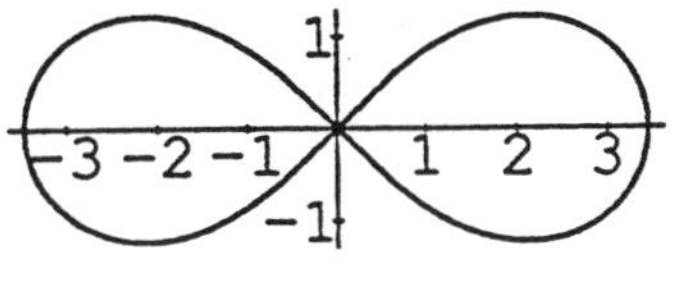

5. a. Implicit differentiation yields $\dfrac{dy}{dx} = \dfrac{2x}{5y^4 - 1}$. When $x = 1$,

$$y^5 - y = y(y^2+1)(y-1)(y+1) = 0,$$

so the curve goes through the points $(1,0)$, $(1,1)$ and $(1,-1)$. The slopes of the tangent lines through these three points are -2, $\frac{1}{2}$ and $\frac{1}{2}$, respectively. Thus equations of the tangent lines are: $y = -2(x-1)$, $y = \frac{1}{2}(x+1)$ and $y = \frac{1}{2}(x-3)$.

b. The tangent line will be vertical when the denominator of the derivative $5y^4 - 1$ is zero (except possibly when the numerator is also zero). Thus, there are vertical tangent lines when $y = \pm\sqrt[4]{\frac{1}{5}}$. The x-coordinates of these points are $x = \pm\sqrt{\pm\frac{4}{5}\sqrt[4]{\frac{1}{5}}+1}$. Thus there are four points on the curve that have a vertical tangent; their approximate coordinates are $(-1.239, -.669)$, $(1.239, -.669)$, $(.682, .669)$ and $(-.682, .669)$.

c. The point $(0, -1.1673)$ is approximately the y-intercept asked for.

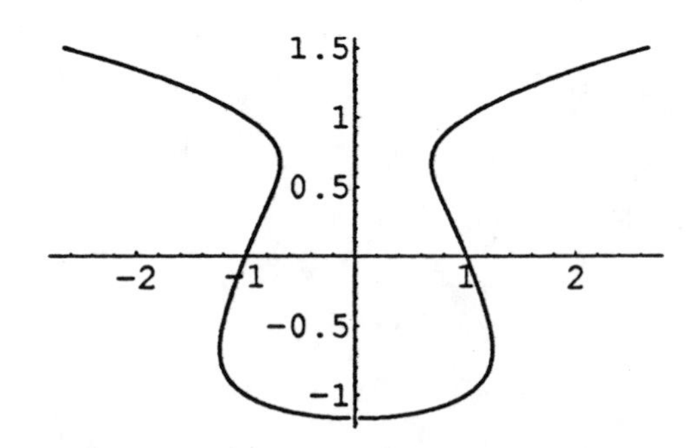

[MN]

6. Let $y = f(x) = x^3 + x$, so $h(y) = x$. Differentiating the first equation with respect to y gives $\dfrac{dx}{dy} = \dfrac{1}{3x^2+1}$. When $y = 2$, $x = 1$, so $h'(2) = \frac{1}{4}$.

[AP]

Section 9: Derivatives of Trigonometric, Log, and Exponential Functions

1. Differentiating both sides of $\sin x = e^y$ gives $\dfrac{dy}{dx} = \dfrac{\cos x}{e^y} = \cot x$. This problem could also be done by taking logarithms of both sides and differentiating explicitly.

[AP]

2. a. $f(-x) = \dfrac{-x + \sin(-x)}{\cos(-x)} = \dfrac{-x + (-\sin x)}{\cos x} = -f(x)$, so f is an odd function.

   b. $f'(x) = \dfrac{1 + \cos x + x \sin x}{\cos^2 x}$

   c. $y = 2x$

[AP]

3. a, b, d all satisfy this equation.

[AP]

4. $f'(e) = \lim\limits_{\Delta x \to 0} \dfrac{e^{e+\Delta x} - e^e}{\Delta x} = \lim\limits_{\Delta x \to 0} \dfrac{e^e e^{\Delta x} - e^e}{\Delta x} = \lim\limits_{\Delta x \to 0} e^e \dfrac{e^{\Delta x} - 1}{\Delta x}$, so the only correct answer is (e).

[AP]

5. There are several ways of attacking this problem; most involve looking at the derivative of f. Perhaps the most straightforward way is to note that $f'(x) = 2(\cos 2x - e^{-2x})$, so $f'(0) = 0$. The only graph with this property is b. Another method is to observe that $f'(x) < 0$ when $x < 0$, which eliminates a. The derivative of the function graphed in part c seems almost constant for positive x, which is clearly not true of $f'(x)$.

[TT]

6. As with the previous problem, this problem is most easily solved by looking at the derivative of f. $f'(x) = \cos x + \dfrac{5}{x+1}$. Hence $f'(0) = 5$, but for large x, $f'(x)$ can be at most slightly bigger than 1. Part a is the only graph that satisfies this property. Also, part b can be eliminated by noting that its derivative is never negative, while $f'(x)$ is clearly negative for some x. Part c has a negative derivative for some $x < 5$, which is not true of $f(x)$, and the oscillations in part c are getting larger, which again does not happen for $f(x)$.

7. Let $f(x) = e^x$. Then $\lim\limits_{x \to 0} \dfrac{e^{3+x} - e^3}{x} = f'(3) = e^3 \approx 20.086$.

8. All three graphs have the same general shape and are always increasing (see the graphs below). If $b_1 > b_2 > 1$, b_1^x increases more rapidly since its derivative is always greater for any fixed value of x. Thus $b_1^x > b_2^x$ for $x > 0$ and $b_2^x > b_1^x$ for $x < 0$.

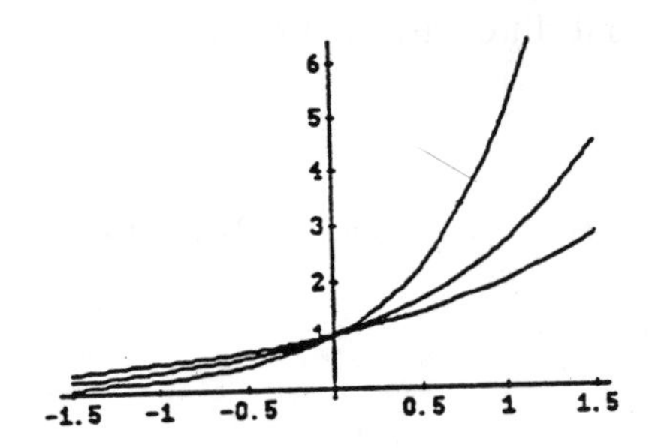

Graph for Problem 8

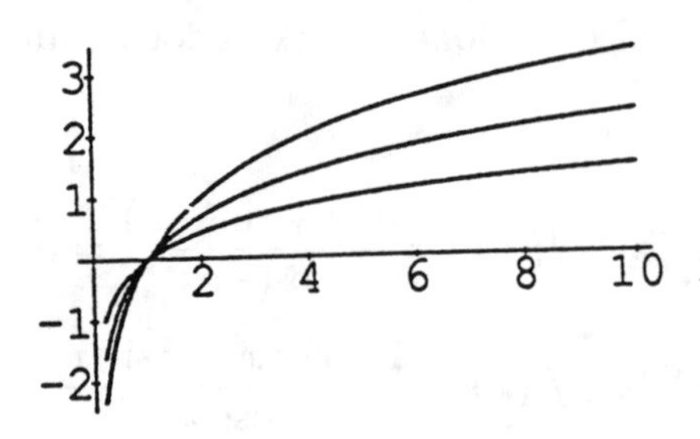

Graph for Problem 9

9. The geometry of the problem is contained in the hint (which you may choose to suppress). If $y = a^x$, then $y' = a^x \ln a$ and in view of the hint, $a^b \ln a = 1$ where (b, b) is the common point of tangency. Similarly if $y = \log_a x$, then $y' = \dfrac{\log_a e}{x}$ and $\log_a e/b = 1$, so $b = \log_a e$. Substituting this value of b in the expression $a^b \ln a = 1$ yields $e \ln a = 1$ and so $a = e^{1/e}$. $b = \log_a e = 1/\ln a = e$. This might be a good place to remind your students that e is a particularly good base for an exponential function since the graph of such a function has slope 1 at $x = 0$.

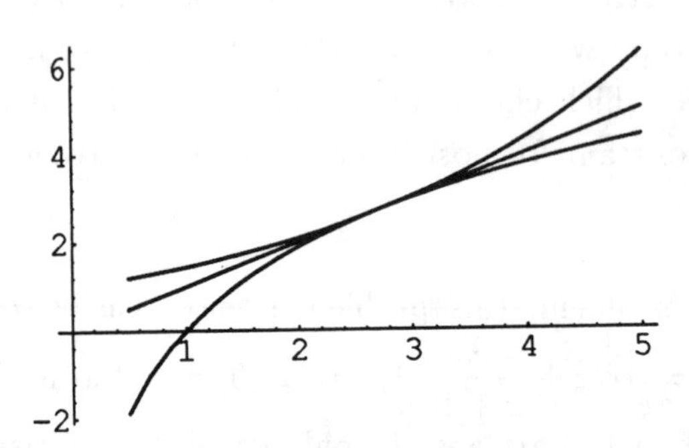

10. Again, all three graphs have the same general shape and are always increasing (see the graph above). If $b_1 > b_2 > 1$, then $\log_{b_1} x$ increasing more slowly than $\log_{b_2} x$ since its derivative is smaller for any given value of x. Hence $\log_{b_1} x < \log_{b_2} x$ for $x > 1$ and $\log_{b_1} x > \log_{b_2} x$ for $0 < x < 1$.

11. Taking natural logarithms of both sides is a valid step since $\ln x$ is an increasing function. Since $f'(x) = \dfrac{1 - \ln x}{x^2}$, f is increasing when $x < e$ and decreasing when $x > e$. Since $f(e) = \dfrac{1}{e}$, this shows that our inequality is satisfied for all positive $x \neq e$. It might be worthwhile to ask the more ambitious students to generalize this problem and solve the inequality $x^a < a^x$.

[MN]

12. The goal of this problem is to have students actually try their hand in calculating the derivative of an inverse function, using the same techniques they have seen used for the inverse trigonometric functions. In fact, f is really just a translation and scaling of the cosine function. Part a is straightforward, since $f'(x) = \cos x - \sin x$ is positive on the domain of f. As for part b, we will find the derivative of $y = f^{-1}(x)$ by differentiating $f(y) = x$ implicitly. Doing this gives $\dfrac{dy}{dx} = \dfrac{1}{\cos y - \sin y}$. Using the given formula yields $\cos y - \sin y = \sqrt{2 - (\cos y + \sin y)^2} = \sqrt{2 - x^2}$. Thus $\dfrac{dy}{dx} = \dfrac{1}{\sqrt{2 - x^2}}$.

Section 10: Root Finding Methods

1. Here is one set of possible graphs:

a.

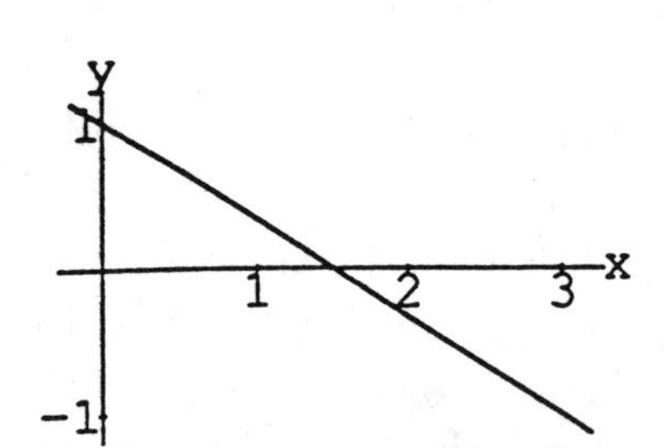

b.

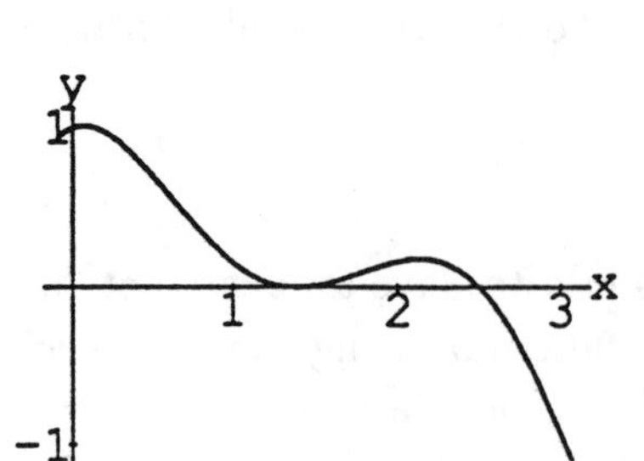

c.

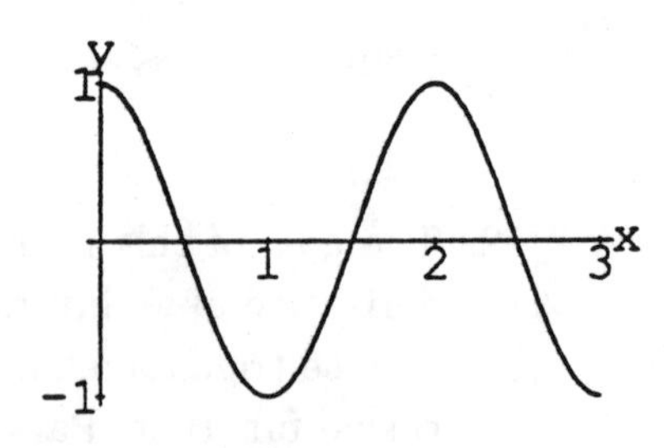

2. a. Since $f'(x) \neq 0$ for all x, f is a monotonic function and therefore can have at most one zero (and may have none). Alternatively one can apply Rolle's Theorem as is done below.

   b. By Rolle's Theorem, the equation $f'(x) = 0$ must have at least one solution between any two solutions to $f(x) = 0$. Therefore f can have at most 2 zeros.

   c. Arguing as in part *b*, we find that f may have at most $n + 1$ zeros.

3. a. For sufficiently large x, $f(x)$ and $f(-x)$ must have different signs and hence $f(x) = 0$ must have at least one solution by the Intermediate Value Theorem.

   b. No. For example, given any even integer n, $f(x) = x^n + 1 \geq 1$ for all x and hence $f(x) = 0$ has no solutions.

4. Let $f(x) = x^3 - 3x + b$. Then $f'(x) = 3x^2 - 3 < 0$ for all $x \in (-1, 1)$. Hence f is (strictly) decreasing on $[-1, 1]$ and thus the equation $f(x) = 0$ can have at most one solution on this interval.

[TA]

5. Let $f(x) = \cos x + x \sin x - x^2$. Now $f(0) = 1$ and $f(\pi) = f(-\pi) = -1 - \pi^2$. (in fact, f is an even function.) Thus the Intermediate Value Theorem implies that there must be at least one solution to $f(x) = 0$ in the interval $(0, \pi)$ and another in the interval $(-\pi, 0)$. Therefore the equation $f(x) = 0$ has at least two solutions. On the other hand, since $f'(x) = -\sin x + \sin x + x \cos x - 2x = x(\cos x - 2)$, the equation $f'(x) = 0$ has only one solution. Hence the equation $f(x) = 0$ has at most two solution by Rolle's Theorem (see Problem 2*b*). Putting the two parts together gives that $f(x) = 0$ must have exactly two solutions.

[TA]

6. It appears from the graph given below that $x_1 = 3$ and $x_2 \approx 2.2$. The purpose of this problem is to help students understand how Newton's method works, so the exact numbers obtained are not as important as drawing the tangent lines used in obtaining them.

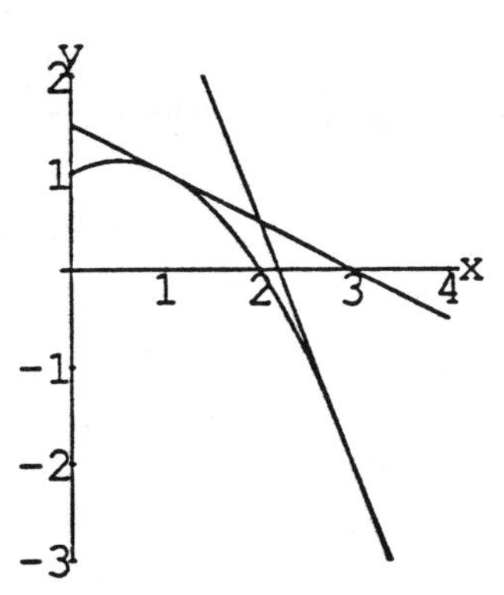

7. Since f is concave up, its tangent lines always lie below the graph of f. Hence any line tangent to f at a point $(x, f(x))$, where $x > z$, will intersect the x-axis to the right of z. Therefore, if x_n is greater than z, x_{n+1} must also be greater than z.

8. a. By inspection, $f(1) = 500$. This says that the hare passes the midway point after exactly one minute. After one minute the tortoise has gone only $g(1) = 350$ meters (still pretty fast for a tortoise), so the hare easily passes the halfway point first. Alternatively, one can solve the equations $f(t) - 500 = 0$ and $g(t) - 500 = 0$ using Newton's method (the latter equation can also be solved by hand). The solutions are $t = 1$ for the first equation as above and $t \approx 1.72$ for the second equation.

   b. We need to solve the equation $f(t) - g(t) = 0$, which can be approximately solved by Newton's method. The only solution for positive t is when $t \approx 4.53$ minutes at which time both runners have traveled about 985.45 meters. Anticipating the answer to part (c), we find that the tortoise passes the hare at this time, less than 15 meters from the finish line!

   c. To find out who crosses the finish line first, we need to solve the equations $f(t) - 1000 = 0$ and $g(t) - 1000 = 0$. Approximate solutions (found by Newton's method) are $t \approx 4.681$ minutes for the first equation and $t \approx 4.624$ for the second equation. Thus the tortoise wins the race, crossing the finish line a mere 3.4 seconds ahead of the hare.

Section 11: Related Rates

The advice given here is a rather loose amalgam of game plans that can be found in numerous texts. The teacher should not inundate students with advice, as they must learn to realize that there is no guaranteed procedure for solving story problems. Nonetheless, a few helpful hints may get students off square one, and start them down the road.

1. Many students are too impatient to read a problem through to its conclusion. They may miss important facts and may also not clearly understand which quantity or quantities are being sought.

2. In any problem with geometric overtones, it is important that students take the initiative to draw their own pictures.

3. This, at least, gets the students started. Encourage the use of mnemonically helpful variables, rather than just "x" and "y".

4. Relationships between variables occur in several different ways. Various forms of the verb "to be," such as "is" and "are," often provide fruitful sources of equations. In a picture, relationships will often occur from the use of triangles. Particular emphasis should be placed on obtaining relationships from either the Pythagorean Theorem or similar triangles. Since triangles are such an important source of relationships, students should be encouraged to draw auxiliary lines to create triangles, as this is often a good way to attack a problem.

5. A common error students often make is to adopt an "all or nothing" attitude. One way to encourage students to get in the habit of breaking up a difficult problem into a succession of easier steps is to stress that *partial credit* will be given on story problems on exams. Breaking down a problem into subgoals often results in significant amounts of progress. Long before Mao Tse Tung, problem solvers realized that a journey of a thousand miles begins with a single step.

6. Students will often get stuck during a problem. A common reaction is that the student abandons the problem, feeling that it is too hard. This piece of advice, as well as the next one, are the story-problem equivalent of "just one more rep" in weightlifting. The adage of "no pain, no gain" applies to story problems as well.

7. Checking a problem is not an onerous task, but it is a way to insure inner tranquility—nothing is quite so reassuring as to *know* that one has obtained the right answer! If that is not easily done, then a plausibility check to see if the answer is in the right ball-park is often possible.

1. Here is another opportunity to use both the Pythagorean Theorem and implicit differentiation. Let H denote the distance of the runner from home plate, S the distance of the runner from second base (see the figure below). The question asks us to evaluate dS/dt when $H = 45$, given that $dH/dt = 20\,\text{ft/sec}$. Clearly

$$S^2 = 90^2 + (90 - H)^2$$

Differentiating, we obtain

$$2S\frac{dS}{dt} = 2(90 - H)\left(-\frac{dH}{dt}\right)$$
$$\frac{dS}{dt} = \frac{(H - 90)dH/dt}{S}$$

When $H = 45$, $\dfrac{dS}{dt} = -4\sqrt{5} \approx -8.944$ ft/s. The negative sign indicates that the distance is decreasing. (i)

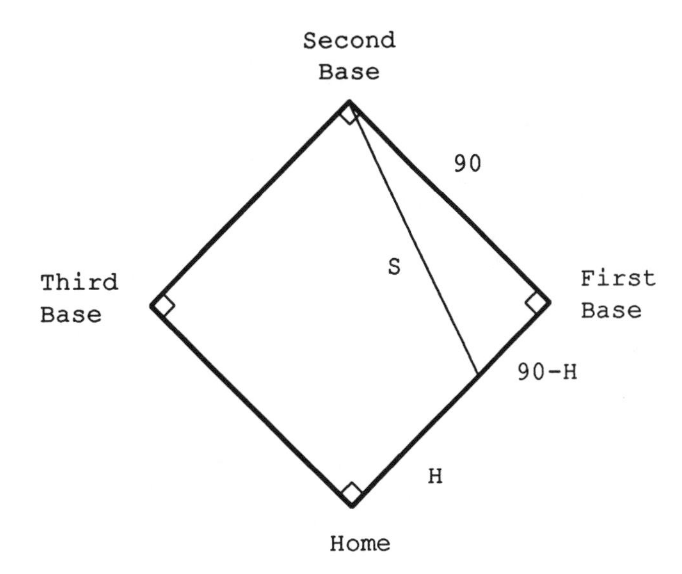

2. This question is not difficult, but once again students must ask what quantity is being sought, what is constant, and what is varying. Differentiating the basic equation, we obtain

$$P(kV^{k-1}\frac{dV}{dt}) + V^k\frac{dP}{dt} = 0$$

Factoring out V^{k-1} yields

$$kP\frac{dV}{dt} + V\frac{dP}{dt} = 0$$

$$k = -\frac{V\ dP/dt}{P\ dV/dt} = \frac{-10(-14)}{25(4)} = 1.4$$

Notice how the fact that differentiation lowers the exponent by 1 enables us to "luck out" and not have to solve a nonlinear equation for k. This is even more apparent if students can differentiate logarithmic functions, as can be seen by taking the logarithms of both sides of the original equation.

3. A good challenge problem! Let k be the proportionality constant, D be the liquid depth, and R the liquid radius. By similar triangles, $R = D/5$, so the volume V equals $\pi R^2 D/3 = \pi D^3/75$, and $\dfrac{dV}{dt} = \dfrac{\pi D^2 dD/dt}{25}$. If L is the slant height, the surface area is equal to $\pi RL = \pi R\sqrt{R^2 + D^2}$. We obtain another expression for dV/dt, as it is the difference between the liquid added at the top and the liquid oozing out. Equating the two yields

$$\frac{\pi}{25} D^2 \frac{dD}{dt} = 1 - k\pi R\sqrt{R^2 + D^2}.$$

Substituting $D = 4$, $dD/dt = -.1$, and $R = \frac{4}{5}$, and solving for k, we obtain $k = (125 + 8\pi)/(80\pi\sqrt{26})$. If we add liquid at a rate r which precisely counteracts the rate of oozing, the liquid level will remain unchanged. This rate of oozing is

$$k\pi R\sqrt{R^2 + D^2}.$$

Using the value above for k and substituting $D = 4$ and $R = \frac{4}{5}$, we obtain a value of $(125 + 8\pi)/125\,\text{cm}^3/\text{min}$ for the rate at which liquid must be poured to maintain a constant liquid level of 4 cm.

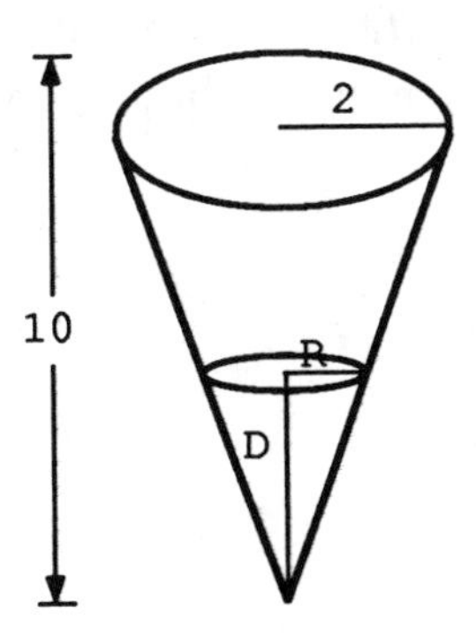

4. Let R be the radius, T the thickness, and V the volume of the slick. Then

$$V = \pi R^2 T$$

Differentiating, we obtain

$$\frac{dV}{dt} = \pi(R\frac{dT}{dt} + 2RT\frac{dR}{dt}).$$

When $R = 500$, $T = .01$ and $dT/dt = -.001$. Of course, $dV/dt = -5$. Therefore,

$$-5 = \pi(-250 + 10\frac{dR}{dt}).$$

So $dR/dt = 25 - 1/(2\pi)$. Since $A = \pi R^2$, we therefore have

$$\frac{dA}{dt} = 2\pi R\frac{dR}{dt} = 1000\pi(25 - \frac{1}{2\pi}) \approx 78{,}040\text{ft}^2/\text{hr}.$$

5. Students may be stumped by this problem, because at this stage of a calculus course they cannot compute the volume of a section of a sphere (hence the hint). By the Pythagorean Theorem,

$$(5 - D)^2 + R^2 = 5^2,$$

Thus $dV/dt = \pi(10D - D^2)dD/dt$. Since $dV/dt = 3$ and $D = 2$, we obtain $dD/dt = 3/(16\pi)$in/sec. This problem will help prepare students for volumes by slices.

6. There are several approaches to this problem. We shall use a coordinate plane, with origin at the center of the track and with the spectator at $(300, 0)$. Let A be the angle between the radius to the runner and the x-axis, let D be the distance from the runner to the spectator, and let s be the distance covered by the runner (arc length). See the figure below. Since $s = 100A$, we see that $ds/dt = 5 = 100dA/dt$, so $dA/dt = .05$. By the distance formula,
$$D^2 = 100^2((3 - \cos A)^2 + \sin^2 A).$$
Simplifying and differentiating yields $2D\dfrac{dD}{dt} = 60{,}000 \sin A \dfrac{dA}{dt}$. If $D = 250$, solving for $\cos A$ yields $\cos A = \frac{5}{8}$. Therefore, we see that
$$\frac{dD}{dt} = \frac{30{,}000 \sin A\ (dA/dt)}{D} = \frac{30{,}000(-\sqrt{39/64})(.05)}{250} = \frac{-3\sqrt{39}}{4}\ \text{m/sec}.$$

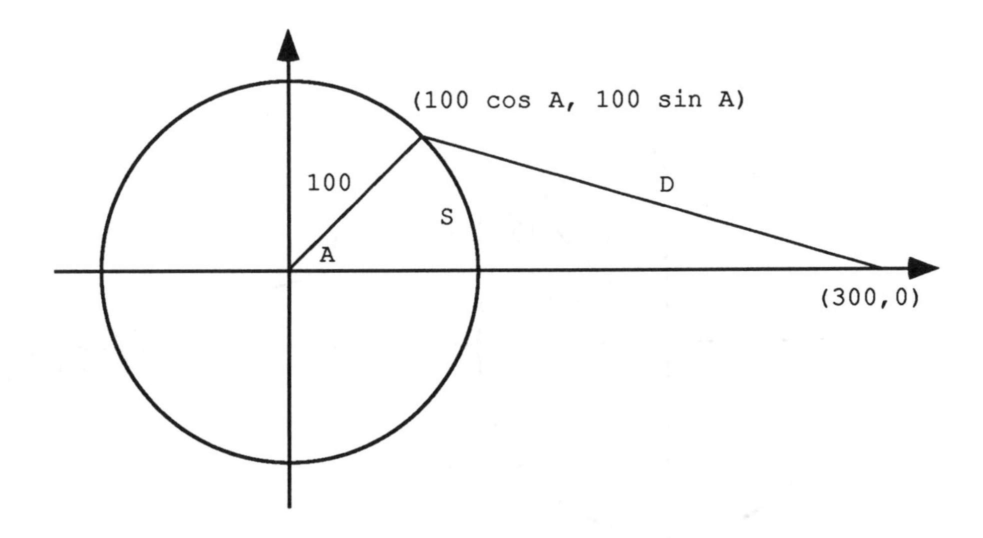

7. In a sense this problem is dual to the previous one. Using the notation in the diagram below, $x = 10\cos A + 30\cos B$, and $10\sin A = 30\sin B$. Thus

$$\frac{dx}{dt} = -10\sin A\frac{dA}{dt} - 30\sin B\frac{dB}{dA}\frac{dA}{dt}.$$

B reaches a maximum when A is an odd multiple of $\pi/2$, so $dB/dA = 0$ at the specified configuration. The speed of the piston at that moment is the absolute value of dx/dt, which is therefore seen to be $10\,dA/dt = 80\pi$ cm/sec.

The solution presented here uses some geometric insight to show that $dB/dA = 0$ at the moment that the right triangle is formed. It can also be done by simply solving for $\cos B$ in terms of $\sin A$ and grinding it out, *apropos* of which you may want to ask your students to solve for x in terms of t and to use a graphics package to graph the result:

$$x = 10\cos(8\pi t) + 30\sqrt{1 - \frac{\sin^2 8\pi t}{9}}.$$

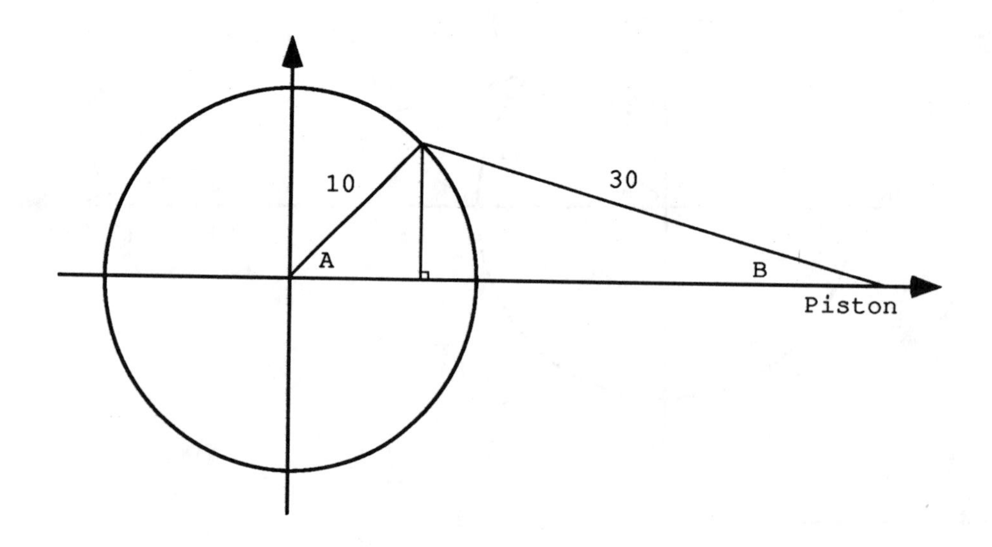

Chapter III: Extreme Values

Section 1: Increasing and Decreasing Functions and Relative Extrema

1. Most textbooks only give problems with polynomials of at most fourth degree. These functions quickly become predictable and boring. These are examples of higher order polynomials which can be done rather easily.

   a. $f'(x) = 72x^7(x-1)$, so f is increasing on $(-\infty, 0]$ and $[1, \infty)$ and is decreasing on $[0, 1]$. There is a relative maximum at 0 and a relative minimum at 1.

   b. $f'(x) = 63x^6(x - \sqrt{2})(x + \sqrt{2})$, so f is increasing on $(-\infty, -\sqrt{2}]$ and on $[\sqrt{2}, \infty)$ and decreasing on $[-\sqrt{2}, \sqrt{2}]$. It has a relative maximum at $-\sqrt{2}$ and a relative minimum at $\sqrt{2}$.

   c. $f'(x) = 30x^3(x+2)(x-1)$, so f is decreasing on $(-\infty, -2]$ and on $[0, 1]$, and is increasing on $[-2, 1]$ and $[1, \infty)$. It has relative minima at -1 and 2 and a relative maximum at 0.

2. Here is the graph of one such function:

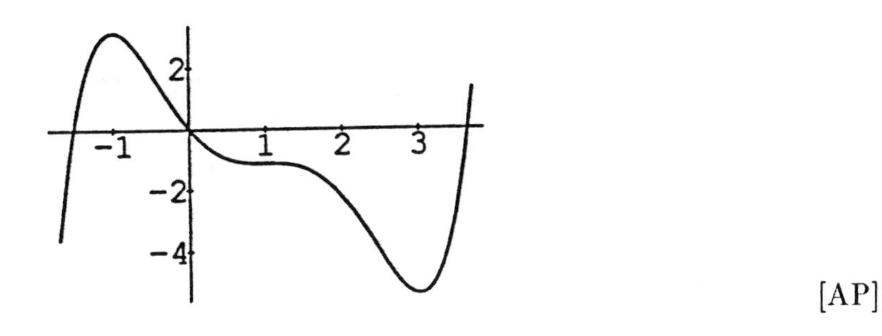

[AP]

3. a. $f'(x) = 4x - 2000$, so f is decreasing on $(-\infty, 500]$ and increasing on $[500, \infty)$.

   b. By part a, $1000^2 = f(0) > f(2) = 998^2 + 2^2$.

   c. $f'(x) = -n(c-x)^{n-1} + nx^{n-1}$. Hence, $f'(x) < 0$ when $c - x > x$, so f is decreasing on $(-\infty, \frac{c}{2}]$ and increasing on $[\frac{c}{2}, \infty)$. As in part *b*, this shows that 10000^{100} is larger than $9000^{100} + 1000^{100}$.

4. Let $h(x) = f(x) - g(x)$. Then h has a relative maximum at c. Thus $0 = h'(c) = f'(c) - g'(c)$, so the slopes of the tangent lines of f and g at c are the same.

[TF]

5. a. Since $f'(x) \leq 2$, f can increase by at most 6 on an interval of length 3. Thus $f(3) \leq -1 + 6 = 5$.

   b. Let $g(x) = f(x) - ax$. Then $g'(x) \leq 0$ for all real numbers, so $b = g(0) \geq g(c) = f(c) - ac$. Hence $f(c) \leq b + ac$.

6. There are two points where f' is not defined: -2 and 1. The derivative of $|x - a|$ is -1 if $x < a$ and 1 if $x > a$. Thus we can compile the following table for the derivative of f:

interval	$(-\infty, -2)$	$(-2, 1)$	$(1, \infty)$
$f'(x)$	-2	0	2

Thus f is increasing on $(-\infty, -2]$, constant on $[-2, 1]$ and increasing on $[1, \infty)$. The graph of f is given below.

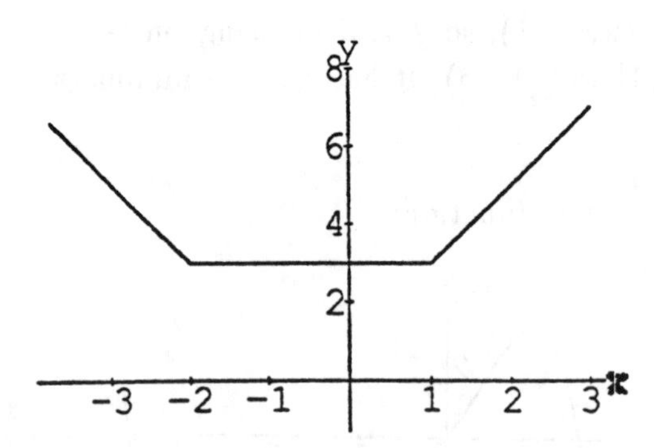

7. A table of the values of f' is given below:

interval	$(-\infty,-2)$	$(-2,0)$	$(0,1)$	$(1,\infty)$
$f'(x)$	-3	-1	1	3

Thus f is decreasing on $(-\infty,0]$ and increasing on $[0,\infty)$. The graph of f is given below.

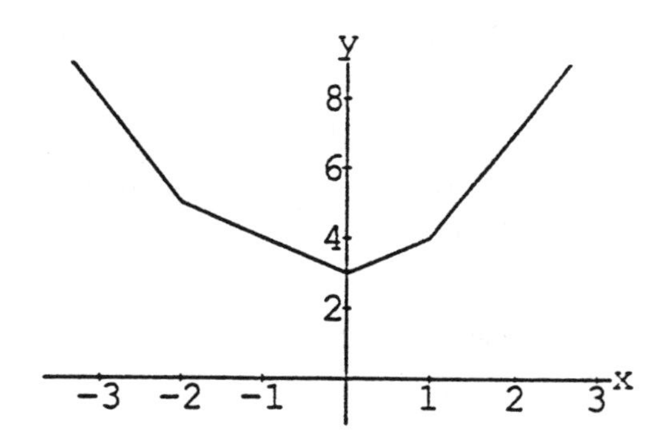

8. Sometimes common sense can save a lot of work. Since $f(x) \geq 0$ and $\cos^2 x = \frac{1}{3}$ for some real number x by the Intermediate Value Theorem, the minimum value of f is 0. Since the range of $\cos^2 x$ is $[0,1]$, the maximum value of f is $\frac{2}{3}$.

[DA]

9. $f'(x)$ is positive on $(-\infty,-2)$, on $(0,2)$ and on $(2,\infty)$, and negative on $(-2,0)$. Hence:

a. f is increasing on $(-\infty,-2]$ and on $[0,\infty)$; it is decreasing on $[-2,0]$.

b. f has a relative maximum at -2 and a relative minimum at 0.

c.

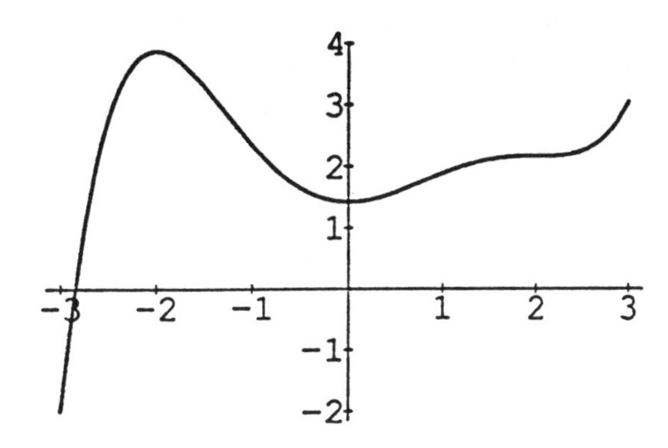

[AP]

10. a. Since $f'(x) > 0$ for all x, f is always increasing so the equation $f(x) = 0$ can have at most one solution.

b. $g(x)$ is decreasing when x is negative and increasing when x is positive. Hence there can be at most one non-positive solution and one non-negative solution to $g(x) = 0$. Thus there can be either 0, 1, or 2 solutions.

c. If there are two solutions, it follows from the discussion of part b that one is negative and one is positive. They appear, in fact, to be symmetric across the y-axis.

11. a. Yes, g is increasing for all x. This can be seen either by noting that $g'(x) = f'[f(x)]f'(x) > 0$ or by noting that if $b > a$, then $f(b) > f(a)$ and hence $g(b) = f(f(b)) > f(f(a)) = g(a)$.

b. No. In fact, g is increasing for all real numbers x. This can be seen by using an argument analogous to that used in part a.

12. a. True.

b. False. For example, take $f(x) = x^3$ on the interval $(-1, 1)$.

c. True. If this were not the case, then $f'(x) = 0$ for all x in (a, b), which implies that f is constant on that interval.

d. False. f could be constant on an interval.

e. True.

f. True since f is increasing on the left of b and decreasing to the right of b.

g. True by the definition of a relative maximum.

h. True. In fact, if f has relative maxima at distinct points a and b, then it must have a relative minimum in the interval (a, b). To see this, note that f has an absolute minimum in the interval $[a, b]$; since this minimum cannot occur at the endpoints (unless f is constant on $[a, b]$), it must occur at a point c in (a, b), so f must have a relative minimum at c as well.

i. True. For example take $f(x) = -e^{-x}$ or $f(x) = \arctan x - 3$. Alternatively, it is reasonable to simply sketch the graph of such a function.

13. a. This is clearly NEVER true; the maximum value of $f(|x|)$ is at most 5.

b. This MIGHT be true.

c. This MUST be true.

d. This can NEVER be true. By the Intermediate Value Theorem, there must be a number c such that $f(c) = 0$. Hence the minimum value of $|f(x)|$ is zero.

[AP]

14\. a. Since $f'(x) = e^x \geq e > 2 = g'(x)$ for all $x \geq 1$, f is increasing faster than g for all $x \geq 1$. Since $f(1) = e > g(1) = 2$, it follows that $f(x) > g(x)$ for all $x \geq 1$.

b. Since $f(0) = 1$ and f is an increasing function, $f(x) \geq 1$ for all $x \geq 0$. On the other hand, $h(x) < 1$ on $[0, 1)$ so the inequality $f(x) > h(x)$ holds for all x in $[0, 1)$. By part a, $f'(x) = e^x > h'(x) = 2x$ for all $x \geq 1$. Since $f(1) = e > h(1) = 1$, this implies that $f(x) > h(x)$ for all x in $[1, \infty)$ as well.

15\. The two marked points are the two relative extrema and hence the two critical points of f. Now $f'(x) = 3ax^2 + 2bx + c$, so $f'(x) = 0$ when $x = \dfrac{-2b \pm \sqrt{4b^2 - 12ac}}{6a} = \dfrac{-b \pm \sqrt{b^2 - 3ac}}{3a}$.

[JM (from Dec. 1972 issue of *The Mathematics Teacher*)]

Section 2: Concavity and the Second Derivative

1. Here is a possible graph:

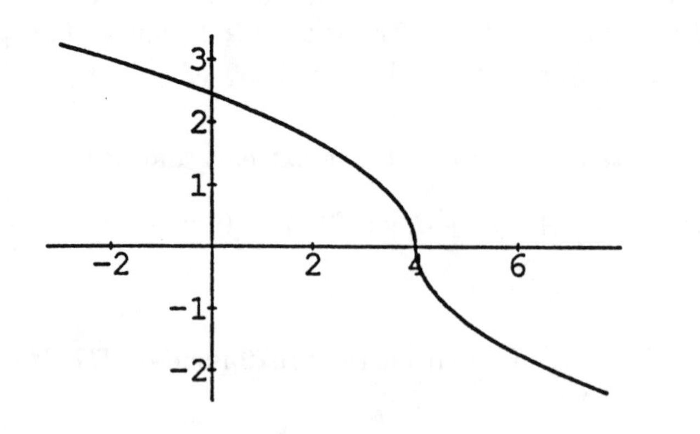

[AP]

2. a. f has a relative maximum at $x = 1$; it has no relative minimums in $[-3, 3]$.

   b. The points of inflection are where the concavity changes, and this happens at $x = 0$ and $x = 1$.

   c.

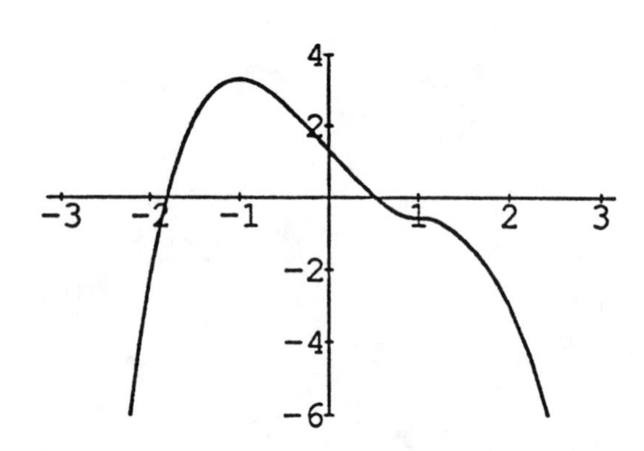

[AP]

3. a. f' is positive at A and D.

   b. f'' is positive at C and D.

[AP]

4. Perhaps the most straightforward way of approaching this problem is to sketch the derivatives of each of the three functions. When one does this, it is seen that (iii) is the graph of the derivative of the function graphed in (i), and (i) is the graph of the derivative of the function graphed in (ii). Thus (ii) is the graph of f, (i) is the graph of f', and (iii) is the graph of f''.

5. a. f is increasing on $[2, 6]$, $[8, 10)$.

   b. f is decreasing on $(0, 2]$, $[6, 8]$.

   c. The relative minima of f are at $x = 2$ and $x = 8$.

   d. The only relative maximum of f is at $x = 6$.

   e. f is concave up on $(0, 4]$ and $[7, 9]$.

   f. f is concave down on $[4, 7]$ and $[9, 10)$.

   g. The inflection points of f are at $x = 4$, $x = 7$, and $x = 9$.

6. a. See the graph below.

   b. $g'(x) = 2xf'(x^2)$. Since $f'(x^2) > 0$ for all $x \neq 0$, $g'(x) < 0$ if $x < 0$ and $g'(x) > 0$ if $x > 0$. Hence the only relative minimum of g occurs when $x = 0$.

   c. $g''(x) = 2f'(x^2) + 4x^2 f''(x^2) > 0$ for all real numbers $x \neq 0$. Hence f is concave up for all real numbers x.

   d. See the graph below.

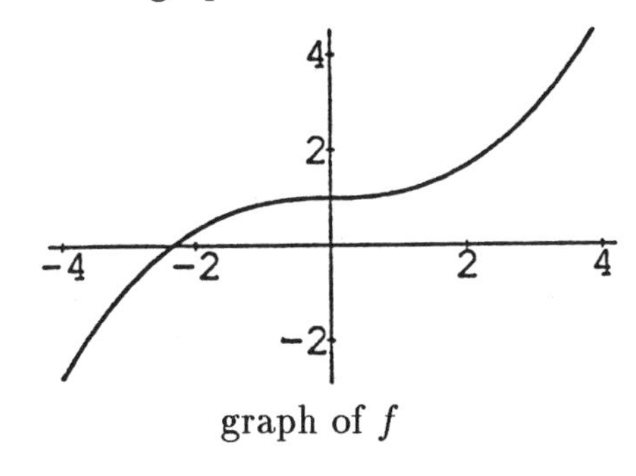

graph of f

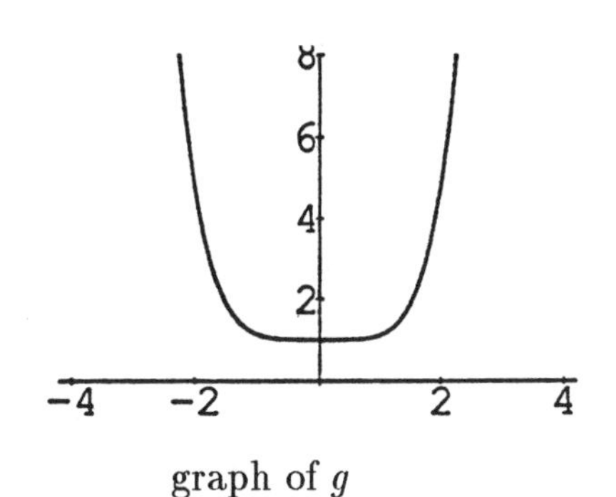

graph of g

[AP]

7. Since $f'''(c) > 0$, f'' is increasing in an interval containing c. Since $f''(c) = 0$, this implies that f'' changes sign at c, so f has a point of inflection at c.

8. We need to find the maximum value of $\dfrac{dy}{dx}$. Differentiating yields $\dfrac{dy}{dx} = -3x^2 + 6x$ and $\dfrac{d^2y}{dx^2} = -6x + 6$. Thus $\dfrac{d^2y}{dx^2}$ is positive when $x < 1$ and negative when $x > 1$. Hence the slope of the tangent line is the largest at $x = 1$ where it is equal to 3.

[CNC]

9. a. This is impossible since for all $x \geq 0$, $f'(x) \geq f'(0)$ and hence $f(x) \geq f'(0)x + f(0)$, which shows that f must eventually be positive.

   b. This is possible. For example, $f(x) = e^{-x}$ satisfies these conditions.

   c. We will show that this is impossible by considering three cases:
   Case 1: $f'(0) > 0$. Then as in part a, for all $x \geq 0$ we have $f(x) \geq f'(0)x + f(0)$ so $f(x)$ would be non-negative for sufficiently large x, contrary to assumption.
   Case 2: $f'(0) < 0$. Then for all $x < 0$, $f(x) > f'(0)x + f(0)$, so $f(x) > 0$ for x sufficiently small, i.e. for $x << 0$, which again is contrary to assumption.
   Case 3: $f'(0) = 0$. Then $f'(1) > f'(0)$ since $f''(x) > 0$. The argument of case 1 thus applies with 1 instead of 0.
   Note that, in the light of Part c, the requirement in Part a that $f'(x) > 0$ is irrelevant.

Section 3: Max-Min Story Problems

1. More points are probably lost on story problems for this reason than for any other.

2. Encourage students to draw their own pictures. This is one of the most useful skills, and it lies at the heart of much of the mathematical modeling that characterizes science and engineering. There are benefits to be derived from "thinking visually" in mathematics, and getting students who are inclined to prefer algorithms to visual methods (see, for example, Theodore Eisenberg and Tommy Dreyfus *On the Reluctance to Visualize in Mathematics*, MAA Notes Number 19, pp 25-37) to draw pictures will propel them in the right direction. Teachers will render a service to mathematics, engineering, and civilization as we know it by continually reinforcing the admonitions cited in these first two points.

3. This helps the student to keep his/her eye on the ball, as well as helping to prevent transpositional errors. One is very unlikely to write a formula such as $R = \pi A^2$.

4. Two good sources of relationships between variables are the Pythagorean Theorem and similar triangles. One cannot be too thin, too rich, or emphasize the value of triangles too strongly.

5. Although many problems can be done without the use of trigonometry, it is important to encourage the incorporation of trigonometric approaches. In addition to improving the students' facility with trigonometry, use of trigonometric variables simplifies many problems, either because differentiation is simpler, or the need to differentiate disappears entirely because the max-min properties of sine and cosine functions are familiar.

1. Although the problem can be solved by using the length of one of the sides of the rectangle as a variable, the use of the angle θ is more efficient. The area is then given by $A = (R\sin\theta)(2R\cos\theta) = R^2\sin 2\theta$ which achieves a maximum value of $A = R^2$ when $\theta = \pi/4$. Note that this problem demonstrates the value of using trigonometric variables, as differentiation was not necessary.

2. Even though this problem is not especially challenging, it does give some experience in solving max-min problems with exponential functions.

   a. $N'(t) = 1000(1 - t/20)e^{-t/20}$. $N'(t) > 0$ if $t < 20$ and $N'(t) < 0$ if $t < 20$, so $N(20) \approx 32{,}360$ is a maximum (the Second Derivative Test should be avoided wherever possible). Since $N(0) = 25{,}000$ and $N(100) > 25{,}000$, the minimum number of bacteria present is 25,000.

   b. $N''(t) = 50(t/20 - 2)e^{-t/20}$. If $t < 40$, $N''(t) < 0$, and if $t > 40$, $N''(t) > 0$, so $N'(t)$ is a minimum when $t = 40$. This problem uses the sign of the second derivative (the derivative of the first derivative) to locate a minimum for the first derivative. Note that it is not necessary to check to see whether the endpoints of the interval yield a minimum, but it would be necessary to check the interval endpoints to see if they yield a maximum.

3. This problem is useful as an elementary max-min problem using exponential functions, and has obvious real-world ramifications. $C'(t) = \dfrac{K(-be^{-bt} + ae^{-at})}{a - b}$. Setting $C'(t) = 0$, we find that $t = \dfrac{\ln(a/b)}{a - b}$. Since $C(0) = 0$, $C(t) > 0$, and $C(t)$ approaches 0 asymptotically, this must be a maximum. Note that, since a and b have units min^{-1}, the answer makes dimensional sense.

4. a. The domain of R should be $[0, B]$. B appears to be the maximum allowable dose of the drug. The constant A is some sort of scale factor, which depends upon the individual being given the drug. Note that the constant B also is likely to depend upon the individual being given the drug. (Heavier people can generally tolerate larger doses.) This type of "formula analysis," while not strictly part of calculus, is an important part of the mathematical consciousness-raising experience.

   b. $R'(x) = Ax(2B - 3x)$ and $R'(x) = 0$ if $x = \frac{2}{3}B$. This is seen to be a maximum by looking at the sign of the first derivative.

   c. $R(\frac{2}{3}B) = \frac{4}{27}AB^3$.

   d. $S'(x) = R''(x) = 2A(b - 3x)$. $S'(x) = 0$ when $x = B/3$. Inspection of the sign of S' makes it clear that this must be a maximum. Note that this problem involves a first derivative of S which is actually the second derivative of R. Students often set equal to zero the first derivative of the first function they get their hands on. This problem requires them to look more carefully at exactly what this process entails.

   e. It is the rate of change of the reaction with respect to the size of the dose. A large change in the amount of the reaction with a small increase in dose size would be described in English by saying that the patient is very sensitive to the dose.

5. Some students are likely to differentiate this function, which is not necessary (and may even be counterproductive) since the maximum value of $\sin 2\theta$ occurs when $\theta = \pi/4$. This represents a good opportunity to hammer home the importance of using trigonometric variables when possible. This problem should be appreciated by anyone who has ever thrown a baseball, or a forward pass in football.

6. The key to this problem is to draw the correct picture, realize that the goal is to maximize the angle α in the figure below, and use the trigonometry of right triangles and calculus to complete the solution. $\tan(\alpha+\beta) = 72/D$; $\tan\beta = 18/D$; $\alpha = (\alpha+\beta) - \beta = \arctan(72/D) - \arctan(18/D)$; $\frac{d\alpha}{dD} = 18\left(\frac{-4}{D^2+72^2} + \frac{1}{D^2+18^2}\right)$.

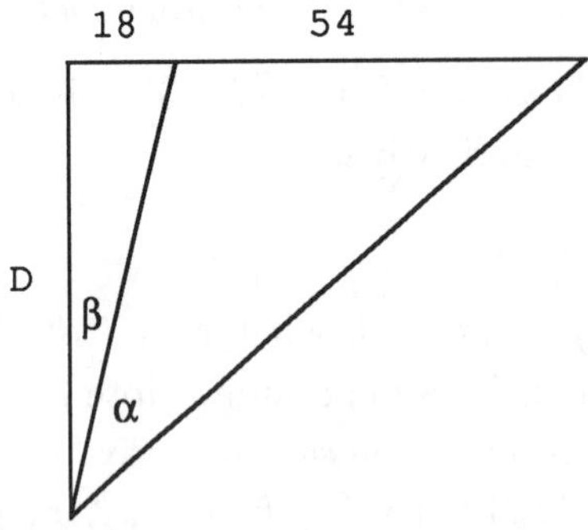

a. Setting $\frac{d\alpha}{dD} = 0$ has a solution $D = 36$. From physical considerations, it is easy to see that this must be a maximum, since as D approaches either 0 or infinity, α approaches 0.

b. One assumes that the passenger is lined up with the right side of the road.

c. This assumption is reasonable if the passenger is in the right-hand seat of a car in the right lane. They might be unreasonable if the distance of the passenger to the right side of the road is non-trivial, as might be the case in the fast lane of a multi-lane highway, or in Great Britain where the passenger is on the left-hand side.

d. It is the ratio of the width of the billboard to its distance from the nearest point on the road which is crucial to this problem. If the ratio is 3:1 (as it is here), then the maximum angle of viewing occurs at twice the distance from the billboard to the nearest point on the road.

Many problems become mechanically easier if we make the solution more algebraically general to begin with. This one is an example of that. Let A be the distance from the road to the near edge of the billboard (18 ft. here), let B be the width of the billboard (54 ft. here) and let $C = A+B$. Then $\alpha = \arctan(\frac{C}{D}) - \arctan(\frac{A}{D})$.

$$\frac{d\alpha}{dD} = \frac{-C}{D^2+C^2} - \frac{-A}{D^2+A^2}$$

Setting this equal to zero, we get $C(D^2+A^2) = A(D^2+C^2)$. $(C-A)D^2 = (C-A)AC$ or $D = \sqrt{AC}$. The optimal viewing distance is the *geometric mean* of the distances from the road to near and far edges of the billboard.

7. This classic problem has many points of interest. Useful themes to reiterate are the importance of rates, the totals principle, and the role of trigonometry. In addition, this problem can be construed as Snell's Law.

   a. $T = \text{time rowing} + \text{time walking} = \dfrac{\sqrt{1+x^2}}{R} + \dfrac{\sqrt{1+(2-x)^2}}{W}$.

   b. $\dfrac{dT}{dx} = \dfrac{x}{R\sqrt{1+x^2}} - \dfrac{2-x}{W\sqrt{1+(2-x)^2}} = \dfrac{\sin\alpha}{R} - \dfrac{\sin\beta}{W}$.

      Setting $\dfrac{dT}{dx} = 0$ yields the desired result.

   c. This is where a CAS might be used. The required equation results from doing standard algebra on setting the expression for $\dfrac{dT}{dx}$ equal to 0 with the substitutions $R = 2$ and $W = 4$.

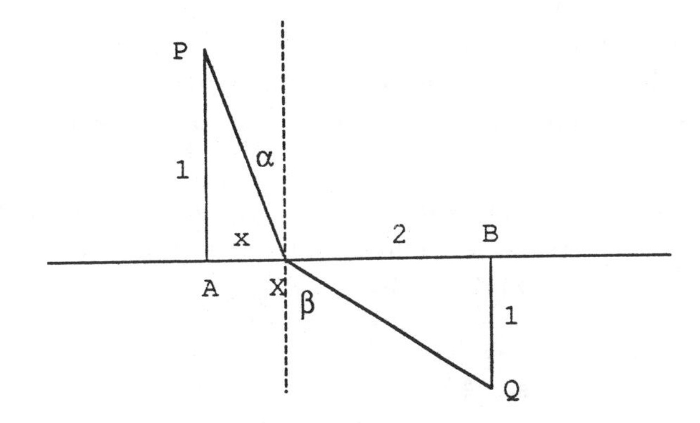

8. The correct approach to this problem is to let R be the rowing speed, and to compute the minimizing point x as in the diagram below.

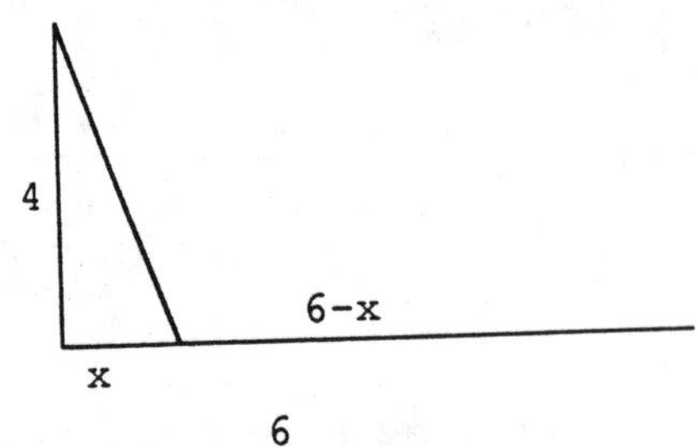

T = time rowing + time walking $= \dfrac{\sqrt{16+x^2}}{R} + \dfrac{6-x}{3}$. $T'(x) = \dfrac{x}{R\sqrt{16+x^2}} - \frac{1}{3}$. In order to have $T'(x) = 0$ when $x = 6$, we must have $\frac{1}{3} = \dfrac{6}{R\sqrt{52}}$, and so $R = \dfrac{9}{\sqrt{13}}$ ≈ 2.496 MPH. Note that if R is greater than this minimum value, $T'(x)$ is never 0 for x between 0 and 6, and so the quickest trip would be at one of the endpoints. By choosing an extremely high value of R (e.g. the speed of light), we can see from physical considerations that the quickest trip must occur by walking as little as possible, and so the quickest trip would be to row directly to the restaurant.

An alternative approach. If the previous problem is about Snell's Law, then this problem is about the critical angle for total internal reflection. In particular, for the optimal path from lake onto shore to continue directly down the shoreline, we must have:

$$\frac{\sin\alpha}{\sin(\frac{\pi}{2})} = \frac{R}{W} \text{ or } \sin\alpha = \frac{R}{W}$$

Call this value of α the critical angle. If the angle that the all-rowing path makes with the perpendicular to the shore, i.e. $\arcsin(\dfrac{6}{\sqrt{6^2+4^2}})$, is greater than this critical angle, then the optimal path will reach shore to the left of the restaurant. If the direct-path angle is less than or equal to the critical angle, then the optimal path is the direct path. Hence the criterion for the direct path to be the optimal path is that:

$$\frac{R}{W} \geq \frac{6}{\sqrt{6^2+4^2}} \text{ or } R \geq \frac{18}{\sqrt{52}} = \frac{9}{\sqrt{13}} \approx 2.496\,\text{MPH}.$$

Note that the direct path may turn out to be the optimal path even if rowing is a little slower than walking.

9. a. $20x+10 < 110 \Longrightarrow x \leq 5$. $10(x^2-8x+22) \leq 310 \Longrightarrow -1 \leq x \leq 9$, so $0 \leq x \leq 5$.

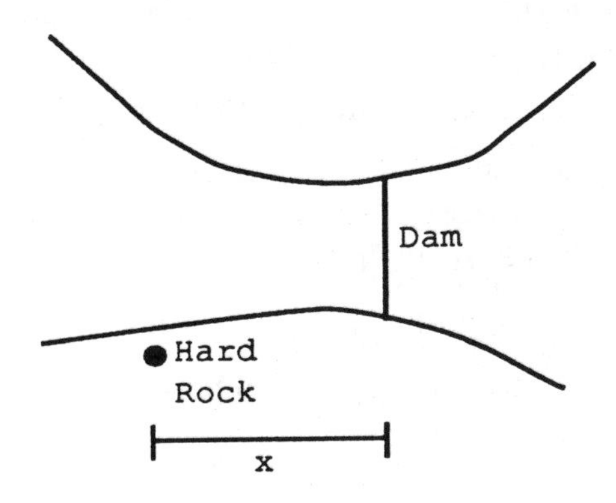

b. $W'(x) = 20x - 80$, so $x = 4$ gives a minimum width of 60 feet. Also, $W(0) = 220$ and $W(5) = 70$, so the maximum width is 220 feet.

c. $C(x) = K(x^2-8x+22)(2x+3) = K(2x^3-13x^2+20x+66)$, K a positive constant. $C'(x) = 2K(3x-10)(x-1)$. Examining the sign of the derivative shows that the cheaper dam (which is not the shortest) occurs at $x = 10/3$. The cost of this dam is approximately \$ 62.30 K whereas the cost of the dam built at Hard Rock ($x = 0$) is \$ 66 K. Students should certainly check the endpoints of the interval which is the domain of the cost function, even though as this problem is formulated, it is still cheaper to locate the dam downstream.

[PF]

10. Clever students may observe that \$2000 is the least cost to be incurred if the pipe crosses the lot, and since this *is* the cost of laying the pipe around the edges of the lot (from A to B to C), the most economical thing to do is to lay the pipe along the edges. If students insist on using calculus, then the following approach is possible: Let $\theta = \angle BAP \leq \frac{\pi}{4}$. Then the cost

$$\begin{aligned} C(\theta) &= \text{cost } AP + \text{cost } PC \\ &= 20(\text{length } AP) + 10(\text{length } PC) \\ &= 2000 \sec\theta + 1000(1 - \tan\theta) \\ &= 1000(2\sec\theta + 1 - \tan\theta). \end{aligned}$$

$C'(\theta) = 1000\sec\theta(2\tan\theta - \sec\theta)$. $C'(\theta) = 0$ when $\theta = \frac{\pi}{6}$. $C(\frac{\pi}{6}) = \$2732.05$. At the endpoint, $C(\frac{\pi}{4}) = \$2828.43$. If $\theta = 0$, the cost cannot be figured by the formula above, since the pipe always goes along one of the sides. In this case, the cost is (length)(\$10) = (200)(\$10) = \$2000, so this is the cheapest way to do it.

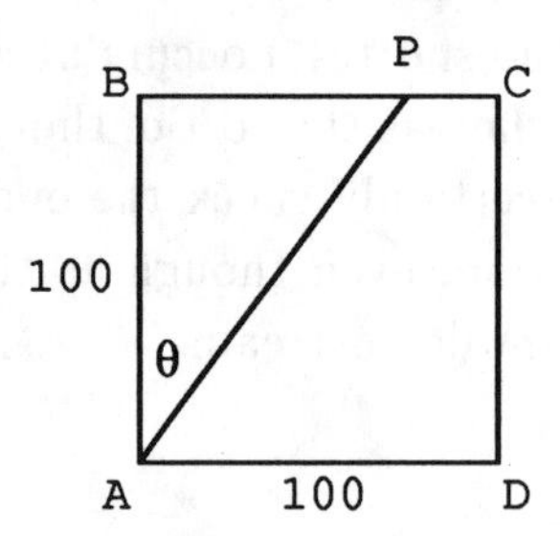

Note that this is an example of a story problem involving a function (the cost, in this case) which is naturally discontinuous.

[RH]

11. a. If v_0 represents initial velocity, the solution to the differential equation is $v(t) = v_0 e^{-t/10}$. If the distance of the throw is D feet, then we must have $D = \int v(t)\,dt$. Letting s be the duration of the throw, we obtain $D = 10v_0(1 - e^{-s/10})$. Therefore $s = -10\ln\left(1 - \dfrac{D}{10v_0}\right)$. The first throw is made with an initial velocity of 100 ft/sec and a distance of $300 - x$ feet, and the second with an initial velocity of 110 ft/sec and a distance of x feet, so the total time taken is

$$T(x) = -10\ln(.7 + x/1000) - 10\ln(1 - x/1100) + R.$$

b. $T'(x) = \dfrac{-10}{700 + x} + \dfrac{10}{1100 - x} = 0$ if $x = 200$. The infielder should therefore be stationed 200 feet away from home plate, and the total time for the two throws plus the relay is approximately $3.06 + R$ seconds. Physical considerations show that this is a minimum.

c. If the infielder is not involved in the play, the throw will reach home plate in $-10\ln(.7+0/1000) - 10\ln(1-0/1100)$ seconds. This is approximately 3.57 seconds, so the relay must be made in under half a second if it is to be a superior line of play.

x 300-x

Home Plate Infielder Outfielder

This extremely attractive problem is due to Roland Minton, of Roanoke College. An interesting project, especially during World Series time, might be to compute the needed relay time as a function of the initial velocities of both throws and the distance of the outfielder from the plate. This project is a good opportunity to use graphic displays on a computer.

L'envoi: the only manager in baseball capable of solving this problem was Davey Johnson of the Mets, who (a) has a degree in mathematics from Trinity College, and (b) compiled the best record in baseball during the past six years. Despite this, he was fired, yet another example of the undervaluation of mathematics in contemporary American society.

[RH]

12. Here is yet another classic type of problem. Let S denote the speed in MPH of the truck, $C(S)$ the total cost of running the truck at S MPH. $C(S)$ = fixed costs + fuel costs + driver costs (the totals principle in action again). Gas mileage is given by the formula $4 - .1(S - 50) = 9 - .1S$, so the number of gallons used on the trip is $260/(9 - .1S)$, and hence the cost of the fuel in dollars is $(260)(0.89)/(9 - .1S)$. At S MPH, the trip takes $260/S$ hours, and so fixed costs + driver costs = $(260)(38.83)/S$. Therefore, $C(S) = 260(0.89/(9 - .1S) + 38.83/S)$ and $C'(S) = 260[.089/(9 - .1S)^2 - 38.83/S^2]$. Setting $C'(S) = 0$ and letting $K = \sqrt{38.83/.089}$ yields $S = 9K/(1 + .1K) = 60.86$ miles per hour. From physical considerations, this must minimize costs, as values of S close to 90 would produce arbitrarily large fuel costs, while values of S close to 0 would produce arbitrarily large driver costs. Arguments such as these are valuable because they reduce the need to do tedious evaluations at the endpoints and help familiarize the student with the behavior of functions.

    Notice that driver costs plus fixed costs can actually be grouped as a single subtotal and that the length of the trip appears as a scale factor in the final answer. Thus the optimum speed is independent of the length of the trip. It might be useful for students to consider under what circumstances the optimum operating speed would depend on the length of the trip, and how such situations might occur in the real world. (One possibility is that drivers are paid at different rates for overtime.)

    [RH]

13. The solution presented here, which uses a trigonometric variable, is an alternative to the conventional approach to this problem. If $\theta = \angle PAB$, then the area $A = \frac{1}{2}(AB + 5)(2 + DE)$. But $AB = 2\cot\theta$ and $DE = 5\tan\theta$. $A(\theta) = (2\cot\theta + 5)(5\tan\theta + 2)/2 = (20 + 25\tan\theta + 4\cot\theta)/2$. $A'(\theta) = 12.5\sec^2\theta - 2\csc^2\theta$. Therefore, $A'(\theta) = 0 \Longrightarrow \tan\theta = \frac{2}{5}$, and so the minimum area of the triangle is $20\,\text{cm}^2$. Positioning the burn at 2.1 cm and 5.1 cm will result in an area of $(2)(2.1)(5.1) = 21.42\,\text{cm}^2$, $1.42\,\text{cm}^2$ more than the original example.

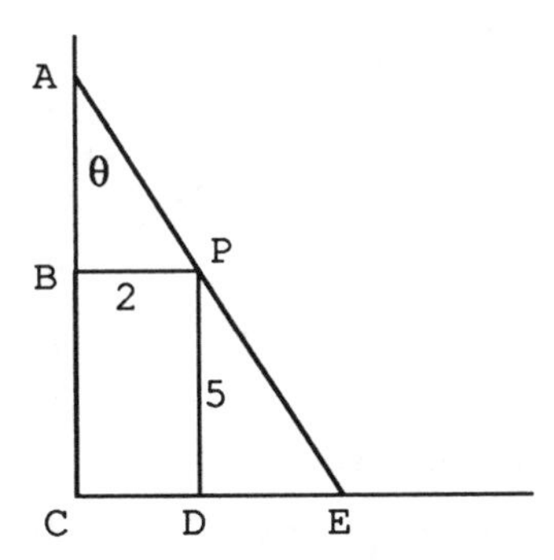

You may want to consider the general case of this problem: Suppose the burn is located at a distance u from one edge and at a distance v from the other edge. Then $A(\theta) = (u\cot\theta + v)(v\tan\theta + u)/2 = (2uv + v^2\tan\theta + u^2\cot\theta)/2$. $A'(\theta) = (v^2\sec^2\theta - u^2\csc^2\theta)/2$. Setting $A'(\theta)$ equal to 0 yields $\tan^2\theta = (\dfrac{u}{v})^2$ or $\tan\theta = u/v$. From this, we deduce that the minimum value of A is $2uv$, as it was in the original statement of the problem.

[RH]

14. Let F be the total number of floors contemplated, $C(F)$ the total cost of the building, and K the cost of the first floor. Then $C(F) =$ initial costs + floor 1 cost + $\cdots$ + floor F cost $= 450K + K + 2K + \cdots + FK = K(450 + \frac{1}{2}F(F + 1))$. Then the average cost $A(F)$ per floor is $K(450/F + \frac{1}{2}F + \frac{1}{2})$, and $A'(F) = K(-450/F^2 + \frac{1}{2})$. Setting $A'(F) = 0$ yields $F = 30$ as the number of floors.

Here is one of those instances in which the Second Derivative Test makes life easier, as it is clear that the second derivative is positive for all positive values of F.

It is worth discussing how to handle this problem if the minimizing number of floors comes out not to be an integer, as this situation arises in many problems in which the variable is allowed to assume only integer values.

[RH]

15. The total $T = \sqrt{16+x^2} + \sqrt{(12-x)^2+25}$. If $Y = \sqrt{(12-x)^2+25}$ and $Z = \sqrt{16+x^2}$, then $T'(x) = \dfrac{x}{Z} - \dfrac{12-x}{Y}$. Setting $T'(x)$ equal to 0 and solving yields $x = \frac{48}{9}$.

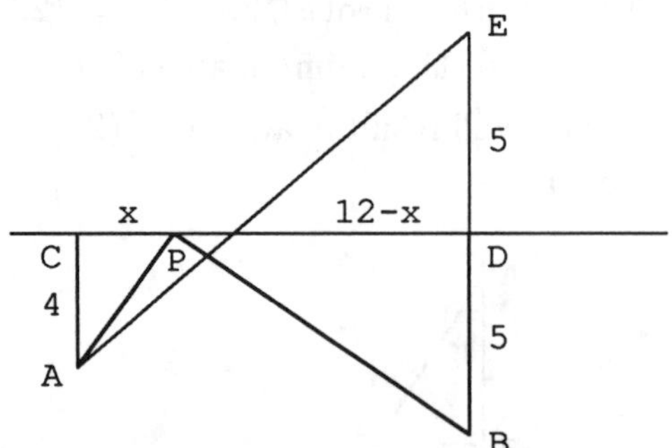

For those students who do not use a CAS to solve $T'(x) = 0$, you can help them appreciate the value of geometry by constructing the auxiliary line segment DE, which is 5 units in length. Draw the line AE. The point where this line intersects the line CD is the point which minimizes the total length of the pipelines. This can easily be seen because $PE = PB$, and therefore the sum of the lengths ofthe pipelines is $AP + PE$. This total is minimized by the line AE, as a straight line is the shortest distance between two points.

If you show students this proof, make sure that they realize that the techniques of calculus are generally more powerful than those of geometry, and that nice solutions such as this tend to be the exception rather than the rule. You might also point out that like Problems #7 and 8 in this section, this one has an optical interpretation. (Here we are describing reflections in a mirror.)

[RH]

16. If F denotes the fuel costs as a function of v, then $F = Kv^2$ for some constant K. Since $300 = K10^2$, $K = 3$. If the ship travels D km at a speed of v km/hr, the trip will take D/v hours. Therefore the total cost $C(v)$ = fuel costs + fixed costs = $(D/v)(3v^2) + 1200D/v = 3Dv + 1200D/v$. $C'(v) = 3D - 1200D/v^2$. Setting this equal to 0 yields $v = 20$ km/hr, which is independent of D. This is clearly a minimum since slow speeds generate exorbitant fixed costs and high speeds generate exorbitant fuel costs.

[RH]

17. The length L of the pencil is $L = PQ + QR = S\sec\theta + B\csc\theta$, and so $L'(\theta) = S\sec\theta\tan\theta - B\csc\theta\cot\theta$. If $L'(\theta) = 0$, then some trigonometry shows that $\tan^3\theta = \frac{B}{S}$, or $\theta = \arctan\left(\frac{B}{S}\right)^{1/3}$. Use this value of θ to compute the length of the largest pencil, which is $(b^{\frac{2}{3}} + s^{\frac{2}{3}})^{\frac{3}{2}}$.

18. Let N be the number of shipments. Then the number of CD players per shipment is $2400/N$, and the yearly fee (in dollars) is $4800/N$. The shipment fees total $50N$ so the total cost $C = 50N + 4800/N$. Differentiating yields $C'(N) = 50 - 4800/N^2$, and so $C'(N) = 0 \Longrightarrow N = 4\sqrt{6} \approx 9.8$. Since one cannot make 9.8 shipments, the alternatives are to make either 10 shipments or 9 shipments. 10 shipments result in a shipping cost of $C(10) = \$980$. Note that one cannot make 9 shipments, as the shipments would not contain an equal number of CD players, so the next candidate would be 8 shipments, and $C(8) = \$1000$. Therefore, the solution is to make 10 shipments of 240 CD players each.

    There is a difficulty if one chooses as a variable the number P of CD players per shipment. The equation for C is simple: $C = 2P + 120{,}000/P$, so $C'(P) = 2 - 120{,}000/P^2$. Setting this equal to 0 gives $P \approx 244.95$ players. In this case, the allowed number of players per shipment are 240 and 300, which are some distance away from 244.95. There is a tendency for students to overlook the fact that one must examine only the solutions which are allowed.

19. This problem uses exponential functions, and also helps to expose the difficulties which arise when one tries to use the same symbol to denote two different variables within the same equation. Let (a, e^{-a}) be a point on the curve in the first quadrant. Since the slope of the tangent to the curve at this point is $-e^{-a}$, the equation of the tangent is $y - e^{-a} = -e^{-a}(x - a)$. The x-intercept is $a + 1$, and the y-intercept is $e^{-a}(1 + a)$. The area A of the triangle is $\frac{1}{2}e^{-a}(1 + a)^2$. $A'(a) = \frac{1}{2}e^{-a}(1 - a^2)$. If $A'(a) = 0$, then $a = 1$ is the only solution with $a > 0$. Examining the sign of the first derivative shows that $a = 1$ is a maximum, and the area $A(1) = \frac{2}{e}$.

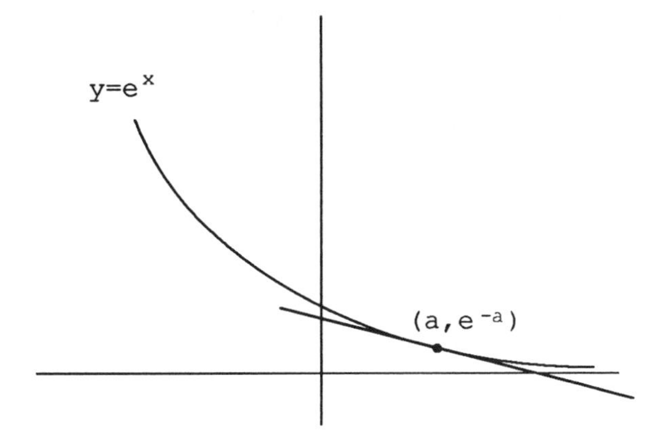

20. This problem is particularly good for the "hands-on" approach. It is clear that the slant height of the cone is R, but until students have actually constructed the cone, the values of the other parameters are not obvious. The circumference of the cone is $2\pi R - R\theta$, and if r is the radius of the cone, $2\pi r = R(2\pi - \theta)$; so $r = R(1 - \frac{1}{2\pi}\theta)$ and h satisfies the equation $h^2 + R^2(1 - \frac{1}{2\pi}\theta)^2 = R^2$. If $Y = 1 - \frac{1}{2\pi}\theta$, then the volume $V = \frac{\pi}{3}R^3Y^2\sqrt{1-Y^2}$. We now obtain $V'(Y) = \dfrac{\pi R^3 Y(2-3Y^2)}{3\sqrt{1-Y^2}}$, so $V'(Y) = 0$ implies that $Y = \sqrt{\frac{2}{3}}$, and $\theta = 2\pi\left(1 - \sqrt{\frac{2}{3}}\right)$. As values of θ near 0 or 2π give small values of the volume, this must be a maximum. Geometric considerations also indicate that θ should be independent of R. The introduction of Y makes differentiation easier, but students should be aware that a "change of variable" may entail an application of the Chain Rule.

21. Although your students may have already seen a standard variation of this problem, it is worthwhile because of the required modeling. $V(x) = x(47-2x)(95-4x)/2$. Solving $V'(x) = 0$ yields $x \approx 7.875$ as the only acceptable answer. This value of x gives a box of dimensions 31.25 x 31.75 x 7.88 cms. (Note that this box is not quite square.)

    This problem also points out that students should be aware that max-min problems pop up unexpectedly in real life. We get maximum volume for $x \approx 8.75$ which is far removed from Pizza Hut's value $x = 4.25$. Of course, Pizza Hut is interested in making boxes which fit pizzas, *not* in boxes which maximize volume.

22. Let R be the radius of the circle, θ the angle in radians. Then the area of the sector is $\frac{1}{2}\theta R^2 = 100$. C = total cost = tulip cost + rose cost $= 15R\theta + 20(2R) = 15R\left(\frac{200}{R^2}\right) + 40R = \frac{3000}{R} + 40R$. $C'(R) = \frac{-3000}{R^2} + 40$. $C'(R) = 0 \implies R = 8.66$ meters. Since $C''(R) > 0$ for $R > 0$, $C(8.66) = \$692.82$ is the minimum cost.

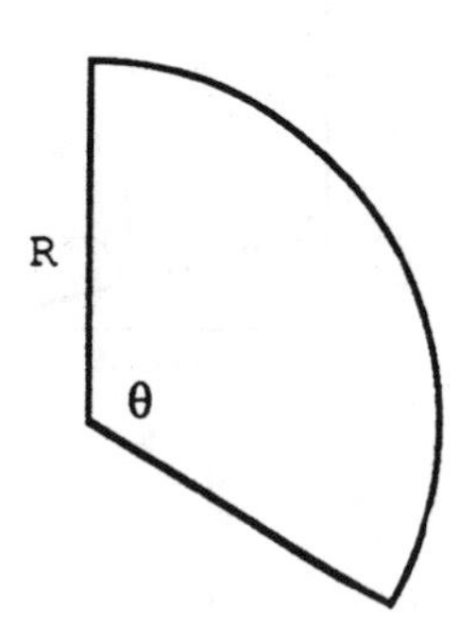

23. Ladder length $L = AB + BC = H\csc\theta + D\sec\theta$. Thus $L'(\theta) = -H\csc\theta\cot\theta + D\sec\theta\tan\theta$. $L'(\theta) = 0 \Longrightarrow H\cos\theta/\sin^2\theta = D\sin\theta/\cos^2\theta \Longrightarrow \tan^3\theta = \frac{H}{D}$, and $\theta = \arctan\left(\frac{H}{D}\right)^{1/3}$. Ask your students to compare this problem with Problem 17, and see if they can find a way to transform one problem into the other.

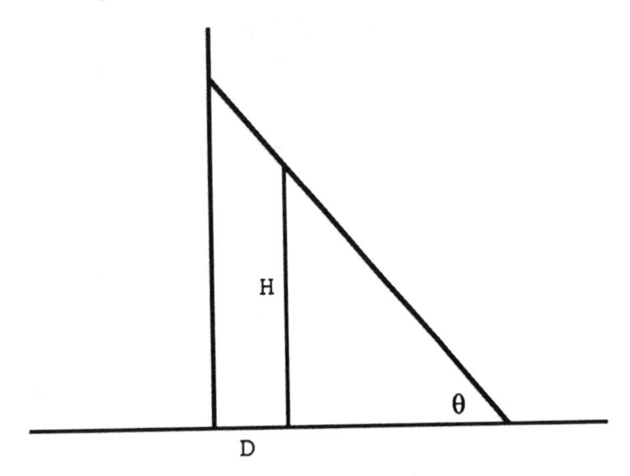

24. Let L = gutter base. Cross sectional area $A = L^2(\sin\theta + \sin\theta\cos\theta)$.
$A'(\theta) = L^2(2\cos^2\theta + \cos\theta - 1)$. The relevant root is $\cos\theta = \frac{1}{2}$, or $\theta = \frac{\pi}{3}$. Computing $A(0)$ and $A(\frac{\pi}{2})$ shows that $A(\theta)$ is a maximum if $\theta = \frac{\pi}{3}$.

This problem has a nice generalization which will give your students a taste of multivariable calculus. Let the base be a fixed fraction r of the total length. Determine area A as a function of θ and r. Then, treating r as a parameter, maximize A in terms of θ. This yields θ as a function of r. Upon substitution of this function in the expression for A, maximize A in terms of R. This reduces a 2-variable maximization to two successive single-variable problems.

25. A judicious parameter choice makes this a piece of cake. Following the hint, we see that $AB = L\sin\theta$, $BC = W\cos\theta$, $AD = L\cos\theta$, and $DE = W\sin\theta$. After some trigonometric simplification, the area A is given by $LW + (L^2 + W^2)\sin\theta\cos\theta = LW + \frac{1}{2}(L^2 + W^2)\sin 2\theta$. This achieves a maximum value of $LW + \frac{1}{2}(L^2 + W^2)$, a result obtained without differentiating a function, setting it equal to 0, and solving!

26. a. If S is the side of the base and H the height of the box, then $100 = 8S + 4H$, and $V = S^2H = S^2(25 - 2S) = 25S^2 - 2S^3$. $V'(S) = 50S - 6S^2$, and so $S = \frac{25}{3}$, $H = \frac{25}{3}$, and the box is a cube of volume $15{,}625/27 \approx 578.7\text{cm}^3$. Geometrical considerations make it clear that this is a maximum.

b. The surface area $A = 2S^2 + 4SH = 2S^2 + 4S(25 - 2S) = 100S - 6S^2$. Therefore $A'(S) = 100 - 12S$, and once again $A'(S) = 0$ if $S = \frac{25}{3}$. This is a cube whose surface area is 1250/3 square centimeters. In this case, the Second Derivative Test is easier to use than an examination of geometrical considerations.

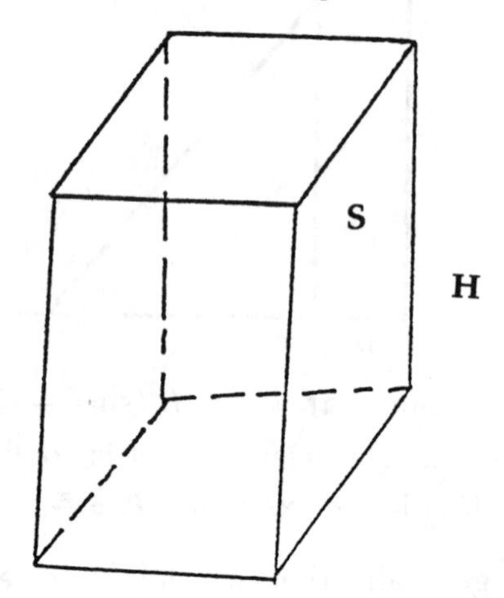

Other variations of this problem involve constructing pyramids with different types of bases. One can also construct the skeleton of an Indian tepee (in the form of a cone) by using some of the wire for the circular base, and three or four supporting wires which meet at the vertex of the cone.

27. Let r be the radius and h the height of the inscribed cone. The volume of the inscribed cone is $V = \frac{1}{3}\pi r^2 h$. By similar triangles, $\dfrac{r}{R} = \dfrac{H-h}{H} = 1 - \dfrac{h}{H}$. Therefore $V = \frac{1}{3}\pi r^2 \dfrac{H}{1 - r/R}$. $V'(r) = 0$ if $r = \frac{2}{3}R$. It follows that $h = \frac{1}{3}H$. Geometrical considerations show that this corresponds to a maximum, and the volume of the inscribed cone is $\frac{4}{27}$ of the volume of the original cone. Other inscription problems also make extensive use of the Pythagorean Theorem and similar triangles.

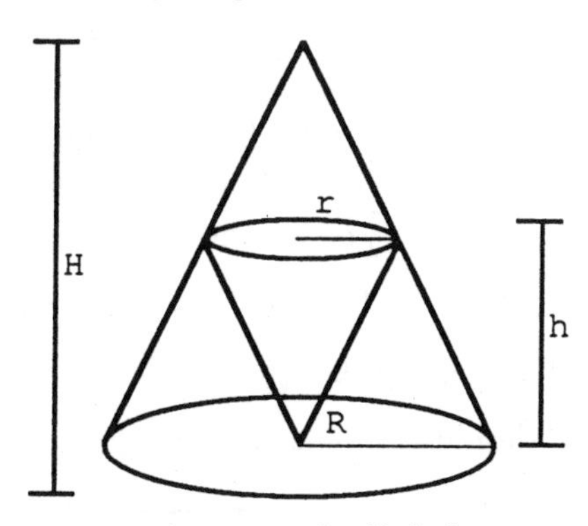

28. Let D be the distance from the shore to the lighthouse, and let θ be the angle $\angle PLA$. It doesn't matter from which point on the shore we choose to measure the position of the light beam, so let $u = PA$. Then $u = D\tan\theta$, so $u'(t) = D(\sec^2\theta)\left(\dfrac{d\theta}{dt}\right)$. Since $\dfrac{d\theta}{dt}$ and D are constants, the problem reduces to finding the angle θ that minimizes $\sec^2\theta$. This is $\theta = 0$, which corresponds to the point on shore nearest the light. This explains why, if you stand at this point, the light beam seems to pause and hit you in the eye before moving on.

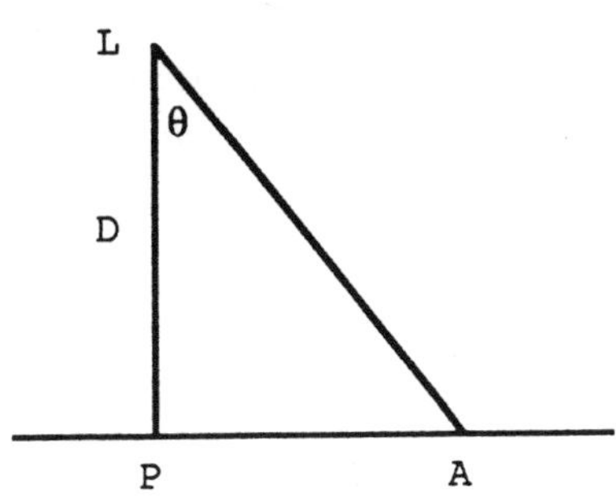

29. The total cash flow C is the rental income less the service and repair charges, so $C = x[65 - \frac{1}{10}(x - 300)] - 30[65 - \frac{1}{10}(x - 300)] = -\frac{1}{10}x^2 + 98x + 30N - 1950 - 900$. $C'(x) = -\frac{1}{5}x + 98$, and $C'(x) = 0 \Longrightarrow x = \$490$. The Second Derivative Test shows that this is a maximum.

30. This is a basically straightforward problem in differentiation and trigonometry. $T'(\theta) = C\left(\frac{b\csc^2\theta}{R^4} - \frac{b\csc\theta\cot\theta}{r^4}\right)$. Setting this equal to 0 yields the desired result. Although the Second Derivative Test can be used to verify that this value for θ yields a minimum, simply factoring the derivative shows that $T'(\theta) = bC\csc^2\theta\left(\frac{1}{R^4} - \frac{\cos\theta}{r^4}\right)$. The last factor is negative for small values of θ and positive for values of θ larger than the critical angle.

[JS]

31. The problem involves maximizing the area of a circular sector with arc length S that is cut off by a chord. Let R be the radius of the circle, and θ the angle of the sector. The area of the sector is $\frac{1}{2}R^2\theta$, and the area of the triangle at the base of the sector is $2(\frac{1}{2}(R\cos\frac{\theta}{2})(R\sin\frac{\theta}{2}))$. Thus $A = \frac{1}{2}R^2(\theta - \sin\theta)$. Since $S = R\theta$, $A = \frac{1}{2}(S/\theta)^2(\theta - \sin\theta)$. Thus $A'(\theta) = -\frac{1}{2}S^2\left(\frac{1+\cos\theta}{\theta^2} - \frac{2\sin\theta}{\theta^3}\right)$, and $A'(\theta) = 0$ if $\theta = \pi$. Therefore the sheet should be bent so that the cross-section is a semi-circle.

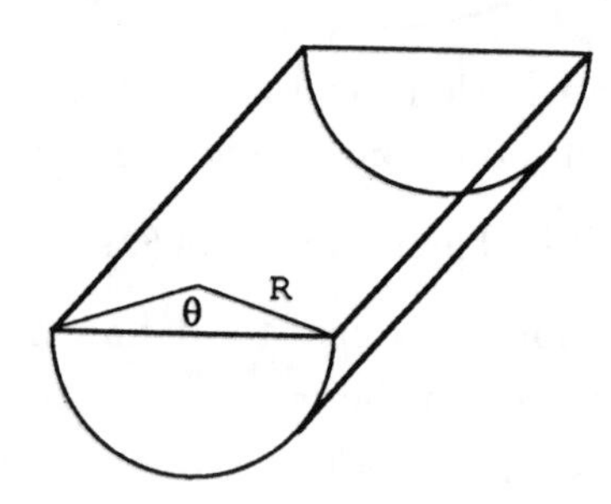

32. This problem is most easily done by using parametric coordinates. Let θ be the angle between the x-axis and a line from the origin to a point on the ellipse. The perimeter $P = 4(a\cos\theta + b\sin\theta)$, so $P'(\theta) = 0$ if $\tan\theta = b/a$. The maximum perimeter is $4\sqrt{a^2+b^2}$.

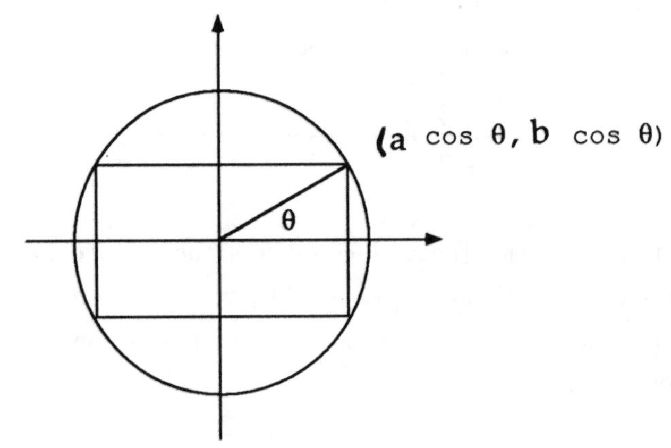

33. A graphics package will help your students avoid a classic trap set for them here. If no such package is available. a hint is clearly in order. Let A be the point with coordinates $(-1, 0)$ and B the point with coordinates $(1, 0)$. If the radial angle is θ, the distance formula shows that the distance swum is $\sqrt{2 + 2\cos\theta}$, and the distance walked is θ. The time T is therefore given by $T = \frac{1}{2}\sqrt{2 + 2\cos\theta} + \frac{1}{5}\theta$. T is actually minimized for $\theta = \pi$, i.e. the man should walk all the way around the pond. This takes $\frac{\pi}{5} \approx .628$ hours. If your students take the derivative of T and set it equal to 0, they should get: $T'(\theta) = \dfrac{-\sin\theta}{2\sqrt{2 + 2\cos\theta}} + \frac{1}{5}$. $T'(\theta) = 0$ yields $\dfrac{\sin^2\theta}{2 + 2\cos\theta} = \frac{4}{25}$. This reduces to the equation $25\cos^2\theta + 8\cos\theta - 17 = 0$. Therefore, $\cos\theta = 0.68$, $\theta = 0.823$ radians. The corresponding time for the trip is approximately 1.08 hours. This *maximizes* the time to go from A to B!

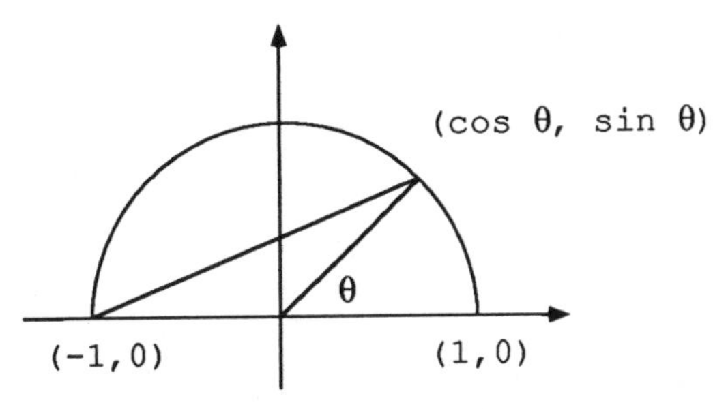

34. This problem has been included to help reinforce the idea that one must pay attention to what is actually being maximized or minimized when a derivative is set equal to zero.

   a. $h(0) = 375$ feet.

   b. $h(5) = 0$. A good opportunity to use Eisenstein's criterion for rational roots of a polynomial.

   c. Velocity $v(t) = h'(t) = 4t^3 - 24t^2 = 4t^2(t - 6)$. $v(t) = 0$ if $t = 0$ or $t = 6$; since the flight ends at $t = 5$ and $v(t) \leq 0$ throughout this period, the rocket is always headed down. Its maximum altitude occurs at $t = 0$; $h(0) = 375$ feet.

   d. Acceleration $a(t) = 12t^2 - 48t = 12t(t - 4)$. $a(t) = 0$ at $t = 0$ and $t = 4$. Since $a(t) < 0$ if $t < 4$, the velocity is decreasing for $0 < t < 4$ and increasing for $4 < t < 5$. The maximum upward velocity occurs at launch, and is $0\,\text{ft/sec}$. (This might be considered a trick question.)

   e. The maximum downward velocity occurs at $t = 4$, and is $-128\,\text{ft/sec}$.

   f. $a'(t) = 24t - 48$, so the acceleration is decreasing for $0 < t < 2$ and increasing for $t > 2$. Since $a(0) = 0$ and $a(5) = 60$, the maximum upward acceleration is $60\,\text{ft/sec}^2$.

   g. $a(2) = -48\,\text{ft/sec}^2$.

Chapter IV: Antiderivatives and Differential Equations

Section 1: Antiderivatives

1. C. The two choices are the graph of an antiderivative and its negative. Students cannot just look to see where the derivative is zero. They must look at the slope of the tangent line elsewhere.

2. C. Encourage students to comment about how they proceed even though they are not asked to do so in this problem.

   [AP]

3. Students can't get this one wrong, but that probably makes it harder for them! Each graph is that of an antiderivative of the given function because they differ only by a constant.

4. In both parts, f and g differ only by a constant, so $f' = g'$.

5. E. Do students comment about how they proceed even though they are not asked to do so in this problem?

   [AP]

6. The following explains why the graphs are not those of f.

   a. The derivative of this function is the constant function -1, so it cannot be the graph of f.

   b. This graph is similar to the graph of f, but the slope of f at -1 and $+1$ must be zero.

   c. This graph is that of $-f$.

   d. The graph below is the graph of one possible f.

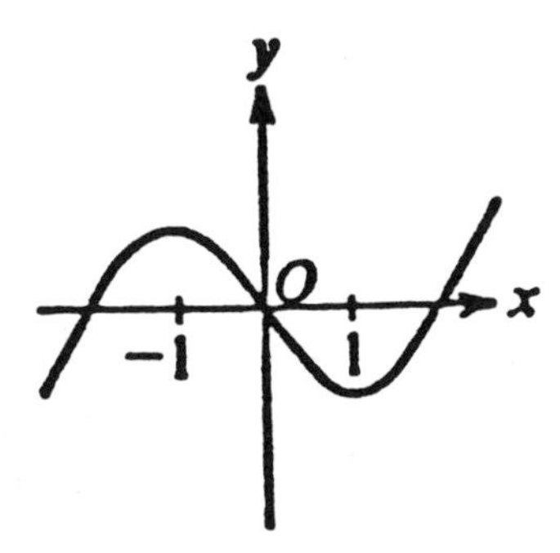

7. This problem is built upon the trigonometric identity $\cos 2x = 1 - 2\sin^2 x$. Since $-\frac{1}{2}\cos 2x$ and $\sin^2 x$ differ by a constant, they have the same derivative.

8. This question deals with the method of undetermined coefficients. The algebra involved in solving for the coefficients (solving linear systems of equations) can be done by a CAS, especially if you want to extend this question to functions like $x^n e^x$, with $n > 2$.

   a. $F'(x) = Axe^x + (A + B)e^x$. By equating coefficients, $A = 1$ and $A + B = 0$. Therefore $A = 1$ and $B = -1$.

   b. $F'(x) = Axe^x + (A + B)e^x$. By equating coefficients, $A = 2$ and $A + B = 0$. Therefore $A = 2$ and $B = -2$.

   c. Let $F(x) = Axe^{2x} + Be^{2x}$. Then $F'(x) = 2Axe^{2x} + (A + 2B)e^{2x}$. By equating coefficients, $2A = 1$ and $A + 2B = 0$. Therefore $A = \frac{1}{2}$ and $B = -\frac{1}{4}$.

   d. Let $F(x) = Ax^2e^x + Bxe^x + Ce^x$. Then $F'(x) = Ax^2e^x + (2A + B)xe^x + (B + C)e^x$. By equating coefficients, $A = 1$, $2A + B = 0$ and $B + C = 0$. Therefore $A = 1$, $B = -\frac{1}{2}$ and $C = \frac{1}{2}$.

9. Although antiderivatives are not exactly found this way, this problem gives students an opportunity to recognize antiderivatives of some basic functions.

   a. $f(x) = x^2$ and $g(x) = e^x$, $F(x) = x^2e^x$.

   b. $f(x) = \ln x$ and $g(x) = \sin x$, $F(x) = \ln x \sin x$.

   c. $f(x) = e^x$ and $g(x) = \sin x$, $F(x) = e^x \sin x$.

   d. $f(x) = e^x$ and $g(x) = -\cos x$, $F(x) = -e^x \cos x$.

   e. $f(x) = \ln x$ and $g(x) = x$, $F(x) = x \ln x$

10. This problem is designed to get students to think about the flow or direction of a function as given by its derivative. Obtaining functions from flow (direction) fields is a natural extension of this problem.

    a. The graphs are given below.

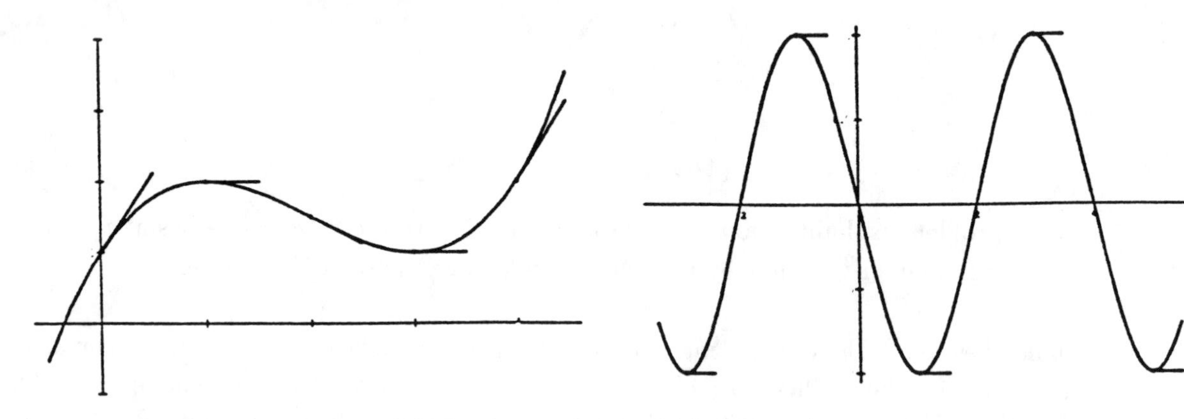

    b. The minimal degree is 3. The polynomial used to produce this graph is $1 + 2.25x - 1.5x^2 + .25x^3$.

    c. $-\sin(\frac{\pi x}{2})$ works!

    d. This part may be going one step too far. Be that as it may, the graphs of the functions and their derivatives are given below.

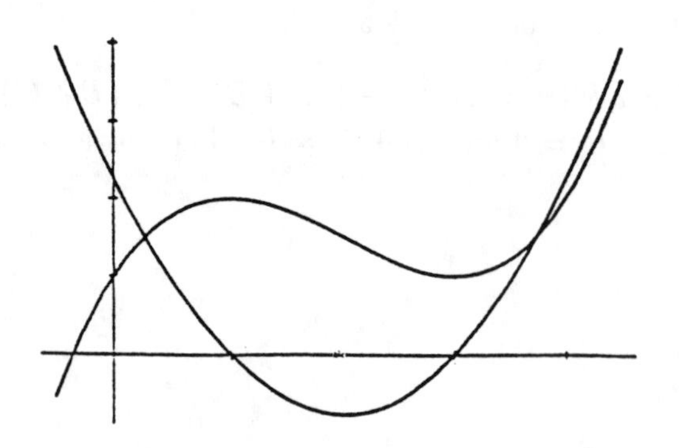

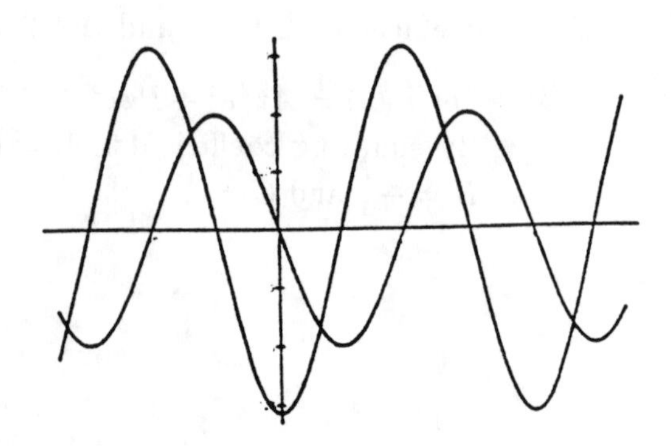

11. The required graphs are given below.

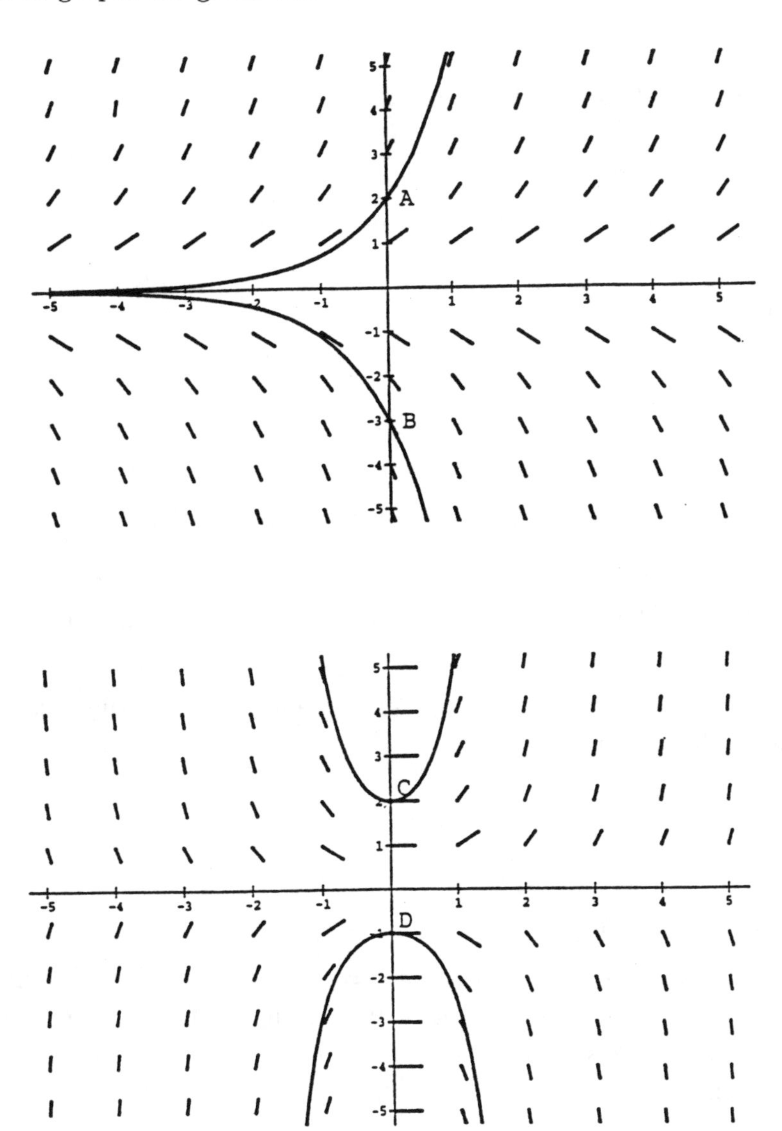

12. The function f is actually the exponential function, and students' graphs of F should look approximately like the graph of f although, of course, the displacement of their graphs is not determined by the question.

13. The graph of g is just one unit higher than that of f. This problem can be used to convey the fact that two functions with the same derivative differ only by a constant.

14. B. This problem is similar to the previous one. The concept of an initial condition, introduced in the next section, is foreshadowed by the correct answer here.

Section 2: Introduction to Differential Equations

1. $\dfrac{dT}{dt} = k(T - T_m)$. In this form k is a negative constant.

2. a. 11:59 am, which is just one minute before it is full!

   b. Since $\dfrac{1}{128} = \dfrac{1}{2^7} < 1\%$, for all but the last 7 minutes the bottle will be "almost empty." Hence for $\dfrac{53}{60} = 88\%$ of the time the bottle is "almost empty."

   Note: Students may use a formula like $A(t) = 2^{t-60}$ where $A(t)$ is the fraction of the bottle filled by bacteria at time t in minutes. Solving $.01 = 2^{t-60}$ gives $t = 53.35$ min. Therefore the bottle is "almost empty" for $\dfrac{53.35}{60} = 88.9\%$ of the time.

   c. Because the doubling time is 1 minute, the bottle will be full at 12:02 pm, or in just two more minutes.

   This problem gets students to consider carefully doubling time and the nature of exponential functions.

3. a. $y = e^x$

   b. $y = e^{-x}$

   c. $y = \cos x$ and $y = \sin x$

   d. $y = e^x$ and $y = e^{-x}$ or $y = \sinh x$ and $y = \cosh x$ The word "different" is meant to suggest that the functions be linearly independent. If, in part c, students answer $y = \sin x$ and $y = -\sin x$ or something similar, recommend that they try to find two functions which are not scalar multiples, one of the other.

4. a. Substitution leads to the following system of equations: $a+b = -1$ and $4a+2b = -1$, which has a solution of $a = \frac{1}{2}$ and $b = -\frac{3}{2}$.

   b. Substitution leads to the following system of equations: $0 + b(x) = -x$ and $2a(x) + 2xb(x) = -x^2$ which has a solution of $a(x) = \frac{1}{2}x^2$ and $b(x) = -x$.

5. A. The point $(8, 0)$ belongs to the curve, and $y' = 24x^2e^{x^3} = 3x^2y$.

[AP]

6. a. The graph below, which shows the data given in the problem, suggests an exponential growth in population. In fact, the population is growing at an even faster rate. (See part *c*.)

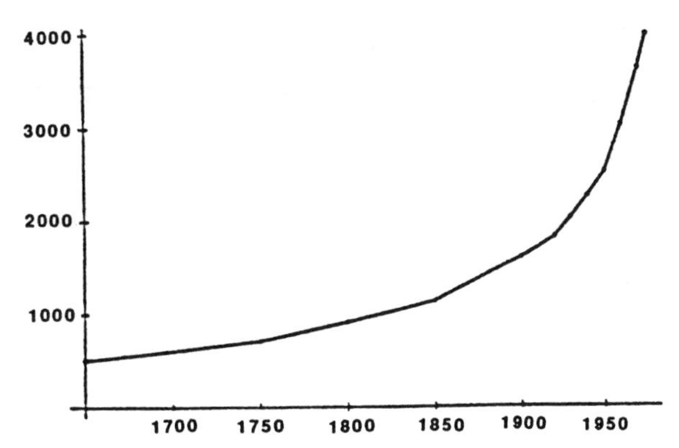

b. $L'(t) = \dfrac{f'(t)}{f(t)} = k.$

c. The line drawn is the line of best fit. It appears that the growth is more rapid than exponential growth.

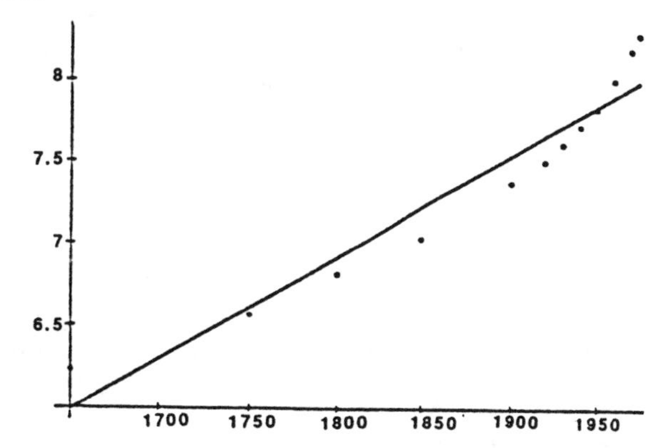

d. Students may cite wars, famine, decreasing infant mortality, or increased life expectancy as reasons.

7. a. $k = -\dfrac{\ln(2)}{6}$ and therefore $y(t) = (1{,}000{,}000)e^{-\frac{\ln(2)}{6}t} = (1{,}000{,}000)2^{-\frac{1}{6}t}$.

b. Using the differential equation, we see that

$$\frac{dy}{dx} = ky = -\frac{\ln(2)}{6}(600{,}000) = -69{,}300\,\text{gallons/year.}$$

Students can answer this part even if they cannot get part a.

c. $t = -6\,\dfrac{\ln(.05)}{\ln(2)} \approx 26\,\text{years.}$

8. a. $\dfrac{y^2}{2} = \dfrac{x^2}{2} + C$

   b. (i) $C = -\frac{3}{2}$. This leads to the solution $y = \sqrt{x^2 - 3}, \quad x \geq \sqrt{3}$ where we take the positive branch because $y(2) = 1$.

   (ii) $C = -\frac{3}{2}$. This leads to the solution $y = -\sqrt{x^2 - 3}, \quad x \geq \sqrt{3}$ where we take the negative branch because $y(2) = -1$.

   (iii) $C = -\frac{3}{2}$. With solution $y = -\sqrt{x^2 - 3}, \quad x \leq -\sqrt{3}$.

   c. The graphs are given below.

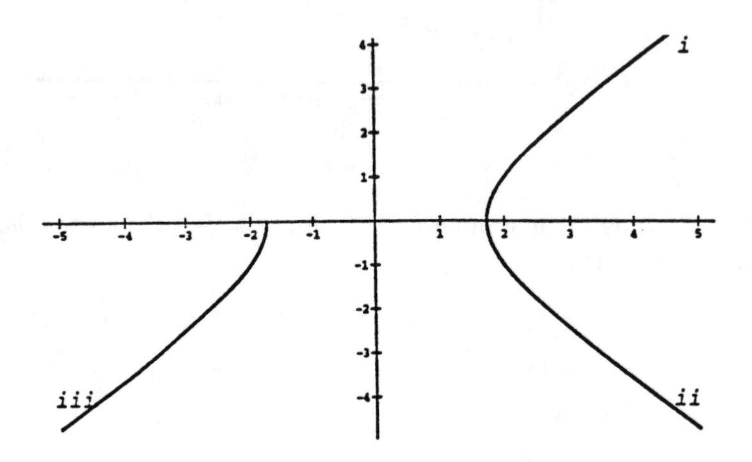

[MN]

9. a. By separating variables, students can get the general solution: $-\dfrac{1}{y} = x^2 + C$

   b. (i) $C = -2$, and therefore $y = -\dfrac{1}{x^2 - 2}$.

   (ii) $C = 0$, and therefore $y = -\dfrac{1}{x^2}$.

   (iii) Substituting into the general solution gives an undefined term of $-\dfrac{1}{0}$. However, if $y(x) \equiv 0$, then $\dfrac{dy}{dx} = 0$, and therefore this constant function is a solution of the differential equation. It is, in fact, the unique solution to satisfy the initial condition $y(0) = 0$.

[MN]

10. a. $y = x^3$

    b. Substitution shows that, as defined, f satisfies the differential equation.

    This problem shows that this type of differential equation can have an infinite number of solutions satisfying $y(0) = 0$.

[MN]

11. Leibniz, who discovered the Product Rule in 1675, first conjectured that $(fg)'$ *was* equal to $f'g'$. He thought that he had an example to confirm this in which $f(x) = cx + d$, $g(x) = x^2 + bx$, where b, c, d are constants, but he quickly realized that this was not the case.

    a. Let $y = g(x)$. $(e^{3x}y)' = (e^{3x})'y'$ implies that $3e^{3x}y + e^{3x}y' = 3e^{3x}y'$. Therefore $2y' = 3y$, which has the solution $y = Ae^{\frac{3}{2}x}$.

    b. $(e^{x^2}y)' = (e^{x^2})'y'$ implies that $2xe^{x^2}y + e^{x^2}\dfrac{dy}{dx} = 2xe^{x^2}\dfrac{dy}{dx}$. This reduces to the differential equation $2xy\,dx + dy = 2x\,dy$, which can be solved by separating the variables. After several steps (which include dividing to show that $\dfrac{2x}{2x-1} = 1 + \dfrac{1}{2x-1}$), the answer can be written in the form $y = Ae^x\sqrt{|2x-1|}$.

[MN]

12. a. The lower and upper bounds are obtained by looking at the simple interest generated if the initial balances were $A(t)$ and $A(t + \Delta t)$ respectively.

    b. Divide the inequalities in part a by Δt and take the limit as Δt approaches 0.

    c. $A(t) = A_0 e^{rt/100}$.

[MN]

13. a. y must satisfy the differential equation: $-\dfrac{dy}{dx} = \dfrac{x-2}{y-0}$. Therefore $(x-2)^2 + y^2 = 2$ is the solution that passes through $(1, 1)$.

[MN]

    b. The graph is a circle with center $(2, 0)$ and radius $\sqrt{2}$. Point out to your students that the specifications in the problem are, in effect, a statement of a well-known property of circles.

14. Using the condition that P is the midpoint of RS leads to the differential equation $y = -x\frac{dy}{dx}$ since slope$(RS) = \frac{dy}{dx} = -\frac{2y}{2x} = -\frac{y}{x}$. See the figure below. The solution through the point $(1,1)$ is $y = \frac{1}{x}$.

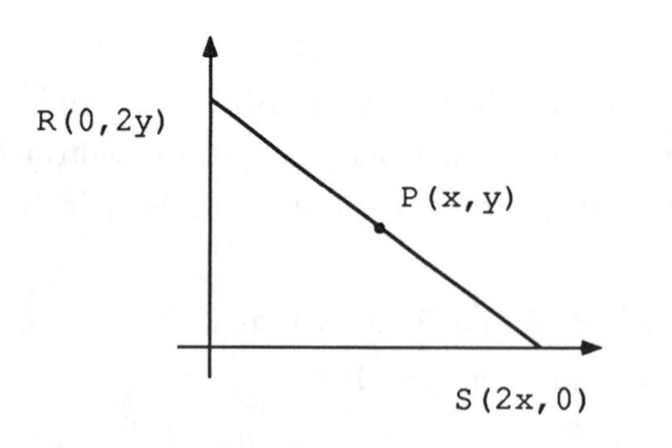

[MN]

15. a. $2 = (1 + r/100)^t$ implies that $\ln(2) = t\ln(1 + \frac{r}{100}) \approx t\frac{r}{100}$. Thus the doubling formula yields $t = \frac{\ln 2}{\ln(1 + \frac{r}{100})}$.

b.

$r\%$	2%	4%	8%	12%	18%
"rule of 72"	36	18	9	6	4
"exact"	35.00	17.67	9.01	6.12	4.19

c. $t = \frac{100\ln(2)}{r} \approx 69/r$.

d.

$r\%$	2%	4%	8%	12%	18%
"rule of 72"	36	18	9	6	4
"rule of 69"	34.50	17.25	8.625	5.750	3.833
"exact"	34.66	17.33	8.664	5.776	3.851

Using the "rule of 72" makes dividing easier.

16. a. The factor $M - P(t)$ causes the rate of learning to decrease as $P(t)$ approaches M.

 b. Solving the differential equation by separation of variables and using the initial condition, we see that $P(t)$ can be written in the form:

$$P(t) = M - (M - P_0)e^{-kt}.$$

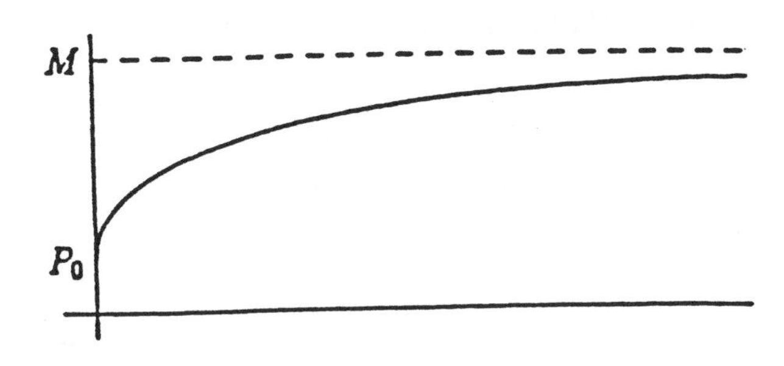

[MN]

 c. Solving $.9M = M - (M - .1M)e^{-.05t}$ for t gives $t = -20\ln(\frac{1}{9}) \approx 44\,\text{hours}$.

17. a. Using separation of variables, we see that $T(t) = T_a + (T_0 - T_a)e^{-kt}$.

 b. The limiting temperature is T_a. The object warms when $T_0 \leq T_a$ and cools when $T_0 \leq T_a$ to the limiting temperature of T_a.

 c. Solving for e^{-k} yields $T(t) = 350 - 290(\frac{23}{29})^{t/2}$. Therefore when $T = 150$, $t = \dfrac{2\ln\frac{20}{29}}{\ln\frac{23}{29}} \approx 3.2\,\text{hours} = 3\,\text{hours}\ 12\,\text{minutes}$.

[MN]

18. a. $\frac{d}{dP}(\frac{dP}{dt}) = k((1000 - P) - P)$. Setting $\frac{d}{dP}(\frac{dP}{dt}) = 0$ yields $P = 500$, which is a maximum for $\frac{dP}{dt}$.

b. If $P = 0$, or $P = 1000$ then $\frac{dP}{dt} = 0$. If there are no rabbits or if the population has reached the maximum the island can support, then the rate of change of rabbits is zero.

c. Students' graphs should show an inflection point when $P = 500$, and an upper limit of 1000 similar to the one below. Tell students that this curve is known as the "logistic" curve.

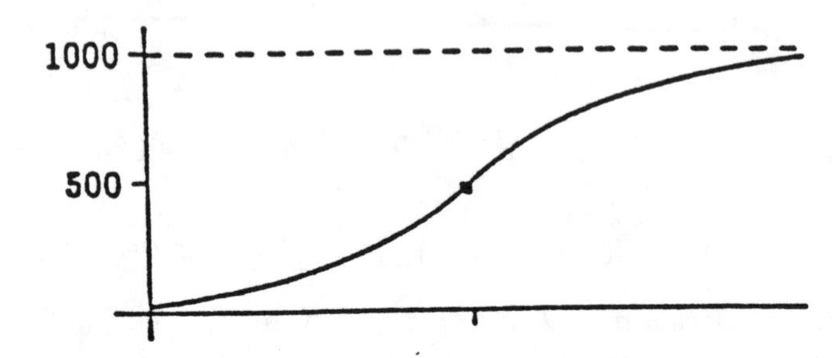

19. a. This term is used to model the number of encounters between predators and prey. Therefore the larger this term is, the faster the predator population and the more slowly the prey population will grow.

b. Solving $\frac{dx}{dt} \geq 0$ gives $y \geq \frac{B}{A}$.

c. Solving $\frac{dy}{dt} \geq 0$ gives $x \leq \frac{C}{D}$.

d. In Region I, $x \geq \frac{C}{D}$ and $y \geq \frac{B}{A}$, which imply that the predator population is increasing while the prey population is decreasing. Similar arguments can be made in the other three regions.

e. The differential equation becomes $(\frac{Ay - B}{y})dy = (\frac{C - Dx}{x})dx$. Integrating and simplifying gives a solution $x^C y^B = Qe^{Ay+Dx}$, Q constant.

[MN]

20. The equation is separable, and integration yields $\arcsin y = x+C$. The great temptation is to "invert" this to $y = \sin(x + C)$, and to produce graphs like the ones below:

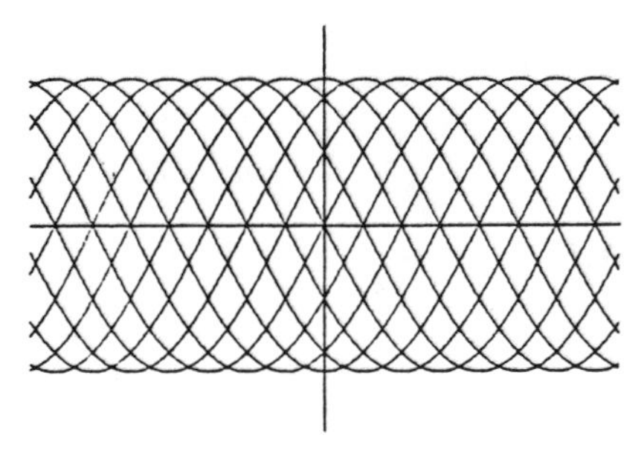

But in answer to part c, the graphs *should not* cross. (This is an informal way of stating the uniqueness of solutions of initial condition problems.) The graphs should always be increasing since $\sqrt{1-y^2}$ is always non-negative. The key revision is that the inverse function to $\arcsin x$ is *not* $\sin x$; it is $\sin x$ restricted to $[-\frac{\pi}{2}, \frac{\pi}{2}]$. Once we recognize this, we should be able to produce the correct set of graphs:

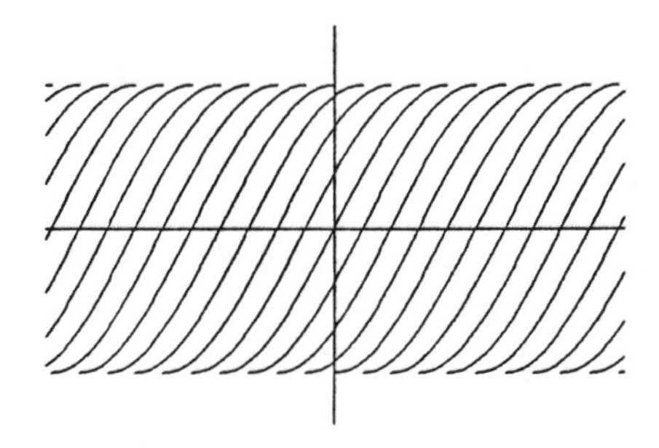

Chapter V: The Definite Integral

Section 1: Riemann Sums

1. a.

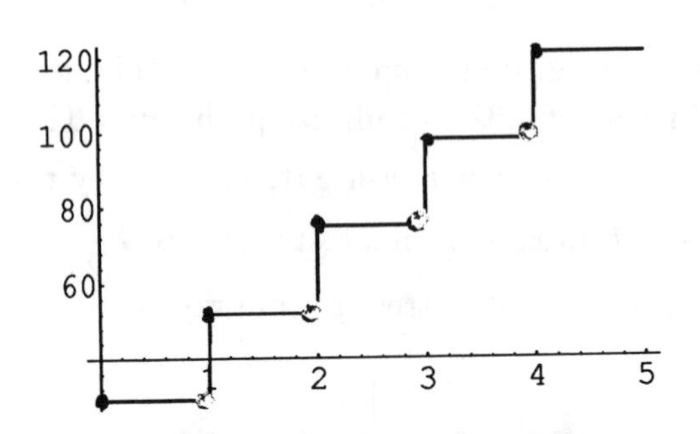

b. \$15.84 since $29 + 52 + 75 + \cdots + 259 = (29)(11) + 23\sum_{i=1}^{10} i = 1584$. This is can be viewed as the area of the region under the graph of the cost function from 0 to 11.

c. \$1,167.50 since $(29)(100) + 23\sum_{i=1}^{99} i = 2900 + 23\frac{(99)(100)}{2} = 116750$.

2. a. $\frac{63}{16}$

b. $\frac{87}{16}$

c. The average of the two answers is $\frac{75}{16}$. This is actually greater than the true area by $\frac{1}{48}$. The midpoint Riemann sum yields $\frac{149}{32}$, which is less than the area by $\frac{1}{96}$.

d. The answer is larger than the actual area. Since the function is concave up, the Trapezoidal Rule (which is the average of the right and left Riemann sums) overestimates the area. The midpoint rule underestimates the area for the same reason.

3. The right endpoint Riemann sum is $\frac{77}{16}$, the left endpoint Riemann sum is $\frac{93}{16}$, and their average (the Trapezoidal Rule) is $\frac{85}{16}$. This average is less than the true area by $\frac{1}{48}$. The midpoint Riemann sum is $\frac{171}{32}$, which is greater than the area by $\frac{1}{96}$. Since the function is concave down, the Trapezoidal Rule underestimates and the Midpoint Rule overestimates the area.

4. a. One estimate might be to take a fraction of the student's height times width at the waist. For example, for a student 6 feet tall and width at the waist of 16 inches, the estimate might be $\left(\frac{2}{3}\right)(6)\left(\frac{16}{12}\right) = \frac{16}{3}\,\mathrm{ft}^2$.

   b. A better estimate might be obtained by lying down and tracing an outline on the floor. Subdivide the region into small squares and/or triangles and add up their areas.

   c. Two ideas that students might discover are: (1) a finer subdivision yields a better answer, and (2) this is a multiplication problem like other applications of integral calculus. It should be noted that formal integration is of no help here.

5. a. The answers obtained by your students will vary depending on how closed spaced they draw their lines. The area of the lake given by the *Encyclopedia Britannica* is 7,550 square miles.

   b. Students could further subdivide the region to get a better estimate. For each square that includes points of the shore, they might estimate the portion of the square that is included in the lake, and add the area of this portion to the lower bound found in part a.

6. The irregularities of the time intervals places this question beyond the pale of the standard Simpson and Trapezoidal Rules. What is called for here is the sort of calculation which lies within the range of a hand calculator.

$$L = \text{lower sum} = \tfrac{1}{7}\left\{\left(10\tfrac{3}{8}\right)\left(\tfrac{72}{60}\right) + (12)\left(\tfrac{53}{60}\right) + \cdots + \left(10\tfrac{1}{4}\right)\left(\tfrac{10}{60}\right)\right\} \approx 13.198.$$

$$U = \text{upper sum} = \tfrac{1}{7}\left\{(12)\left(\tfrac{72}{60}\right) + \left(16\tfrac{3}{4}\right)\left(\tfrac{53}{60}\right) + \cdots + \left(10\tfrac{3}{4}\right)\left(\tfrac{10}{60}\right)\right\} \approx 14.959.$$

$$L < \text{Average Value} < U.$$

7. a. = Average temperature = $\frac{1}{24}\sum_{i=1}^{8} T_i(t)\Delta t_i$ where T_i is the temperature given for the time interval Δt_i.

   b. $\frac{1}{24}(5\cdot42+2\cdot57+2\cdot72+4\cdot84+3\cdot89+3\cdot75+2\cdot66+3\cdot52) = 66$ degrees Fahrenheit. Riemann sums are frequently taught with only regular partitions. This is a good problem to show students what happens when the partition is not regular.

   It may be of interest to students to find out the method the weather service uses to calculate the average temperature for the day.

8. $(30)\frac{15}{60}+(45)\frac{25}{60}+(50)\frac{30}{60}+(25)\frac{15}{60}+(45)\frac{20}{60}+(55)\frac{35}{60}+(50)\frac{40}{60}+(35)\frac{20}{60}+(25)\frac{10}{60} = \frac{9225}{60} =$ 153.75 miles. His average speed is $\dfrac{(153.75)(60)}{210} = 43.9$ MPH.

9. a. The region is bounded above by a quarter circle of radius 4, one unit above the x-axis. For this region, none of the rectangles used in the Riemann sum has zero area.

   b. 16.74

   c. 17.98

   d. 13.98

   e. 15.98. This is not the same as the answer to part *b* because the function is not linear over this interval.

   f. 16.63

   g. This is a forward-looking problem which may generate some surprising conjectures.

   h. The exact area is $4\pi + 4 = 16.57$. Students should see that they obtain a good approximation using eight subintervals, and a fair approximation using just four subintervals and the midpoint of the subinterval.

10. Since each subinterval is 1 minute wide, the distance traveled can be estimated by the Trapezoidal Rule: $(\dfrac{90}{2}+110+125+135+120+100+\dfrac{70}{2})/60 = 11.3$ miles. Students who do not divide by 60 to convert from minutes to hours should realize that their answers are impossibly large.

11. a. $(500+820+\cdots+4820)100 = \$2{,}420{,}000$.

    b. The cost starting from both ends is $2(500+820+1180+1580+2020)(100) = \$1{,}220{,}000$. Therefore there is a savings of \$1,200,000.

[PF]

12. The answer is c. Counting squares shows that the answer lies between 19 and 28.

13. If students use the left endpoints of each subinterval, they will get an upper bound of

$$4 \cdot 1 + 3 \cdot .5 + 2 \cdot 1.5 + 1.5 \cdot 1 = 10.$$

If students use the right endpoints of each subinterval, they will get a lower bound of

$$3 \cdot 1 + 2 \cdot .5 + 1.5 \cdot 1.5 + 1 \cdot 1 = 7.25.$$

Note that the subintervals are not all of the same length.

[CB]

14. a. $.14 + .21 + .28 + .36 + .44 + .54 + .61 + .70 + .78 = 4.06$ is a one estimate, while $.21 + .28 + .36 + .44 + .54 + .61 + .70 + .78 + .85 = 4.77$ is another estimate.

b. Taking an average value over each subinterval yields an estimate of 4.36.

c. If students make the assumption that f is strictly increasing, then an error estimate is the difference between the two estimates found in part a.

15. c. This can be justified by simple area arguments.

16. a. The answer required is a good picture.

b. The key algebra step in establishing the first of the inequalities is:

$$\frac{1}{n}\left[\frac{1}{1+\frac{1}{n}} + \frac{1}{1+\frac{2}{n}} + \cdots + \frac{1}{1+\frac{n-1}{n}}\right] = \frac{1}{n+1} + \frac{1}{n+2} + \cdots + \frac{1}{n+(n-1)}.$$

The second inequality is established in a similar way.

c. The difference is $\dfrac{1}{n} - \dfrac{1}{2n} = \dfrac{1}{2n}$.

d. $\dfrac{1}{2n} < .00005$ implies that $n > 10{,}000$.

Section 2: Properties of Integrals

1. This problem checks properties of integrals.

   a. $\frac{4}{3} + \frac{8}{3} = 4$.

   b. $\frac{11}{3} - \frac{4}{3} = \frac{7}{3}$.

   c. $\frac{11}{3} - 4 = -\frac{1}{3}$. This uses the answer to part a.

2. a. Simple addition: $18 + 5 = 23$.

   b. Simple subtraction: $18 - 5 = 13$.

   c. Calculus is not that simple: no answer can be given.

   d. Simple addition again, this time with a negative answer: $5 - 11 = -6$.

   e. Simple subtraction: $5 - (-11) = 16$.

   f. Again, there is no answer.

   g. Use the answer to part a and subtract: $23 - (-11) = 34$.

3. The statement is true. Suppose that $f(x) \neq 0$ for some x in R. Without loss of generality, we can assume that $f(x) > 0$. Since f is continuous, there is an open interval I on which f is positive. If we choose a and b in I, $\int_a^b f(x)\,dx > 0$ which contradicts what is given.

4. In order to answer this question, students need to use the fact that the integral of a sum is the sum of the integrals and that the integral of a constant is the constant times the length of the interval. Putting these two facts together makes this problem a little harder than it first appears. Answer: $4 + 2(10 - (-5)) = 34$.

5. a. The areas required are the squares of the values of x, i.e. $F(x) = x^2$ in this foretaste of the Fundamental Theorem of Calculus.

   b. In this case, the function is given by $F(x) = x^2 + x$, which is a tad more difficult to discover geometrically unless g is treated as a sum of two functions, one of which yields areas of rectangles and other of which yields areas of triangles.

6. This is a more challenging version of Problem 1, which offers students a bit more to think about.

   a. $\displaystyle\int_0^4 f(x)\,dx = \int_0^2 f(x)\,dx + \int_2^4 f(x)\,dx = 1 + 7 = 8$

   b. $\displaystyle\int_1^0 f(x)\,dx = -\int_0^1 f(x)\,dx = -2$

   c. $\displaystyle\int_1^2 f(x)\,dx = \int_0^2 f(x)\,dx - \int_0^1 f(x)\,dx = 1 - 2 = -1$

   d. The previous answer shows that the integral over this interval is negative. Therefore the function must be negative somewhere in $[1,2]$.

   e. A rectangle of height 3.5 over the interval $[2,4]$ has area 7, and $\displaystyle\int_2^4 f(x)\,dx = 7$. This implies that $f(x) \geq 3.5$ somewhere in $[2,4]$.

7. Graphing the integrand function is important here. If $b > 0$, the value of the integral is b^2 and if $b < 0$, its value is $-b^2$. In both cases, the correct answer is given by (D), $b|b|$. [AP]

8. Counterexamples for I and II can be found easily. Choice III, however, is always true, and the correct answer is C. [AP]

Section 3: Geometric Integrals

1. The greatest possible value is $(4)(2) = 8$, so the correct answer is D.

[AP]

2. I, II, and III are all true, and this is choice E.

[AP]

3. The integral in E is not equal to 0 nsince $\cos^2 x \geq 0$ for all x. The integrands in (A) and (B) are odd functions, and the graph of the cosine function in (C) is symmetric about the point $\pi/2$. The integral in (D) can be expressed as $\int_{-\pi}^{0} \cos^3 x\,dx + \int_{0}^{\pi} \cos^3 x\,dx$ and each of these is 0.

[AP]

4. a. $\int_0^{15} f(x)\,dx = 4 + \frac{9\pi}{4} - 4\pi - 12 = -8 - \frac{7\pi}{4} \approx -13.5.$

   b. $\int_9^{12} f(x)\,dx = -9.$

   c. $\int_0^{15} |f(x)|\,dx = 4 + \frac{9\pi}{4} + 4\pi + 12 = 16 + \frac{25\pi}{4} \approx 35.6.$

5. a. The graphs are straightforward.

   b. The values required are as follows:

   (i) 10 (ii) 22.5 (iii) 30 (iv) 32.5 (v) 65

   c. Students may possibly observe that $\int_0^5 3f(x)\,dx = 3\int_0^5 f(x)\,dx$ or that $\int_0^5 (f(x) + g(x))\,dx = \int_0^5 f(x)\,dx + \int_0^5 g(x)\,dx$. Finally they may observe that: $\int_0^5 f(x)g(x)\,dx \neq \int_0^5 f(x)\,dx \int_0^5 g(x)\,dx$ or even that $\int_0^5 [\![2x]\!]\,dx \neq 2\int_0^5 [\![x]\!]\,dx$.

6. a. Graphs of $f(x) + g(x)$, $f(x)g(x)$ and $\dfrac{g(x)}{f(x)}$ are provided below.

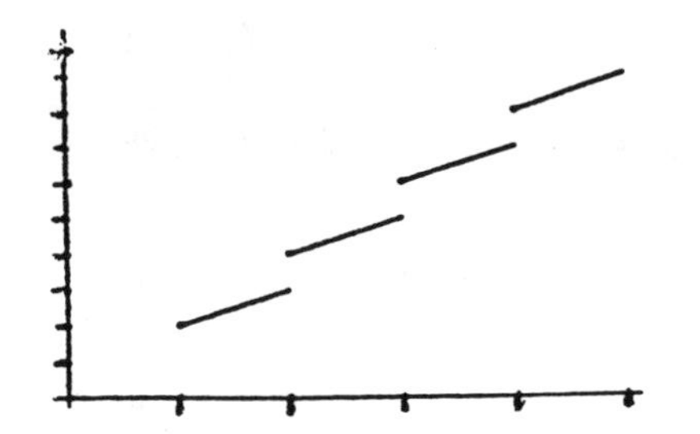

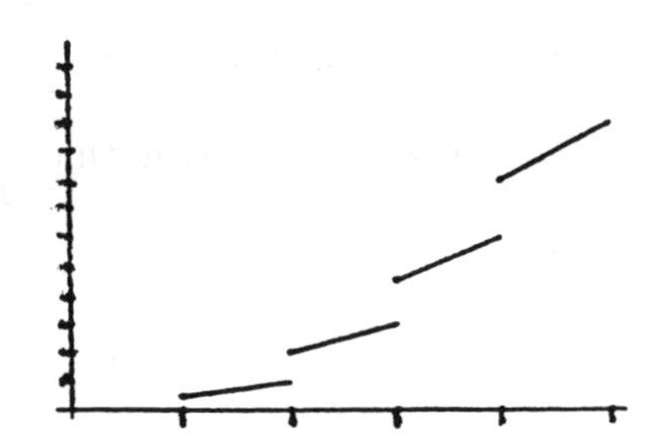

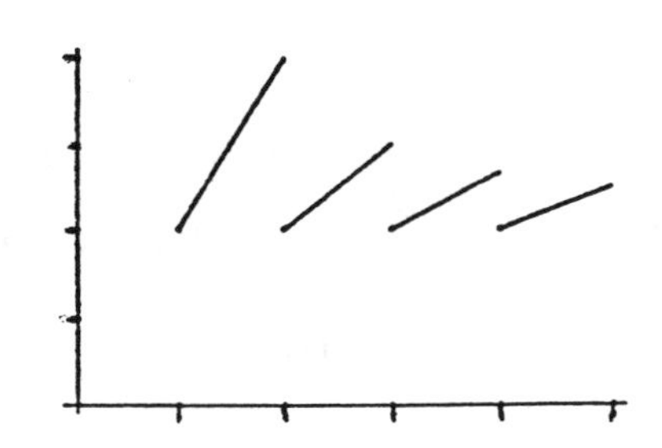

b. The required values follow.

(i) 10 (ii) 12.5 (iii) 22.5

(iv) 35 (v) $\frac{121}{24} = 5.0417$

c. Only the integral in (iii) can be related to integrals of f and g. Students should note, in particular, that:

$$\int_1^5 \frac{g(x)}{f(x)}\,dx \neq \frac{\int_1^5 g(x)\,dx}{\int_1^5 f(x)\,dx}.$$

7. $\displaystyle\int_{-3}^{3} (x+5)\sqrt{9-x^2}\,dx = \int_{-3}^{3} x\sqrt{9-x^2}\,dx + \int_{-3}^{3} 5\sqrt{9-x^2}\,dx$. Since the first integrand is odd, the value of the first integral is zero. $y = \sqrt{9-x^2}$ for $x \in [-3,3]$ is a half-circle, with radius 3, so the value of the second integral (and the value of the integral in the problem) is $\dfrac{45}{2}\pi$.

8. a. If a and b have the opposite parity, then the graph of f is symmetric about the point $(1/2, 0)$ and $f(x) = -f(1-x)$. Clearly $\int_0^1 f(x)\,dx = 0$. If a and b have the same parity, then the graph of f is symmetric about the line $x = 1/2$ and $f(x) = f(1-x)$. It is less clear but still true that $\int_0^1 f(x)\,dx = 0$. See the figures below for two representative cases.

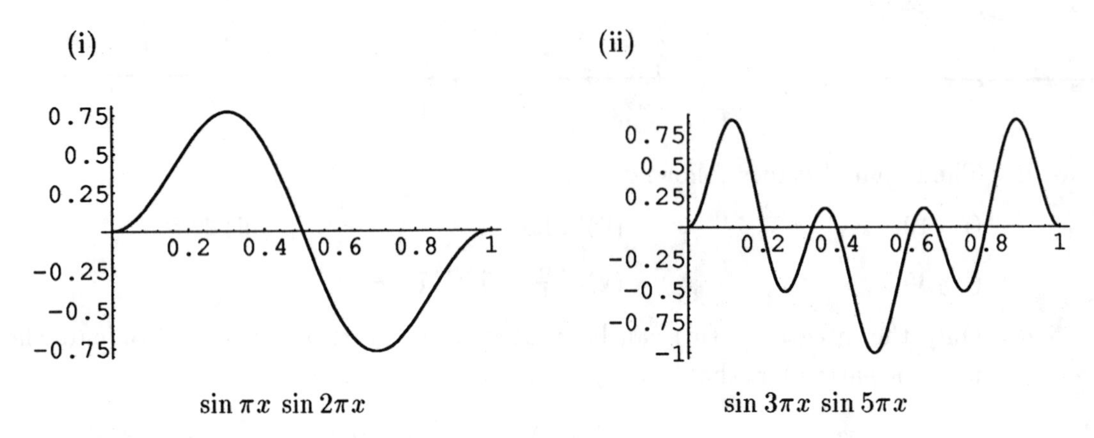

$\sin \pi x \sin 2\pi x$ $\sin 3\pi x \sin 5\pi x$

b. Let S be the rectangle bounded by the lines $x = 0$, $x = 1/2$, $y = 0$, $y = 1$. If $a = b$, then $f(x) = 1 - f(1/2 - x)$, and so the area of the region in S under the graph of f is equal to the area of the region in S above the graph of f. Since the graph of f is symmetric about the line $x = 1/2$, it follows that $\int_0^1 f(x)\,dx = 1/2$. Experimenting with a graphics package may induce students to try to establish the symmetries given above. In any case, stress the point that conjectures, made on the basis of examples, precede proofs.

[AZ]

9. Since $0 \leq \sin^8 x \leq 1$, Jack's answer is too large and Ed's negative answer is too small. Lesley's answer of $\frac{\pi}{2}$ looks good, but an examination of the graph of $y = \sin^8 x$ given below reveals that the area under the curve is less than half of that of the rectangle of height 1 and length π.

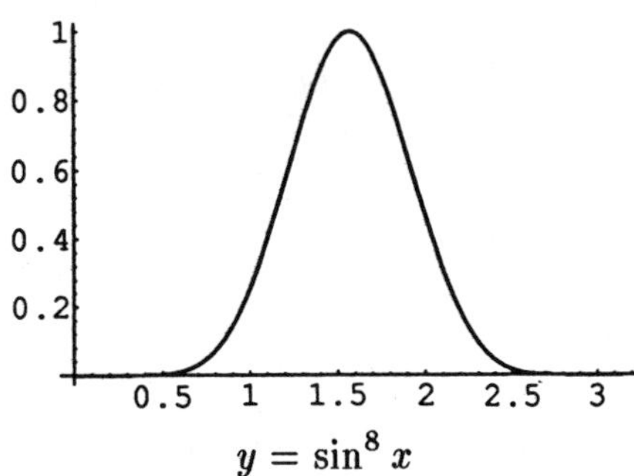

$y = \sin^8 x$

10. In both cases, the functions are inverses of each other.

    a. The area under $y = x^{1/n}$ from 0 to 1 is exactly the area above $y = x^n$ and under $y = 1$ from $x = 0$ to $x = 1$.

    b. Similarly the regions are complementary. Their union is a rectangle of width e and height 1, and thus the sum of their areas is 1.

[MK]

11. The polar region is the union of three subregions: the two triangles ORP and ORQ and the commn region PRQ. The area of the triangle ORP is half the area of the rectangle $PRq'p'$, and the area of the triangle ORQ is half the area of the rectangle $RQqp$. Each of the vertical and horizontal regions contains the common region PRQ, and each contains one of the two rectangles mentioned in the hint. Thus: area(horizontal region) + area(horizontal region) = 2 area(polar region). In other words, the area of the polar region is the mean of the areas of the vertical and horizontal regions.

[GS]

12. A graph of the integrand is given below. An estimate of the integral in the problem makes it clear that the correct answer is C. Its value is approximately 45.744.

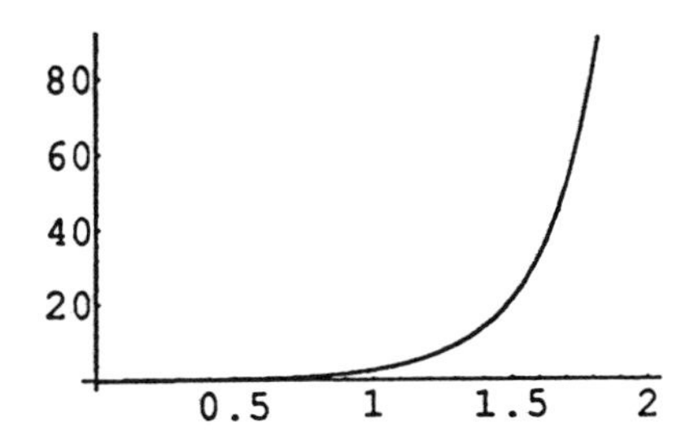

Section 4: The Fundamental Theorem of Calculus

1. a. True.

   b. False. The role of the function f and its antiderivative F are interchanged.

   c. False. The result does not have to be non-negative. It is easy to find examples which show that this is generally false. It could be corrected by assuming that $f(x) \geq 0$ for $a \leq x \leq b$.

   d. True. This is a combination of the Fundamental Theorem of Calculus and the rule that $\int cf(x)\,dx = c\int f(x)\,dx$.

   e. True. This is one form of the Mean Value Theorem.

2. a. $\int_0^2 f''(x)\,dx = f'(2) - f'(0) < 0$

   b. $\int_0^2 f'(x)\,dx = f(2) - f(0) < 0$

   c. $\int_0^2 f(x)dx > 0$. The printer's ink blob makes an accurate estimate of the integral impossible, but the graph of the function does not dip far enough below the x-axis for the integral to be 0 or negative.

3. a. Students have seen the graph of this function.

   b. The lower sum is 4.14626437 while the upper sum is 5.14626437.

   c. The average is 4.64626437, which underestimates the area since the curve is concave down.

   d. The answer is 4.66666667, which is as close to the exact answer as students can get with a calculator.

   e. $\int_1^4 x^{\frac{1}{2}}\,dx = 4\frac{2}{3}$

[PF]

4. a. $g(x) = \frac{2}{3}x^{3/2}$, so $g'(x) = x^{1/2}$. $g'(c) = \dfrac{\frac{2}{3}(2^{3/2}) - \frac{2}{3}(1^{3/2})}{2-1} = 1.21895142$.

   $c^{1/2} = 1.21895142$ so $c = 1.48584256$.

   b. g is an antiderivative of f.

   c. Do the same thing for the interval $[2, 3]$.

   If the author of your calculus text uses this argument in the proof of the Fundamental Theorem of Calculus, then you should consider asking students to look at the proof to see where it appears.

[PF]

5. a. The average values are, respectively, $\frac{1}{2}$, $\frac{1}{3}$, and $\frac{1}{4}$.

   b. The average value is $\frac{1}{n+1}$.

   c. The limit of the average value of f as n approaches ∞ is 0. A graphical explanation of this result involves the fact that, for large values of n, $f(x) = x^n$ is nearly 0 for most of the interval $[0, 1]$. Students might be asked for graphs of $y = x^{10}$, and $y = x^{100}$ to see this.

6. This problem complements the previous one.

   a. The average values are, respectively, $\frac{1}{2}$, $\frac{2}{3}$, and $\frac{3}{4}$.

   b. The average value of $f_n(x)$ over $[0,1]$ is $\frac{n}{n+1}$.

   c. The limit as n approaches ∞ is 1. A graphical explanation of this result involves the fact, for large values of n, $f(x) = x^{1/n}$ is nearly 1 for most of the interval $[0,1]$. You might ask your students for graphs of $y = x^{1/10}$ and $y = x^{1/100}$ in order to make this point.

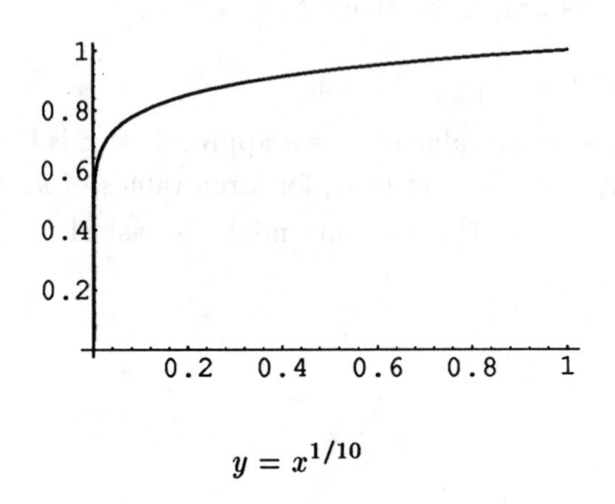

$y = x^{1/10}$

7. The average value, in each case, is 0 because the functions are all odd, and the intervals are all of the form $[-a,a]$.

8. a. $\int_0^1 x^3\,dx = \frac{1}{4}$

   b. $\int_0^4 x^3\,dx = 64$

   c. $\int_0^1 x^5\,dx = \frac{1}{6}$

   d. $\int_1^2 x^3\,dx = \frac{15}{4}$

9. a. $\frac{d}{dx}(5x^3+40) = 15x^3 = f(x)$

b. $\int_c^x 15x^2\,dx = 5x^3 - 5c^3$ and therefore $5x^3+4 = 5x^3-5c^3$. This implies $c^3 = -8$, so $c = -2$.

[AP]

10. a. It is geometrically evident that $G(2) = -G(-2)$ since the integrand is an even function: $G(-2) = \int_0^{-2} \sqrt{16-t^2}\,dt = -\int_{-2}^{0} \sqrt{16-t^2}\,dt = -\int_0^2 \sqrt{16-t^2}\,dt = -G(2)$.

b. $G(0) = 0$

c. $G'(x) = \sqrt{16-x^2}$, so $G'(2) = \sqrt{12}$.

d. $G(4)$ is the area of one quarter of a disk of radius 4, and so $G(4) = 4\pi$. $G(-4) = -G(4) = -4\pi$

[CNC]

11. F is a constant function. Therefore its derivative is 0.

12. a. Clearly $v(0) = 55$, $v(3) = 64$, which is the maximum speed, and $v(6) = 55$.

b. Average speed is $\frac{1}{6}\int_0^6 (55+6t-t^2)\,dt = 61$ MPH. Therefore the fine should be ($3.00)(6)(61-55) = $108.

[PF]

13. a. $\int_0^2 f'(x)\,dx = f(2) - f(0) = 0$.

b. The average value of f' on $[0,2]$ is 0.

14. Position the parabola so that the axis of symmetry is the y-axis:

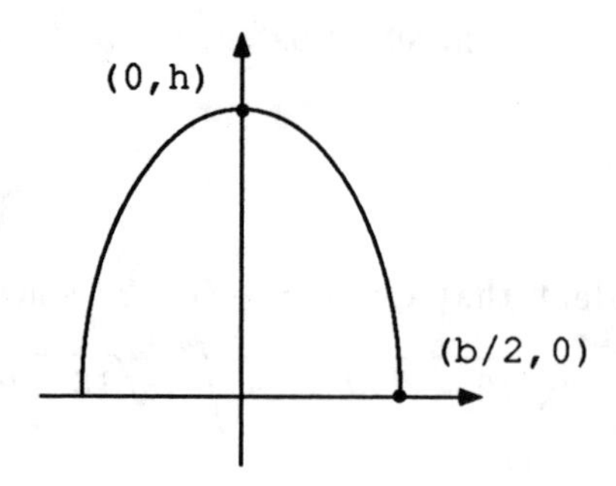

It is not hard to see then that the parabola has the equation $y = -\frac{h}{(b/2)^2}x^2 + h$.

$$\text{Area(arch)} = 2\int_0^{\frac{b}{2}} (-\frac{h}{(b/2)^2}x^2 + h)\,dx = \frac{2}{3}hb$$

This rule was known to Archimedes (c. 287-212 B.C.).

Section 5: Functions Defined by Integrals

1. In part *d*, note that students should get $\int_0^x \sin(t)\,dt = 1 - \cos x$. In part e, they should notice that the integral of the sum of the two parts of f is the sum of the integrals of the two parts.

2. Although this problem expresses a relatively simple idea, it is important that students learn how to modify existing results to solve a wider class of problems.

   a. The answers are:

   (i) $-x^2$ (ii) $-e^x$ (iii) $-\cos x$

   b. $\frac{d}{dx}\int_x^a f(t)\,dt = \frac{d}{dx}\Big(-\int_a^x f(t)\,dt\Big) = -f(x).$

3. a. $\frac{d}{dx}\int_0^{x^2} e^t\,dt = e^{x^2}(2x)$

   b. $\frac{d}{dx}\int_0^{x^2} \cos(t)\,dt = (\cos x^2)(2x)$

   c. $\frac{d}{dx}\int_0^{x^2} f(t)\,dt = f(x^2)(2x)$

   d. $\frac{d}{dx}\int_a^{u(x)} f(t)\,dt = f(u(x))u'(x)$

   If you did not assign Problem 2, then you might not want to assign this one. Students need problems that get at the heart of the chain rule. This one attempts to do just that.

4. Let $F(x)$ be an antiderivative of $\sqrt{x^5+8}$. Then the limit requested equals $\lim_{h\to 0}\frac{F(1+h)-F(1)}{h} = F'(1) = \sqrt{1^5+8} = 3$. The correct answer is C.

   [AP]

5. The derivatives are as follows:

   a. $\dfrac{\sin x}{x}$

   b. e^{-x^2}

   c. $\dfrac{1}{\cos x}(-\sin x) = -\tan x$

   d. 0

   e. $\dfrac{1}{2x^2}2x - \dfrac{1}{2x} = \dfrac{1}{2x}$. This answer can also be obtained from direct calculation.

6. B is the graph of the function, C the graph of f', while A is the graph of $\int_0^x f(t)\,dt$.

7. a. $F(2) = 4$, $F(0) = 0$, and $F(-1) = -\frac{3}{2}$.

   b. Critical points occur where $f(t) = 0$, hence at $x = -\frac{7}{2}$, -2, 2, and 5.

   c. Inflection points occur where $F''(x) = f'(x)$ changes sign. This happens at $x = -3$, -2, 1 and 3.

   d. $\int_{-5}^{5} f(t)\,dt = -\frac{9}{4} + \frac{1}{4} + (2 - \frac{\pi}{2}) + 4 + 0 - 3 = \dfrac{2-\pi}{2}$, and thus the average value of f on $[-5,5]$ is $\dfrac{2-\pi}{20}$.

   e. Since $G''(x) = F'(x) = f(x)$, G will be concave up on those intervals where $f(x) \geq 0$. The interval is $[-3.5, 2]$.

8. $F'(1) = \sqrt[3]{1^2 + 7} = 2$. Furthermore $F(1) = 0$. Therefore the equation of the tangent line is: $y - 0 = 2(x - 1)$.

9. a. $\int_0^2 f(t)\,dt = -4, \quad \int_0^4 f(t)\,dt = -4 + 1 = -3,$

$$\int_2^4 f(t)\,dt = \int_0^4 f(t)\,dt - \int_0^2 f(t)\,dt = -3 - (-4) = 1,$$

$$\int_5^{10} f(t)\,dt = -\tfrac{3}{2} + (-6) + (-3) = -\tfrac{21}{2}, \quad \int_1^7 f(t)\,dt = -2 + 1 + 0 + (-3) = -4$$

b. $F(0) = 0, \quad F(2) = -4, \quad F(5) = -\frac{3}{2}, \quad F(7) = -6.$

c. $$f(x) = \begin{cases} -2 & \text{if } 0 \le x < 2, \\ \frac{5}{2}x - 7 & \text{if } 2 \le x < 4, \\ -3x + 15 & \text{if } 4 \le x < 6, \\ -3 & \text{if } 6 \le x < 8, \\ \frac{3}{2}x - 15 & \text{if } 8 \le x < 10. \end{cases}$$

d. $$F(x) = \begin{cases} -2x & \text{if } 0 \le x < 2, \\ \frac{5}{4}x^2 - 7x + 5 & \text{if } 2 \le x < 4, \\ -\frac{3}{2}x^2 + 15x - 39 & \text{if } 4 \le x < 6, \\ -3x + 15 & \text{if } 6 \le x < 8, \\ \frac{3}{4}x^2 - 15x + 63 & \text{if } 8 \le x < 10. \end{cases}$$

e.

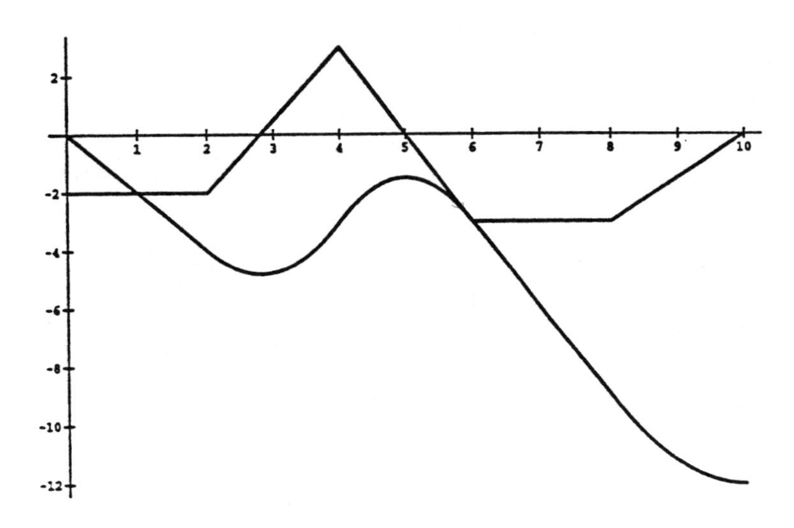

f. F has a local maximum at $x = 5$.

g. F is decreasing where $f(x) < 0$, which is on $[0, \frac{14}{5}]$ and $[5, 10]$.

h. F is increasing where $f(x) > 0$, which is on $[\frac{14}{5}, 5]$.

i. The graph of F is concave up where $f'(x) > 0$, which occurs on $[2, 4]$ and $[8, 10]$.

j. The graph of F is concave down where $f'(x) < 0$, which occurs on $[4, 6]$.

10. a. $F(5) = -\frac{2}{3} - \frac{3}{2} = -\frac{13}{6}$.

b. $F(3) = \frac{3}{2}$, and $F(4)$ is clearly negative. Since F is continuous, the Intermediate Value Theorem guarantees the existence of a zero of F between 3 and 4. Since $F'(x) = f(x)$ is negative between 3 and 4, F is decreasing over $[3,4]$, and the zero between 3 and 4 is unique.

c. Since $F'(3) = f(3) = -3$, the equation of the tangent line is $y - \frac{3}{2} = -3(x-3)$ or $y = -3x + \frac{21}{2}$.

d. The tangent line in part c crosses the x-axis at $x = \frac{21}{6}$.

11. a. The graphs are given below.

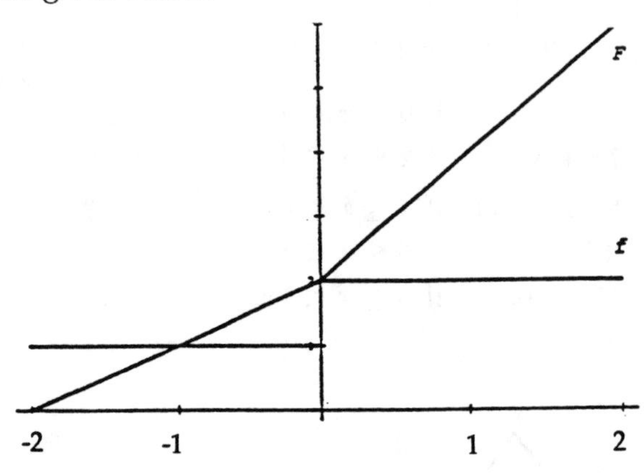

b. $F'(-1) = 1, \quad F'(1) = 2$.

c. F is continuous at zero, but $F'(0)$ does not exist. This is demonstrated by the graph of F.

12. a. $\int_{-1}^{X} g'(x)\,dx = g(X) - g(-1)$; therefore,

$$g(5) = \int_{-1}^{5} g'(x)\,dx + g(-1) = (-\pi + 4) + 2 = 6 - \pi$$

and

$$g(-5) = \int_{-1}^{-5} g'(x)\,dx + g(-1) = -(-\pi + 3) + 2 = \pi - 1.$$

b. $$\int_{-5}^{5} g'(x)\,dx = \int_{-5}^{-1} g'(x)\,dx + \int_{-1}^{5} g'(x)\,dx = -(\pi - 1) + (6 - \pi) = 7 - 2\pi.$$

13. Since $g'(x) = f(x)$, the extrema are located at the zeroes of f.

a. Local maxima occur at $x = 1$, 5, and 9.

b. Local minima occur at $x = 3$ and 7.

c. The absolute maximum is at $x = 1$.

d. The absolute minimum is at $x = 3$.

e. Since $g''(x) = f'(x)$, g is concave up where f' is positive. This occurs on the intervals $(2,4)$ and $(6,8)$.

f. The graph of g is given below:

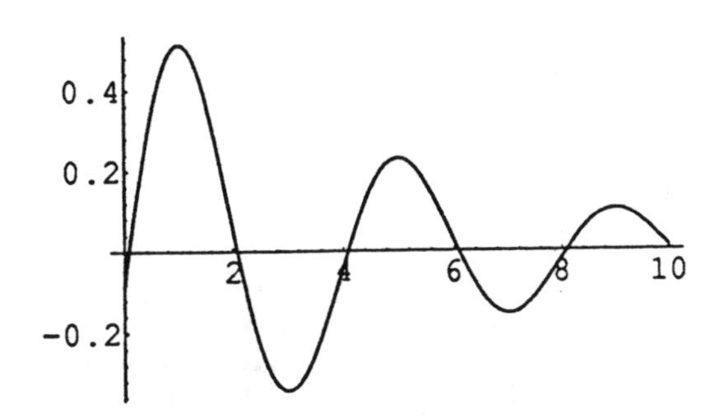

Note: The function used for the graph of this problem is defined by the rule $f(x) = e^{-x/5}\cos(\frac{\pi}{2}x)$.

$$g(x) = \int_0^x f(t)\,dt = \frac{-10}{4+25\pi^2}e^{-x/5}\Big(2\cos\big(\tfrac{\pi}{2}x\big) - 5\pi\sin\big(\tfrac{\pi}{2}x\big)\Big) - \frac{20}{4+25\pi^2}$$

14. The graphs required in part f are provided below. Other parts of the answers can be obtained easily from the graphs.

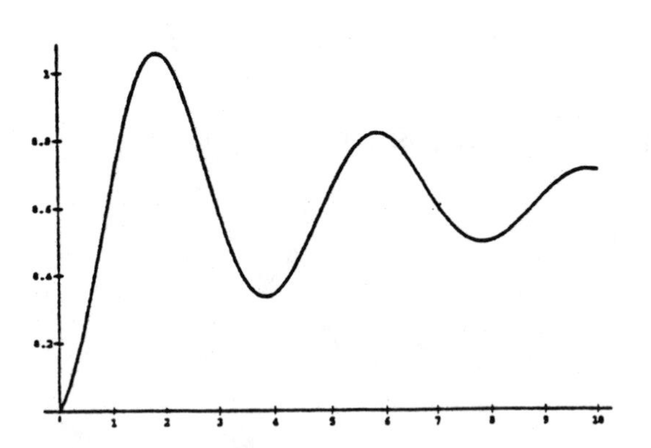

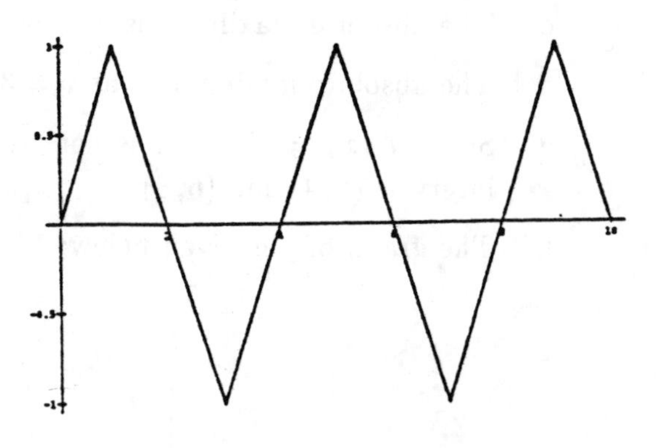

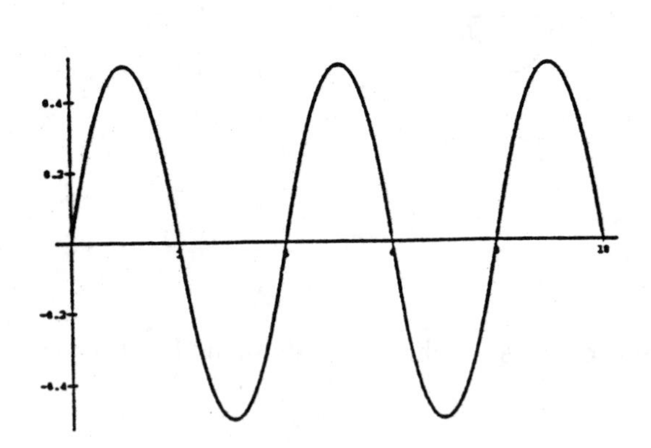

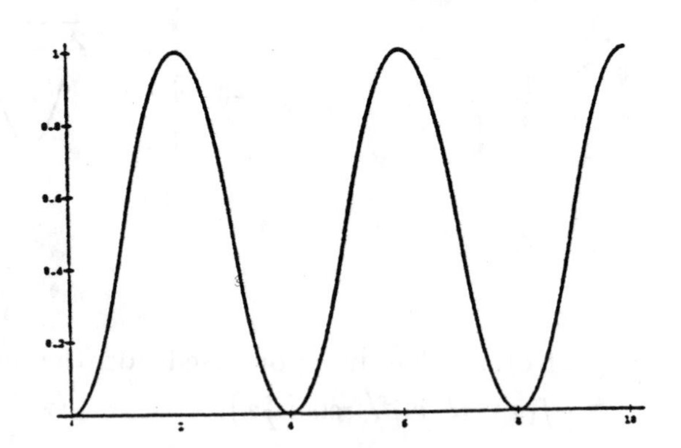

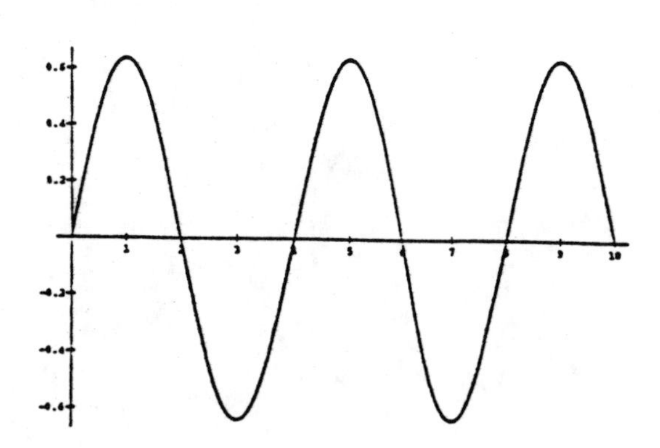

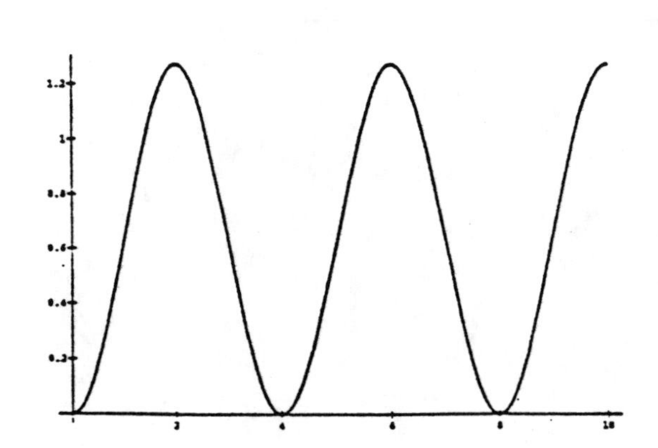

Chapter VI: The Definite Integral Revisited

Section 1: Exact Values from the Fundamental Theorem of Calculus

1. The ACM-GLCA calculus reform project, of which *Calculus Problems for a New Century* is part, developed syllabi for a 2-semester course in calculus in which integration appears in both semesters. This collection is structured along the lines of the proposed syllabi, and this section continues the set of problems of Section V.4.

   If $g' = f$, the conjecture is true. This question will reinforce what should become students' natural inclination on how to use the Fundamental Theorem. The problem, of course, is that so many integrable functions do not have elementary antiderivatives. This question will be explored later in this chapter. $\int xe^{2x}\,dx = \frac{1}{2}xe^{2x} - \frac{1}{4}e^{2x} + C$, so that the third option given is correct.

2. This is a paper and pencil exercise in the use of the Fundamental Theorem and the quotient rule. It also offers a particularly nice example of a reduction rule.

$$\begin{aligned}\int \frac{dx}{(x+1)^3} &= \frac{3}{4}\int \frac{dx}{(x^2+1)^2} + \frac{1}{4}\frac{x}{(x^2+1)^2} \\ &= \frac{3}{4}\left\{\frac{1}{2}\int \frac{dx}{(x^2+1)} + \frac{1}{2}\frac{x}{(x^2+1)}\right\} + \frac{1}{4}\frac{x}{(x^2+1)^2} \\ &= \frac{3}{8}\arctan x + \frac{x(3x^2+5)}{8(x^2+1)^2}.\end{aligned}$$

3. Each of the Advanced Placement Examinations includes a multiple choice section. This question, like a number of others selected from past exams, comes from such a section. The correct answer is (E).

   [AP]

4. Arc length problems which involve do-able integrals are not too common. This is one, and it is a 'natural'.

a.

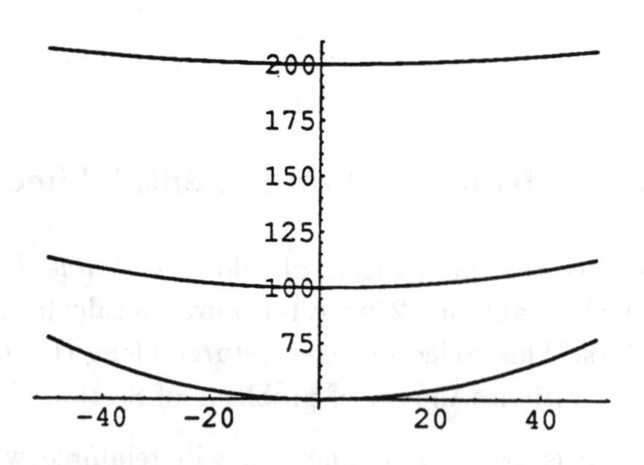

b.

$$\text{arc length} = 2\int_0^{50} \sqrt{1+[f'(x)]^2}\,dx \quad \text{where } f(x) = a\cosh(x/a).$$

$$\text{arc length} = 2\int_0^{50} \sqrt{1+\sinh^2(x/a)}\,dx = 2\int_0^{50} \cosh(x/a)\,dx = 2a\sinh(x/a)\Big|_0^{50}$$

$$= 2a\sinh(50/a)$$

c. For the given values of a, we have:

a	arc length (in meters)	sag required in part d (in meters)
50	117.52	27.154
100	104.219	12.7626
200	101.045	6.2826

d. The distance between the two points of suspension is 100 meters. Students may be puzzled by "sags" which are large in comparison with the increase in length of cable over linear distance. This is a time for a bit of class experimentation. Drape a piece of string over the backs of two chairs, and have your students compare the length of the string and the sag.

5. After introducing polar coordinates, many standard textbooks make no attempt to relate polar and Cartesian coordinate formulations of a problem. There is a batch of problems for which one method works best and another batch for which the other method is preferable. Like Rudyard Kipling's East and West, never shall the twain meet, and students may be left with the impression that area and arc length depend on whatever method is used to measure them. This problem is meant to dispel that notion.

   a. $\theta = 0$ for the side joining $(0,0)$ and $(1,0)$; $\theta = \frac{\pi}{4}$ for the side joining $(0,0)$ and $(1,0)$; and $r = \sec\theta$ for the side joining $(1,0)$ and $(1,1)$.

   b. Area $= \frac{1}{2}\displaystyle\int_0^{\pi/4} \sec^2\theta\, dt = \frac{1}{2}\tan\frac{\pi}{4} = \frac{1}{2}$.

   c. Length $= \displaystyle\int_0^{\pi/4} \sqrt{\sec^2\theta + \sec^2\theta\tan^2\theta}\, d\theta = \int_0^{\pi/4} \sec^2\theta\, d\theta = 1$.

6. a. A sketch produced by one graphics package follows:

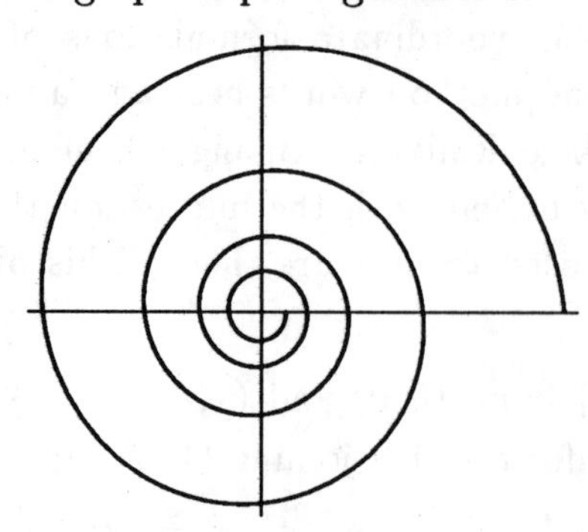

b. The parameter value $t = 0$ gives the point $(1, 0)$. Since

$$\frac{dy}{dx} = \frac{dy/dt}{dx/dt} = \frac{\cos t - .1\sin t}{.1\cos t + \sin t},$$

the slope of the tangent line for $t = 0$ is -10, and the equation of the tangent line is $y = 10 - 10x$.

c. Students who have studied polar coordinates may reformulate the equation of the curve as $r = e^{-.1t}$. For them, the length of the curve from $t = 0$ to $t = 8\pi$ is given by the integral

$$\int_0^{8\pi} \sqrt{r^2 + \left(\tfrac{dr}{dt}\right)^2}\, dt = \int_0^{8\pi} \sqrt{1.01}e^{-.1t}\, dt \approx 9.23581.$$

Students who use the formula

$$\int_0^{8\pi} \sqrt{\left(\tfrac{dx}{dt}\right)^2 + \left(\tfrac{dy}{dt}\right)^2}\, dt$$

will have the same integral to work (and will get the same answer).

d.

$$\text{Area} = \tfrac{1}{2}\int_0^{\pi/2} r^2\, dt = \tfrac{1}{2}\int_0^{\pi/2} e^{-.2t}\, dt = \tfrac{5}{2}[1 - e^{-.1\pi}] \approx 0.673993.$$

Those who use the given equations are best advised to use a Simpson's Rule approximation:

$$\begin{aligned}\text{Area} &= \int_0^1 y\, dx = \int_{\pi/2}^0 e^{-.2t}(-.1\cos t\sin t - \sin^2 t)\, dt \\ &= \int_0^{\pi/2} e^{-.2t}(.1\cos t\sin t + \sin^2 t)\, dt.\end{aligned}$$

Note that the area is less than that of a quarter circle with radius 1, hence the upper bound of $\pi/4$.

7. One hopes that asking students to guess here will lead them to speculate about the effect of multiplying $\sqrt{16-x^2}$ by x for $x < 1$ and $x > 1$.

$$\int_0^4 x\sqrt{16-x^2}\,dx = -\tfrac{1}{3}(16-x^2)^{3/2}\Big|_0^4 = \tfrac{64}{3} \approx 21.3333$$

$$\int_0^4 \sqrt{16-x^2}\,dx = \frac{16\pi}{4} = 4\pi \approx 12.5664$$

8. a. One purpose of this collection of problems is to act as a safe haven for classic calculus problems which have been jettisoned from contemporary texts. The Buffon Needle Problem is one such example. In the figure to the left below, the length of the vertical projection of the needle is $L\sin\theta$. A standard approach to this problem results in a double integral, but this is not necessary. Consider the phase plane in the figure to the right below:

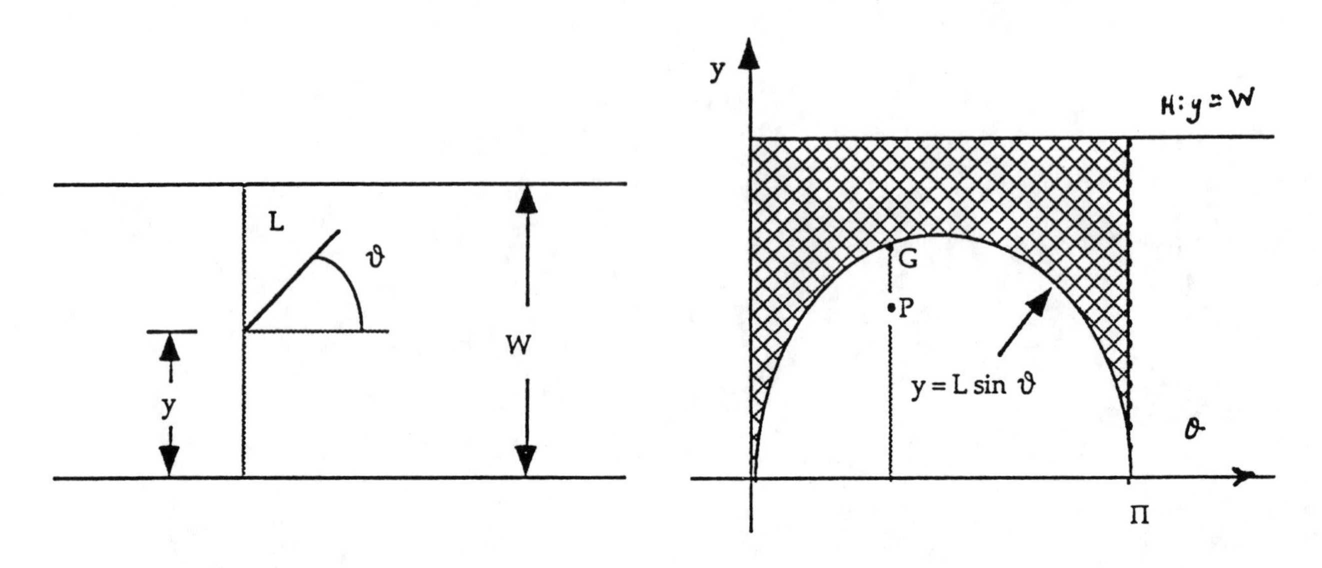

For a fixed θ, the distance from G to H, $W - L\sin\theta$, is the value of y at which the needle just reaches the crack. Points like $P = (\theta, y)$ which lie below G have a distance from H which is greater than this "just-reach" distance, and they are thus associated with outcomes in which the needle crosses a crack in the floor. The probability that the needle stays inside a single plank, which is the (normalized) area of the cross-hatched region in the diagram above, is

$$1 - \frac{\int_0^{\pi} L\sin\theta\,d\theta}{W\pi} = 1 - \frac{2L}{W\pi}.$$

b. $\pi \approx 3.0505$. You may want to extend this part by producing more data for the same needle, subjecting it to 2,000 drops each time for several more times and then estimating π for each simulation. Statisticians might want to explain why so little accuracy has been achieved.

9. Hydrostatic pressure problems can be found in most, if not all, the standard calculus textbooks currently in use, but by no means do all contain this problem or one like it. The thrust of the question is to get students to think about the ideas behind the calculation of force rather than the rote methods of getting an answer.

   a. Most students will argue "from physical principles" or will carry out a calculation to show that the dimensions of the window are irrelevant. In the latter case, if we let w be the width of the window, h its height (so that $wh = 1$), both in meters, then the force f in newtons is given by

   $$f = \rho g w \int_{3-\frac{h}{2}}^{3+\frac{h}{2}} x\,dx = 3\rho wgh = 3\rho g.$$

   Substituting the given values for ρ and g in the expression for f gives a force of 29,400 newtons. Students who point out that this answer assumes that the water level in the pool is higher than the top of the window should get gold stars for the day.

   b. Students who cope with part a will grasp that the symmetry of the window about a vertical axis passing through its center obviates the need to calculate the centroid. A disk-shaped port hole with an area of $1\,\text{m}^2$, for example, would be subject to the same force if its center were 3 feet beneath the surface of the water (and no problem about the top of the window rising above the water level). If you wish to generalize this problem, you may want to ask your students to show that the force on the window is given by $\rho g A x_c$ where A is the area of the window and x_c is the depth of its centroid.

10. a. The equality $\int_{-1}^{c} f(x)\,dx = 2\int_{c}^{1} f(x)\,dx$ leads to $\frac{(x-2)^3}{3}\Big|_{-1}^{c} = \frac{2}{3}(x-2)^3\Big|_{c}^{1}$ whose solution is $c = 2 - \sqrt[3]{\frac{29}{3}} \approx -.130$.

    b. In this case, the same equality yields $\left(x + \dfrac{1}{x}\right)\Big|_{1}^{c} = 2\left(x + \dfrac{1}{x}\right)\Big|_{c}^{2}$ and the (acceptable) solution here is $c = (7 + \sqrt{13})/6 \approx 1.7676$.

11. In this problem, a picture is worth a few hundred words. The area calculation which follows is independent of a: $\int_0^{2a} y\,dx = \left(\frac{x^2}{a^2} - \frac{x^3}{3a^3}\right)\Big|_0^{2a} = \frac{4}{3}$. The locus of the vertices provides a nice bit of geometry. Since $f(x) = 0$ when $x = 0$ and $x = 2a$, the x-coordinate of the vertex is a, and $y = \frac{2}{a} - \frac{1}{a} = \frac{1}{a}$. So the curve determined by the vertices is the hyperbola: $xy = 1$.

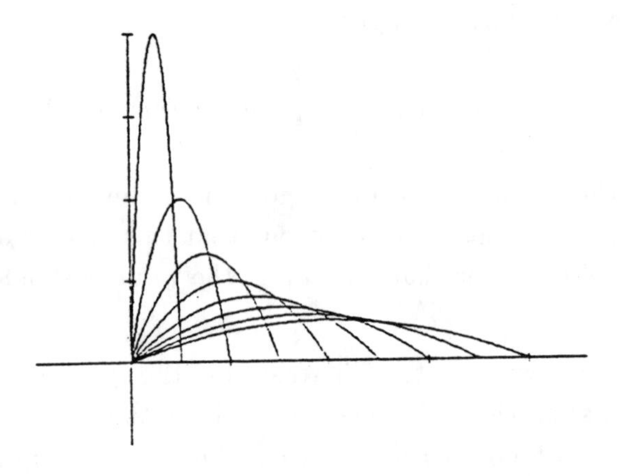

12. a. Both rings have the same volume. We ask students to guess an answer to a question in the hope that it will start them thinking, formulating reasoned opinions, and structuring an analytical attack on the question at hand. In this case, asking them to guess may be something of a dirty trick. Students should value their intuition; it is hard, nonetheless, to resist the appeal of the occasional counterintuitive example.

    b. The shell method gives the following volume of a napkin ring formed from a sphere of radius R by drilling a hole of radius r as specified in the problem.

$$2\int_r^R 2\pi x\sqrt{R^2 - x^2}\,dx = \frac{4\pi}{3}\left(R^2 - r^2\right)^{3/2}$$

    Since $R^2 - r^2 = \left(\frac{h}{2}\right)^2$ for both rings, we conclude that the rings have the same volume, $\frac{1}{6}\pi h^3$.

13. All calculus texts contain center-of-mass examples, but many do not include this one whose principal charm lies in part *c*.

a.

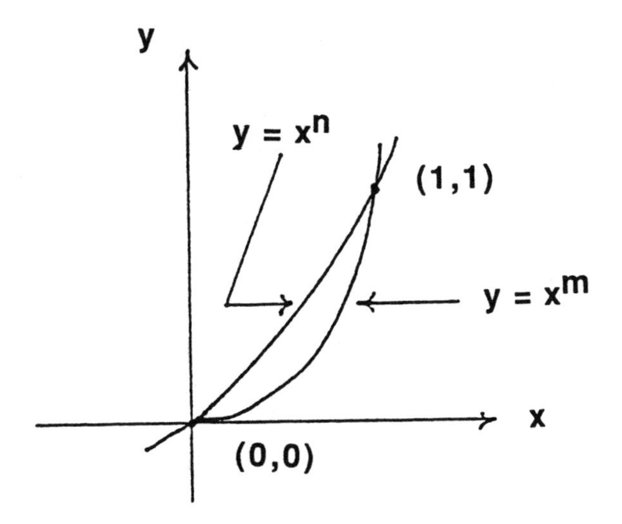

b.

$$\bar{x} = \frac{\int_0^1 x(x^n - x^m)\,dx}{\int_0^1 (x^n - x^m)\,dx} = \frac{(n+1)(m+1)}{(n+2)(m+2)}.$$

$$\text{Similarly } \bar{y} = \frac{\int_0^1 y(y^{1/m} - y^{1/n})\,dy}{\int_0^1 (x^n - x^m)\,dx} = \frac{(n+1)(m+1)}{(2n+1)(2m+1)}.$$

$$\text{Thus } (\bar{x}, \bar{y}) = \left(\frac{(n+1)(m+1)}{(n+1)(m+2)}, \frac{(n+1)(m+1)}{(2n+1)(2m+1)}\right).$$

Access to a CAS will help with the algebra even if the CAS cannot cope with two parameters in the integrand (without being taught how to do this!).

c. The trick is to find values for n and m such that $\bar{x}^n < \bar{y}$. For such values, the center of mass lies above the graph of y^n. There are many possibilities, the simplest of which is $n = 3$, $m = 4$.

$$(\bar{x}, \bar{y}) = \left(\tfrac{2}{3}, \tfrac{20}{63}\right);$$

$$\bar{x}^n = \left(\tfrac{2}{3}\right)^3 = \tfrac{8}{2}7 = \tfrac{56}{189} < \tfrac{60}{189} = \tfrac{20}{63} = \bar{y}.$$

14. This problem is included to encourage students to reflect before they leap into action. If they work with the hint, they will achieve, after combining terms

$$I'(x) = e^{-x}[(a_1 - a_0) + (2a_2 - a_1)x + \cdots + (5a_5 - a_4)x^4 - a_5x^5].$$

Setting $I'(x) = e^{-x}x^5$ yields values for the coefficients in a cascade of falling mathematical dominoes:

$$a_5 = -1 \qquad a_2 = 3a_3 = -60$$

$$a_4 = 5a_5 = -5 \qquad a_1 = 2a_2 = -120$$

$$a_3 = 4a_4 = -20 \qquad a_0 = a_1 = -120$$

You may want to return to this problem later, after your students have seen Taylor series. The polynomial part of $I(x)$ is the 5th degree Maclaurin polynomial for e^x multiplied by $-5!$ To calculate $\int e^{-x}x^n\,dx$ for small natural numbers n provides an example where integration by parts can be an invitation to discover a recursive pattern.

15. This problem can be worked with pencil and paper although it is a good exercise for a CAS. The loaf can be viewed as half a surface of revolution generated by revolving the function defined by the rule $y = \sqrt{r^2 - x^2}$ about the x-axis. The surface area of each slice is then given by the definite integral $\pi \int_{x_0}^{x_0+2r/n} y\sqrt{1+(y')^2}\,dx$. where x_0 is the x-coordinate of the left edge of the slice. (The coefficient '2' which one expects to see in this integral is absent since we are dealing here with only half a loaf, which is still better than no loaf at all.) Calculation gives the answer $2\pi r^2/n$, which indicates that each slice has the same amount of crust as every other slice.

[MP]

Section 2: Techniques of Integration

1. Try a. Note that in these AP multiple-choice questions, the choice of answer reflects the errors students make in working definite integrals by substitution.

   [AP]

2. The aim of this exercise is to relate the geometry of the problem, first to a ballpark feel for the answer and then to the mechanics of solution.

   a.

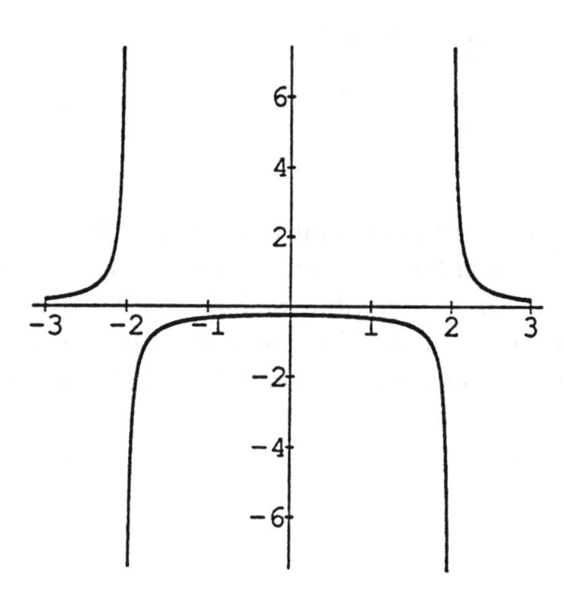

   b. Negative, since the integrand is negative for $x \in (-2, 2)$. Students should be encouraged to use the fact that the function is even and the interval of the type $[-a, a]$.

   c.

$$\frac{1}{x^2-4} = \tfrac{1}{4}\left(\frac{1}{x-2} - \frac{1}{x+2}\right).$$

$$\int_{-1}^{1} \frac{dx}{x^2-4} = \tfrac{1}{4}\ln\left|\frac{x-2}{x+2}\right|\Bigg|_{-1}^{1} = \tfrac{1}{4}\left[\ln\left(\tfrac{1}{3}\right) - \ln 3\right]$$
$$= -\tfrac{1}{2}\ln 3 \approx -.549306.$$

   d. With friends like these, you do not need enemies. Since the secant function never takes values between -1 and 1, x cannot take values between -2 and 2. Yet in this problem, this is just what x does.

3. The right-hand side of the equality suggests that an integration by parts, in which $u = f(x)$ and $dv = \sin x\, dx$, has been carried out. Since $f'(x) = 3x^2$, the correct answer is B.

[AP]

4. This exercise is suitable for students with some knowledge of polar coordinate substitution and the half-angle formula.

$$\begin{aligned}(ds)^2 &= (dr)^2 + (r\, d\theta)^2\\ &= (2a^2)(1+\cos\theta)(d\theta)^2;\\ s &= 2\int_0^{\pi} a\sqrt{2}\sqrt{1+\cos\theta}\, d\theta = 4a\int_0^{\pi} \cos\left(\frac{\theta}{2}\right)\, d\theta\\ &= 8a.\end{aligned}$$

The sketch called for is the well-known valentine shape. The probability of obtaining an antiderivative in closed form for an integrand in an arc length problem is close to zero if the curve in the problem is selected at random. This fact is generally not disclosed to students in elementary calculus. They are exposed to two or three carefully selected problems which lead to "do-able" integrals. What life has in store is hinted at in the next section.

[AL]

5. With a question in Chapter III about making pizza boxes, it seems only appropriate to consider the contents of such a box.

   a. With the center of symmetry of the pizza at the origin of a coordinate system, the pizza rim is given by $x^2 + y^2 = 36$. The area of the right-most slice (and that of the left-most slice which is congruent to it) is given by $2\int_2^6 \sqrt{36 - x^2}\,dx$. The area of the center slice is given by $2\int_{-2}^2 \sqrt{36 - x^2}\,dx$. These integrals yield either to a trigonometric substitution or to a CAS. The area of each of the "outside" slices is approximately 33 square inches; the area of the center slice is approximately 47.1 square inches. The center slice, therefore, contains about 41.6% of the pizza, with each of the other slices containing 29%. You may want to ask your students who have access to a CAS how the pizza is to be sliced if each of the three is to get the same amount of pizza. $\int_{-s}^{s} \sqrt{36 - x^2}\,dx = \int_s^6 \sqrt{36 - x^2}\,dx$ yields $s = 1.58959$.

   b. The hungry student still takes the middle slice.

$$2\int_{-2}^{2} \sqrt{49 - x^2}\,dx = 55.2284$$

$$2\int_{2}^{7} \sqrt{49 - x^2}\,dx = 49.3548$$

Note, however, that the injustice is not so great. The middle slice is about 35.9% of the pizza.

6. Here is an example of how students with access to an appropriate software package can use a CAS to work on pattern recognition and study antiderivatives. Integrals b and d can be worked by the substitution $u = x^2$ to yield answers $-\frac{1}{2}\cos x^2$ and $\frac{1}{2}\sin x^2 - \frac{1}{2}x^2\cos x^2$ respectively. Integrals of this type are "do-able" if the exponent of x in the integrand is odd.

7. a. The sketch follows:

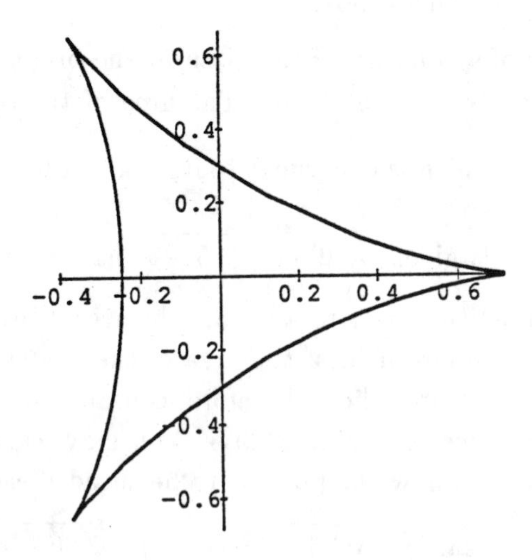

b. arc length $= 3\int_0^{2\pi/3} \sqrt{\left(\frac{dx}{dt}\right)^2 + \left(\frac{dy}{dt}\right)^2}\, dt = \frac{3\sqrt{2}}{2}\int_0^{2\pi/3} \sqrt{1-\cos 3t}\, dt$

$= \frac{\sqrt{2}}{2}\left(-2\sqrt{2}\cos\frac{3}{2}t\right)\Big|_0^{2\pi/3} = 4.$

Using Simpson's Rule, we get:

n	10	20	50
approximation	4.00022	4.00001	4.00000

You may wish to extend this problem by asking students to calculate the area of the region enclosed by the curve. This area, which is $\pi/8$, can be most easily computed by using the formula:

$$\text{Area} = \frac{1}{2}\int_0^{2\pi} \left(x(t)y'(t) - y(t)x'(t)\right) dt.$$

Again Simpson's Rule can be used; $n = 10$ gives an accuracy of 10^{-6}.

8. a. This is a straightforward exercise in the technique of integration by parts.

b. The first of these is a disguised version of part a: Let $w = \ln x$, so that $x = e^w$ and $dx = e^w\,dw$. Then $\int \ln^n x\,dx$ becomes $\int w^n e^w\,dw$. Otherwise you could use integration by parts, with $u = \ln^n x$, $dv = dx$. Then $du = \frac{n}{x}\ln^{n-1} x\,dx$ and $v = x$.

$$\int \ln^n x\,dx = x\ln^n x - n\int \ln^{n-1} x\,dx.$$

For the second, let $u = x^n$, $dv = \sin x\,dx$. Then $du = nx^{n-1}\,dx$, $v = -\cos x$.

$$\int x^n \sin x\,dx = -x^n \cos x + n\int x^{n-1}\cos x\,dx.$$

The integral in the answer can be restored to the sine family by a second integration by parts.

$$\int x^n \sin x\,dx = -x^n \cos x + nx^{n-1}\sin x - n(n-1)\int x^{n-2}\sin x\,dx.$$

9. a.
$$\begin{aligned}\int \sin^n x\,dx &= -\sin^{n-1} x\cos x + (n-1)\int \sin^{n-2} x\cos^2 x\,dx\\ &= -\sin^{n-1} x\cos x + (n-1)\int \left(\sin^{n-2} x - \sin^n x\right)dx\\ &= -\frac{\sin^{n-1} x\cos x}{n} + \frac{n-1}{n}\int \sin^{n-2} x\,dx.\end{aligned}$$

b. A second application of the reduction formula yields

$$\int_0^{\pi/2} \sin^n x\,dx = \frac{(n-1)(n-3)}{n(n-2)}\int_0^{\pi/2} \sin^{n-4} x\,dx$$

and repeated applications yield a Wallis product as the coefficient of $\int_0^{\pi/2} \sin x\,dx$, whose value is 1.

10. The functions f_n are the Chebyshev polynomials encountered in numerical analysis.

    a. Symmetry about the origin implies that the integrals are 0 when n is odd. For n even and non-zero, the integrals are negative.

    b. Students may conjecture that f_n is a polynomial of degree n. They are right, too. The Chebychev polynomials can be defined recursively:

    $$f_{n+1}(x) = 2xf_n(x) - f_{n-1}(x) \quad \text{where} \quad f_0(x) = 1 \text{ and } f_1(x) = x.$$

    c. If n is even, the integral is $\frac{2}{(1-n^2)}$. If n is odd and $n \neq 1$, the integral is 0, as students will probably conjecture from their work in part a. For $n = 1$, it is clear that the integral is also 0.

[AZ]

11. The integral on the right-hand side of the identity in part a reappears in part *b* after students use a trigonometric substitution:

$$\begin{aligned} \int_0^a \arcsin x \, dx &= x \arcsin x \Big|_0^a - \int_0^a \frac{x}{\sqrt{1-x^2}} \, dx \\ &= a \arcsin a - \int_0^{\arcsin a} \frac{\sin y \cos y}{\cos y} \, dy \\ &= a \arcsin a + \sqrt{1-a^2} - 1. \end{aligned}$$

This problem is included because it provides a good geometric interpretation of integration by parts.

12. Implicitly the pattern which students are asked to find is the general formula

$$\int_0^1 x^n(1-x)^m \, dx = \frac{n!m!}{(n+m+1)!}.$$

How they find it—or whether they find it at all—will depend on the software package available to them. In this case, students are better off with pencil and paper and a yen to explore. The substitution should suggest to students that the general formula is symmetric in terms of n and m since it leaves invariant the integral in which it is made.

For many students, this problem may offer an example of the limitations of what is currently available by way of CAS's (although this is certain to change, perhaps soon). It also offers scope for students to experiment and obtain partial answers. The general formula, after all, is not explicitly asked for. This is an exploration-type problem in which obtaining partial answers should be applauded. Teachers will probably find that some of their students—possibly the physics and pre-engineering majors—will get their answers from tables. It might provoke an interesting class discussion to ask such students how such tables relate to the way their CAS works its integrals.

13. We have here one of many problems in which a parametric representation of a curve arises naturally. The only force acting on the projectile is that due to gravity.

$$\begin{aligned} & y'' = -9.8; \\ x' = 4, \qquad & y' = -9.8t; \\ x = 4t, \qquad & y = -4.9t^2 + 100. \end{aligned}$$

The time it takes the projectile to hit the ground is calculated, as usual, by setting $y = 0$ and solving to obtain $t = 10/\sqrt{4.9}$. The length of the trajectory (in meters) is given by the definite integral

$$\int_0^{10/\sqrt{4.9}} \sqrt{16 + (-9.8t)^2}\, dt \approx 102.94.$$

The integral yields to a trigonometric substitution $\left(\tan\theta = \frac{9.8t}{4}\right)$ but this is a good problem for a CAS.

[RH]

14. Pythagoras resolves this dispute. Since $\sec^2 x - \tan^2 x = 1$, the two answers differ by a constant. Perhaps the argument would have been less serious if both students had not forgotten the constant of integration in their antiderivatives.

15. This problem has already appeared in V.3. Including it twice shows how fond we are of it.

    A picture here always helps.

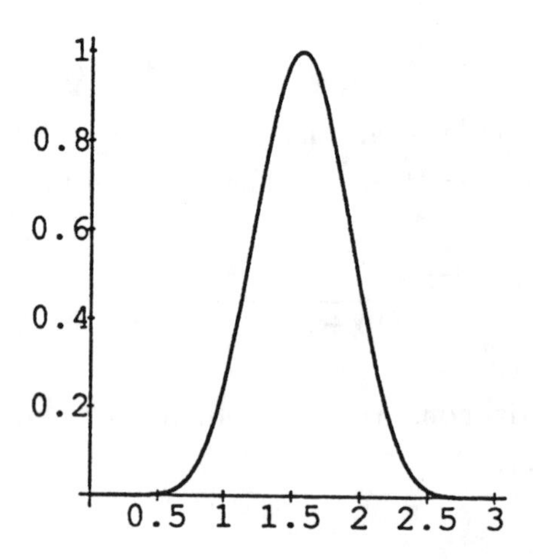

This lets us see that $0 < \int_0^{\pi} \sin^8 x\, dx < \pi$ so that Jack and Ed strike out. To eliminate Lesley, a sharper inequality is called for. Students may find a bounding upper sum for the integral. Another possibility is to observe that

$$\int_0^{\pi} \sin^8 x\, dx < \int_0^{\pi} \sin^2 x\, dx = \tfrac{\pi}{2}.$$

16. This problem nicely illustrates the pitfalls into which incautious substitutions can lead. The difficulty lies in the second calculation. $\cos\theta$ is negative for $\frac{\pi}{2} < \theta < \pi$ and yet $\sqrt{1-u^2}$ is always non-negative. If the integral is split up into two parts, the difficulty is avoided:

$$\begin{aligned}\int_0^{\pi} \cos^2\theta\, d\theta &= \int_0^{\pi/2} \cos^2\theta\, d\theta + \int_{\pi/2}^{\pi} \cos^2\theta\, d\theta \\ &= \int_0^1 \sqrt{1-u^2}\, du + \int_1^0 -\sqrt{1-u^2}\, du \\ &= 2\int_0^1 \sqrt{1-u^2}\, du = 2(\tfrac{\pi}{4}) = \tfrac{\pi}{2}.\end{aligned}$$

Now the answers from both calculations agree.

[MN]

17. Many of the problems developed in the Minnesota Teacher Renewal Project, of which this is one, stress geometry. This problem and others like it prepare students to understand how calculus relates to the subject of probability. What is required here is an area calculation. The region whose area is numerically equal to the desired probability is sketched below.

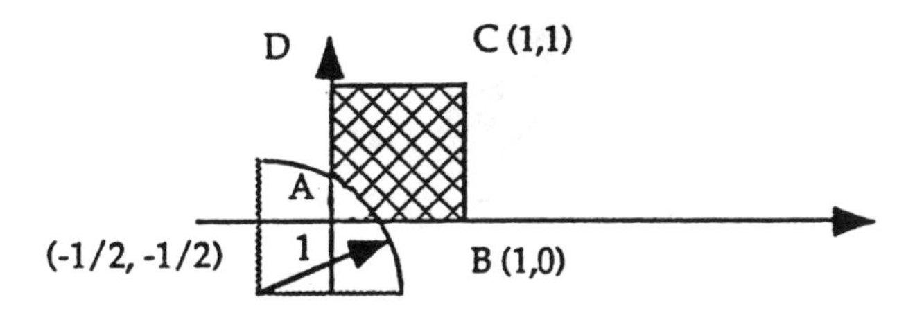

One attack on the problem is to calculate the area of the region (in the unit square) which is the complement of the cross-hatched region above and to subtract it from 1, the area of the square. This can be done without calculus, but we shall not be tempted.

The equation of the circle is $\left(x+\frac{1}{2}\right)^2+\left(y+\frac{1}{2}\right)^2=1$. The circle intersects the x-axis at $(\sqrt{3}-1)/2$ and the integral becomes

$$\int_0^{(\sqrt{3}-1)/2}\left(\sqrt{1-\left(x+\tfrac{1}{2}\right)^2}-\tfrac{1}{2}\right)dx=\int_0^{(\sqrt{3}-1)/2}\sqrt{1-\left(x+\tfrac{1}{2}\right)^2}\,dx-\int_0^{(\sqrt{3}-1)/2}\tfrac{1}{2}\,dx.$$

The second of these integrals presents no problem. The first yields to a double substitution: First $u=x+\frac{1}{2}$ and then $u=\sin\theta$. Students without access to a CAS will get a trigonometric workout. The area is:

$$\int_{\pi/6}^{\pi/3}\cos^2\theta\,d\theta-\tfrac{1}{2}\left(\frac{\sqrt{3}-1}{2}\right)=\frac{\pi}{12}-\tfrac{1}{2}\left(\frac{\sqrt{3}-1}{2}\right)\approx .0788.$$

The probability that the distance is greater than 1 is, therefore, $1-.0788=.9212$.

[MN]

18. This is similar to the preceding problem, but the context is different enough to warrant inclusion. Let x and y be the two given numbers. Then $x^2 + y^2 \leq 9$, and our interest lies in determining "how many" such numbers have a sum greater than 2: $x + y \geq 2$. This boils down to calculating the area of the cross-hatched region in the figure below. The probability sought is obtained by dividing this area by 9π, the area of the disk in the figure.

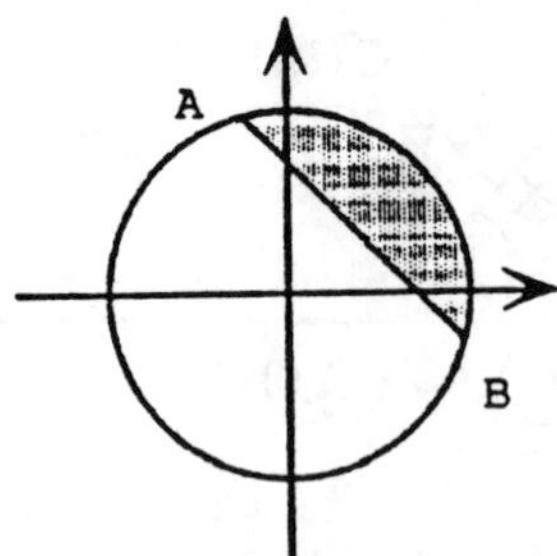

It simplifies matters if the circle is rotated so that AB is horizontal. Note that the line $y = x$ is perpendicular to AB and that the distance of AB from the origin is $\sqrt{2}$. After the rotation occurs, AB is horizontal and both second coordinates of A and B are $\sqrt{2}$, with the first coordinates obtained by substitution in the equation of the circle.

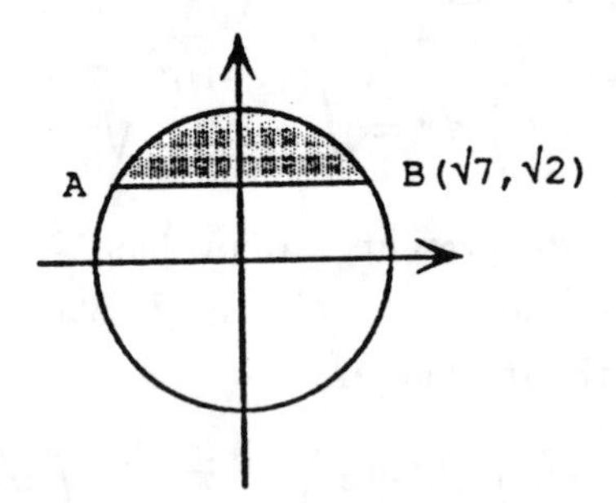

The area of the cross-hatched region is now obtained by calculating a definite integral very much like the integral in the problem before:

$$\text{Area} = 2\int_0^{\sqrt{7}} \sqrt{9 - x^2}\,dx - 2\sqrt{14}$$

where the term $2\sqrt{14}$ is included to eliminate the area of the rectangle under the cross-hatched region. The area is found to be $\sqrt{14} + 9\arcsin(\frac{\sqrt{7}}{3}) - 2\sqrt{14} \approx 5.9776$, and the desired probability is $\dfrac{5.9776}{9\pi} = .2114$.

[MN]

19. This problem will test students' knowledge of polynomials, rational functions, and integration technique, without requiring them to calculate any integrals.

The first two conditions on P require that P be of the form ax^2+1. Since the integral is to be rational, certain ingredients of the partial fraction decomposition of the integrand must be missing:

$$\frac{ax^2+1}{x^3(x-1)^2} = \frac{A}{x^2} + \frac{B}{x^3} + \frac{C}{(x-1)^2} = \frac{Ax(x-1)^2 + B(x-1)^2 + Cx^3}{x^3(x-1)^2}.$$

We analyze, in the usual fashion, the numerators of the left and right-hand sides.

$$\text{Set } x = 0: \qquad B = 1$$

$$\text{Set } x = 1: \qquad C = a+1$$

Since the coefficient of x^3 is 0, we get: $a + C = 0 \Longrightarrow A = -(a+1)$. In like fashion, the coefficient of x is 0: $A - 2B = 0 \Longrightarrow A = 2B = 2$. Equating the two expressions for A in these last expressions allows us to conclude that $a = -3$. A check is provided by looking at the coefficients of x^2.

Section 3: Approximation Techniques and Error Analysis

1. The required graphs are given below. The difficulty with all the definite integrals is that the given functions, each of which is integrable over the given interval, do not have elementary antiderivatives. The Fundamental Theorem cannot be used to calculate the integrals which must be approximated by one of the standard methods.

a.

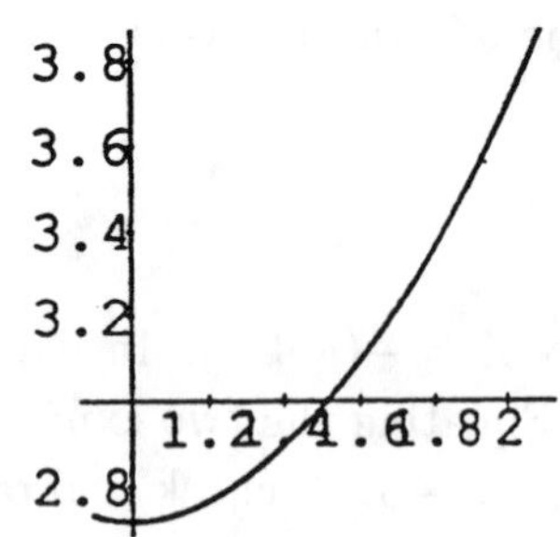

`Plot[Exp[x]/x,{x,0.9,2.1}]`

b.

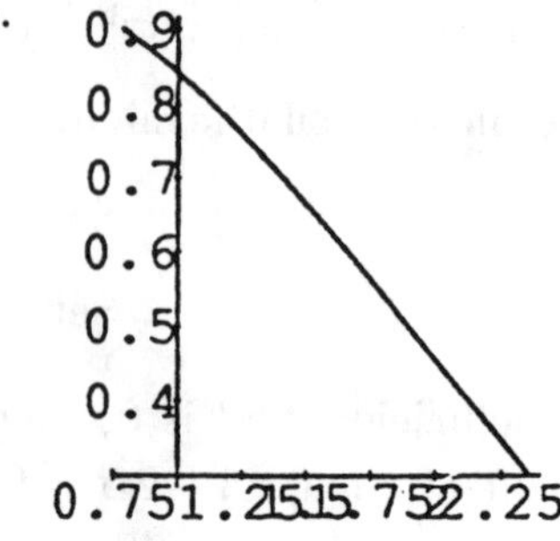

`Plot[Sin[x]/x,{x,Pi/4,3Pi/4}]`

c.

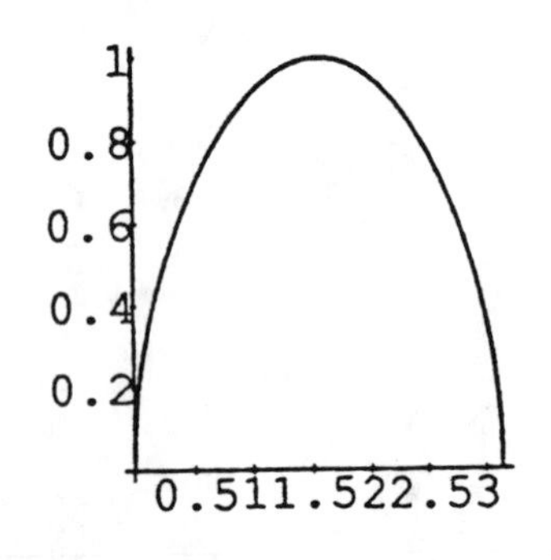

`Plot[Sqrt[Sin[x]],{x,0,Pi}]`

d.

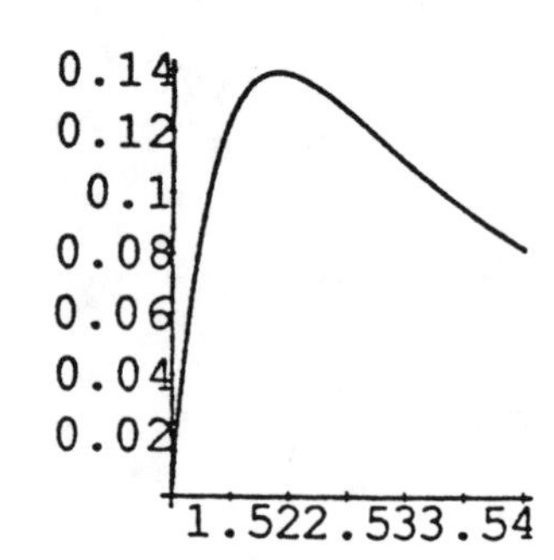

`Plot[Log[x]/(1+x^2),{x,1,4}]`

2. Computer packages come into their own on number crunching exercises like this one. With one such package, the following results were achieved:

n	Trap.	Simp.
4	$.948959\pi$	1.022799π
8	$.987116\pi$	1.000135π
20	$.997943\pi$	π
50	$.999671\pi$	π

The integral is worked by parts:

$$\int x \sin x \, dx = -x \cos x + \int \cos x \, dx = -x \cos x + \sin x + C$$

$$\int_0^\pi x \sin x \, dx = \pi$$

[RH]

3. a. The integrand in the definite integral to be approximated is the function g defined by the rule $g(x) = \pi\big(f(x)\big)^2 = \pi x^2 \sin^2 x$. The required table follows:

n	Trap.	Simp.
20	$1.39492\pi^2$	$1.39498\pi^2$
50	$1.39493\pi^2$	$1.39493\pi^2$
100	$1.39493\pi^2$	$1.39493\pi^2$

b. The integral yields to integration by parts when $\sin^2 x$ is expressed as $\frac{1}{2}(1 - \cos 2x)$:

$$\int_0^\pi \pi x^2 \sin^2 x \, dx = \left[\tfrac{1}{6}x^3 - \left(\tfrac{1}{4}x^2 - \tfrac{1}{8}\right)\sin 2x - \tfrac{1}{4}x \cos 2x\right]\Big|_0^\pi = 1.3949341\pi^2.$$

The best rule of thumb is to pull the plug on the approximation process as soon as you perceive that, beyond a certain point, increasing the number of subdivisions has no significant effect on the answers achieved.

4. This problem continues the investigation of the function $x \sin x$ introduced in the previous two questions, but it is quite independent of them. In part a, students should observe that the integrand is non-negative for $x \in [0,1]$ and for all non-negative integers n. Part *b* is straightforward, and part c prods students to study the shape of x^n for $x \in [0,1]$ and for the given n (the graphs are given below). Students may be asked to push the speculation in part *d* "to the limit," i.e. to look at the value of the integrand for large n. These values will fall only a little short of $\frac{1}{(n+1)}$.

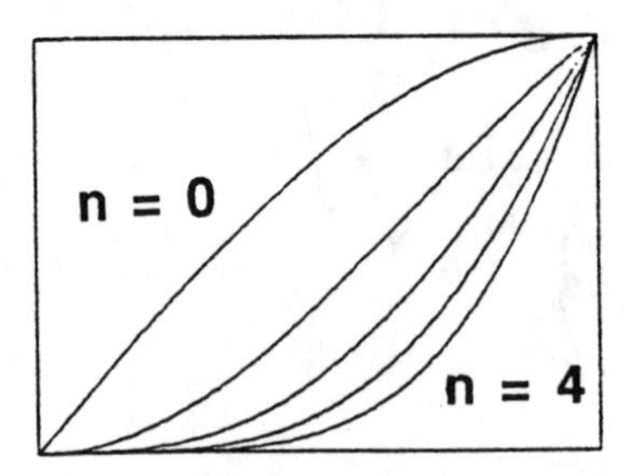

5. In this case, the results obtained are summarized in the table below:

n	Trap.	Simp.
4	7.64382	7.71676
10	7.71348	7.72696
20	7.7239	7.72738
50	7.72686	7.72742
100	7.72727	7.72741

The value of the integral is approximately 7.7274113. As one might expect, Simpson's Rule is better for relatively coarse subdivisions. This problem, like its predecessor, offers instructors with access to a CAS the opportunity to expound on upper bounds for the absolute value of the error due to the Trapezoidal and Simpson's Rules, an explicit consideration of which is being held in reserve (see the next problem, in fact). Students should first form their own opinion of the relative accuracy of the two rules. A CAS is essential in cases where the drudgery of hand-calculating second and/or fourth order derivatives (and how big each of them gets over the interval of integration) daunts both students and teachers.

[RH]

6. a. $I = \ln 3 \approx 1.0986$

b. $T_4 = \left(\frac{1}{2}\right)\left(\frac{1}{2}\right)\left\{\frac{1}{1} + \frac{2}{1.5} + \frac{2}{2} + \frac{2}{2.5} + \frac{1}{3}\right\} \approx 1.1167$

The maximum possible error for T_4 is given by $\dfrac{B(3-1)^3}{(12)(4^2)}$ where B can be taken to be $\frac{2}{1^3} = 2$. Thus the maximum error for T_4 is $\frac{1}{12} \approx .0833$. To achieve a tolerance of .001, we find that $n = 37$ is the least positive integer which satisfies the inequality $\dfrac{(2)(2^3)}{12n^2} < .001$.

c. $S_4 = \left(\frac{1}{2}\right)\left(\frac{1}{3}\right)\left\{\frac{1}{1} + \frac{4}{1.5} + \frac{2}{2} + \frac{4}{2.5} + \frac{1}{3}\right\} = 1.1$

The fourth derivative of the integrand is $24/x^5$, and $\dfrac{24}{x^5} \le 24$ for $x \in [1,3]$. The maximum possible error for S_4 is given by $\dfrac{B(3-1)^5}{(180)(4^4)}$ which becomes $\dfrac{(24)(2^5)}{(180)(4^4)} = \frac{1}{60} \approx .0167$ when B is taken to be 24. To meet the same tolerance as before, we find that $n = 10$ is the least *even* positive integer to satisfy the inequality $\dfrac{(24)(2^5)}{180n^4} < .001$.

7. This problem will dispel the myth that approximations always fall well within the bounds of their theoretical tolerance.

   a. $T_2 = \left(\frac{1}{2}\right)\left(\frac{1}{2}\right)\left\{0 + \frac{2}{2^2} + 1^2\right\} = \frac{3}{8}$. This is too big because the integrand function is concave up. The point of this question, in fact, is to prompt students to think about the graph of the function under consideration and how the approximation of area of a region relates to the shape of that region. The actual error made by T_2 is $\frac{3}{8} - \frac{1}{3} = \frac{1}{24}$. The *maximum* possible error for T_2 is the same: $\dfrac{(2)(1^2)}{(12)(2^2)} = \frac{1}{24}$. You may wish to ask your students to try using a midpoint Riemann sum here, for which the worst-case error estimate is half the estimate for the Trapezoidal Rule with the same number of subdivisions. On this problem, the estimate is

$$\left(\frac{1}{2}\right)\left(\left(\frac{1}{4}\right)^2 + \left(\frac{3}{4}\right)^2\right) = \frac{5}{16}.$$

The error is $\dfrac{5}{16} - \dfrac{1}{3} = -\dfrac{1}{48}$, which again, in absolute value, is the maximum possible error.

   b. Does Simpson perform any better for this integral?

$$S_2 = \left(\tfrac{1}{2}\right)\left(\tfrac{1}{3}\right)\left\{0 + \tfrac{4}{2^4} + 1^4\right\} = \tfrac{5}{24}.$$

   The actual error is $\frac{5}{24} - \frac{1}{4} = \frac{1}{120}$ and this again is the maximum possible error: $\dfrac{(4!)(1^5)}{(180)(2^4)} = \frac{1}{120}$.

8. a. The sketches required are given below. Students with access to a CAS may find it more instructive to obtain these sketches electronically, using a parametric plot command, rather than with pencil and paper.

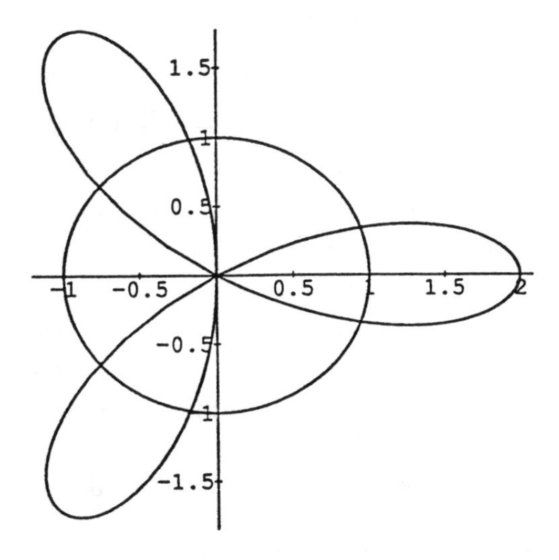

b. Area $= \frac{1}{2}\int_{-\pi/9}^{\pi/9}(4\cos^2 3\theta - 1)\,d\theta = \frac{\pi}{9} + \frac{\sqrt{3}}{6} \approx .637741.$

c. Length $= \int_{-\pi/9}^{\pi/9}\sqrt{4\cos^2 3\theta + 36\sin^2 3\theta\,d\theta} = \int_{-\pi/9}^{\pi/9}\sqrt{4 + 32\sin^2 3\theta}\,d\theta.$ Here we have another elliptic integral arising innocently from a calculation of arc length. You might want to ask your students how they would like to find an antiderivative of the integrand function. While they are thinking that over, a CAS will generate the required approximations:

d.

n	10	20	50	100
approximation	2.44384	2.44406	2.44406	2.44406

9. This problem uses numerical integration to approximate π. The Simpson's Rule approximations called for yield the following results:

n	6	10	20	50
$\int_0^1 4\sqrt{1-x^2}\,dx$	3.11013	3.12701	3.13645	3.14029
$\int_0^1 \frac{4}{1+x^2}\,dx$	3.14159	3.14159	3.14159	3.14159

The secret of success for the second method lies, of course, in the behavior of the derivatives of the function involved. The fourth derivative of the function in part a involves terms with $(1-x^2)$ in the denominator whereas no such difficulties occur with the function in part c. Students without access to a CAS cannot be expected to find the fourth derivatives exactly, but even they should have a qualitative understanding of why the second way of calculating π is superior to the first.

10. The shape of the pool permits us to treat its upper and lower edges as graphs of functions f and g in which case Simpson's Rule can be used to approximate $\int_a^b \{f(x) - g(x)\}\,dx$ where a and b are the points of intersection of the two graphs.

$$\text{Area} \approx \tfrac{3}{3}\{0 + (4)(4.9) + \cdots + (4)(4) + 0\} = 124.7 \text{ (in square feet)}.$$

11. This problem appeared in V.1 as an exercise in Riemann sums. Simpson's Rule can also be used to solve it. Since distance = speed $\times$ time, we can get a ball-park estimate for the distance traveled by the plane

$$\text{Distance (in miles)} = \tfrac{1}{(60)(3)}\{90 + (4)(110) + \cdots + (4)(100) + 70\} \approx 11.3\,\text{miles}.$$

Note that the answer obtained here agrees with the one obtained by approximating the distance as a Riemann sum.

12. The approximation obtained is .6827. The integrand, of course, is the (standard) normal probability density function. The given integral represents the probability that, in a standard normal distribution, the value of the random variable lies between -1 and 1.

13. Having introduced the specter of statistics, we might as well give students the flavor of a statistical query. For $n = 20$, a Simpson's Rule estimate of 0.47 is obtained.

14. a. This can be dispatched elegantly by a CAS, but for students with no more than pencil and paper, it is still a good exercise. The composite quadratic curve through the three points given is parametrized by the two slopes m and n. For $f(x) = ax^2 + bx$, $a = \dfrac{(mX - Y)}{X^2}$ and $b = 2\dfrac{Y}{X} - m$, and $\displaystyle\int_0^X f(x)\,dx = \tfrac{1}{6}X\{4Y - mX\}$. For $g(x) = cx^2 + dx + e$, $c = \dfrac{Z - Y - nX}{X^2}$, $d = n$, $e = Y$, and $\displaystyle\int_0^X g(x)\,dx = \tfrac{1}{6}X\{2Z + 4Y + nX\}$. The area of the total region is $\frac{1}{6}X\{8Y + 2Z + (n - m)X\}$.

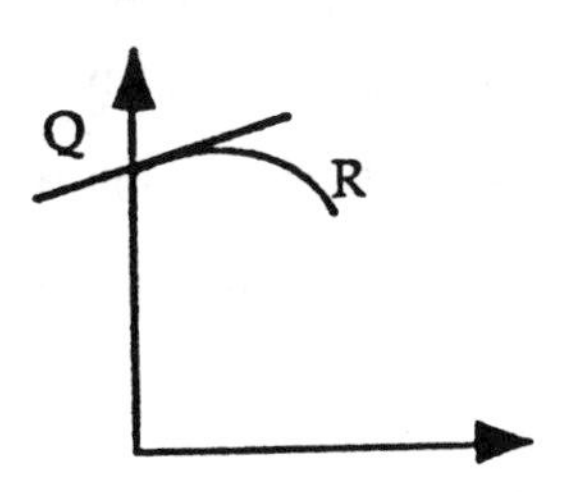

b. If $n = m$, the area becomes $\frac{1}{6}X\{8Y + 2Z\}$ which is the estimate given by Simpson's Rule: $\frac{1}{3}X(0 + 4Y + Z) = \frac{1}{3}X(4Y + Z)$. A moment's reflection should suffice to persuade oneself that if $n = m$, the composite curve is a parabola, and since Simpson uses a parabolic arc to approximate the curve, the answers are the same: A parabola with a vertical axis of symmetry is uniquely determined by three points on its graph.

[MP]

15. One approach to this question is to take $\Delta x = \frac{1}{4}$ and to read the ordinate of each point of intersection of the graph and set of vertical lines over $[0, 4]$. Total amount of oil, in thousands of barrels, produced during the first 4 days is approximately

$$\tfrac{1}{(3)(4)}\{10 + (4)(8.8) + (2)(8.8) + \cdots + (4)(9.3) + 11\} \approx 44.9.$$

This problem, like that of the Simpsons' swimming pool, teaches students that carefully drawn graphs allow them to perform calculations, even when the rule defining a function is unknown.

16. a. Reducing I to a consideration of an integral over $[0, 2\pi]$ is important because it is our intention to approximate I using Simpson's Rule. The error bound is then proportional to the fifth power of the length of the interval over which the integration is carried out. When students have finished with this problem, you might want to ask them to find the value of n which will guarantee an error of less than .001, *without* reducing I.

    b. This is the only part of the problem for which student access to a CAS is psychologicaly (if not literally) necessary. For students without such access, the correctness of the formula of $f^{(4)}(x)$ may be assumed, and students can move on to the rest of the problem, including the upper bound for $|S_n|$.

    c. A hand calculator suffices to produce a value of 124 for n.

    d. $I \approx 795.493$ for $n = 124$. The accuracy is reduced to .001 since our guaranteed accuracy for the coefficient of 100 in the reduction of I is 10^{-5}.

17. This problem leads students through a bread-and-butter exercise in calculating an approximation of a stubborn elliptic integral.

n	Simpson
4	8.6667π
6	8.1101π
8	8.1641π
10	8.1251π
12	8.1305π
14	8.1255π
16	8.1263π
18	8.1255π
20	8.1256π
22	8.1255π
24	8.1255π

18. a. In effect, Simpson's Rule substitutes for the graph of f a sequence of parabolic arcs. If the graph of f is already a parabola, then this substitution yields a carbon copy of the graph of f. If f is linear, then the best-fit parabolic arcs are line segments (the leading coefficients of the polynomials being 0), and again a carbon copy of the graph is obtained. In fact, Simpson's Rule is exact for polynomials of degree 3, regardless of the number of subdivisions of the interval of integration. Students who have been exposed to the formula for the upper bound of the error made by Simpson are aware that the bound involves the fourth derivative of the function whose integral is being approximated. For them, the degree 2 can be strengthened to degree 3.

    b. For n even, $S(f, I, n) = \sum_{m=0}^{n-2} [f(x_m) + 4f(x_{m+1}) + f(x_{m+2})]$,

    and $S(g, I, n) = \sum_{m=0}^{n-2} [g(x_m) + 4g(x_{m+1}) + g(x_{m+2})]$.

    Since the addition of functions is distributive in the sense that $(f + g)(x) = f(x) + g(x)$,

$$S(f + g, I, n) = S(f, I, n) + S(g, I, n).$$

19. Since $\ln 2 = \int_1^2 \frac{dx}{x}$, Simpson's Rule can be used to approximate the integral. For $n = 10$, the approximation is 0.69315, which agrees well with the answer of 0.6931472, given by a hand calculator. Of course, calculating the natural logarithm of any number b can be approximated by Simpson's Rule in much the same manner. For large b, the following ploy avoids the need to subdivide the interval $[1, b]$ many times. The number b can be expressed uniquely as a product of an integral power of 2 and a number c, $b = 2^n c$, where $1 \leq c < 2$. Then $\ln b = n \ln 2 + \ln c$ and the problem of calculating $\ln b$ is reduced to the problem of calculating $\ln c$.

Chapter VII: Sequences and Series of Numbers

Section 1: Sequences of Numbers

1. A rectangle with n by $n+1$ dots is divided by the diagonal into two triangles each with $1+2+\cdots+n$ dots. So $2(1+2+\cdots+n) = n(n+1)$.

[MN]

2. After n days, you have $1+2+\cdots+n = \dfrac{n(n+1)}{2}$ pennies. Choose n so that $\dfrac{n(n+1)}{2} =$ 20,000 (at least). By the quadratic formula, $n = \dfrac{-1 \pm \sqrt{1+160{,}000}}{2} \approx \dfrac{399}{2}$. Rounding up, we see that 200 days are needed. Check by evaluating $\dfrac{n(n+1)}{2}$ for $n = 199$ and $n = 200$.

[MN]

3. All the limits exist. They are:

   a. 1   b. 1   c. e d. 0

   e. 0   f. 0   g. 0

   For part b, you may have to remind your students to put their calculators in radian mode.

4. $\lim\limits_{n\to\infty} a_n = \dfrac{c}{d}$ if $p = q$, and $\lim\limits_{n\to\infty} a_n = 0$ if $p < q$; otherwise $\{a_n\}$ diverges.

5. $$\lim_{n\to\infty} \frac{n^2}{n^2-20}\cdot\frac{1/n^2}{1/n^2} = \lim_{n\to\infty}\frac{1}{1-20/n^2} = 1,$$
   $$\lim_{n\to\infty} \frac{n}{n+1}\cdot\frac{1/n}{1/n} = \lim_{n\to\infty}\frac{1}{1+1/n} = 1, \text{ and}$$
   $$\lim_{n\to\infty} \frac{\sqrt{n}}{\sqrt{n+3}}\cdot\frac{1/\sqrt{n}}{1/\sqrt{n}} = \lim_{n\to\infty}\frac{1}{\sqrt{1+3/n}} = 1.$$
   All three subsequences converge to 1, so $\lim\limits_{n\to\infty} a_n = 1$.

6. $\lim_{n\to\infty}\left(\frac{2n}{n-1}\cdot\frac{1/n}{1/n}\right) = \lim_{n\to\infty}\frac{2}{1-1/n} = 2,$

$\lim_{n\to\infty}\left(\frac{3n}{n+1}\cdot\frac{1/n}{1/n}\right) = \lim_{n\to\infty}\frac{3}{1+1/n} = 3,$

$\lim_{n\to\infty}\left(\frac{n^2}{n^2+10}\cdot\frac{1/n^2}{1/n^2}\right) = \lim_{n\to\infty}\frac{1}{1+10/n^2} = 1.$

Since there are at least two subsequences of $\{a_n\}$ that converge to different limits, the sequence $\{a_n\}$ diverges.

7. a. $0 \quad (x = n)$ b. $1 \quad (x = \frac{1}{n})$ c. $0 \quad (x = n)$

d. $1 \quad (x = n$, apply ln$)$ e. $e \quad (x = \frac{1}{n}$, apply ln$)$ f. $\frac{2}{3} \quad (x = \frac{1}{n})$

8. About 2.85×10^{64}. We could say that the sequence $\{\ln(\ln n)\}$ diverges *slowly*!

9. a. False, as when $a_n = \frac{1}{n}$.

b. True.

c. False, as when $a_n = (-1)^n$.

d. True.

e. False, as when $a_n = -n$.

f. True.

g. False, as when $a_n = \dfrac{(-1)^n}{n}$.

h. True.

10. a. $\dfrac{4^n}{n!} \geq \dfrac{4^{n+1}}{(n+1)!}$ if and only if $(n+1) \geq 4$ or $n \geq 3$. $\lim_{n\to\infty}\dfrac{4^n}{n!} = L$ for some L since the sequence is eventually decreasing and is bounded below by 0.

b. $L = \lim_{n\to\infty}\dfrac{4^{n+1}}{(n+1)!} = \lim_{n\to\infty}\left(\dfrac{4}{n+1}\cdot\dfrac{4^n}{n!}\right) = 0 \cdot L = 0$

[MN]

11. Let a_n denote the nth exam score and A_n the average of $a_1, \ldots, a_n$. Since $a_{n+1} \geq A_n$,

$$\begin{aligned} A_{n+1} &= \frac{1}{n+1}(a_1 + \cdots + a_n + a_{n+1}) \\ &\geq \frac{1}{n+1}(a_1 + \cdots + a_n + A_n) \\ &= \frac{1}{n+1}(nA_n + A_n) = A_n. \end{aligned}$$

Therefore $\{A_n\}$ converges since it is monotonic and bounded above by 100.

12. a. If $a_n \to L$, then taking the limit of both sides of the defining equation yields $L = \frac{1}{2}\left(L + \frac{2}{L}\right)$ from which it follows that $L = \sqrt{2}$.

b. Yes, the terms do get closer to $\sqrt{2} = 1.4142\ldots$, and the value of a_1 does not affect the limit, provided $a_1 > 0$. If $a_1 < 0$, $L = -\sqrt{2}$.

13. $a_n \to \sqrt{B}$ if $a_1 > 0$ and $a_n \to -\sqrt{B}$ if $a_1 < 0$.

14. a. Clearly $a_n < a_{n-1}$ and all terms are bounded below by 0; so $\{a_n\}$ has a limit.

b. The limit is $\frac{1}{2}$. In fact, it can be shown that $a_n = \frac{1}{2}(1 + \frac{1}{n})$.

15. a. If $a_n \to L$, then $L = \sqrt{L}$ and so $L = 0$ or $L = 1$.

b. The sequence always converges to 1.

c. You may have to explain to your students what is meant by a "web diagram". Mark (a_1, a_1) on the line $y = x$, and draw a vertical line through it. Where the line hits the graph of $y = \sqrt{x}$, draw a horizontal line; where this hits $y = x$ is (a_2, a_2). Continue in this fashion. The points (a_n, a_n) will approach $(1, 1)$.

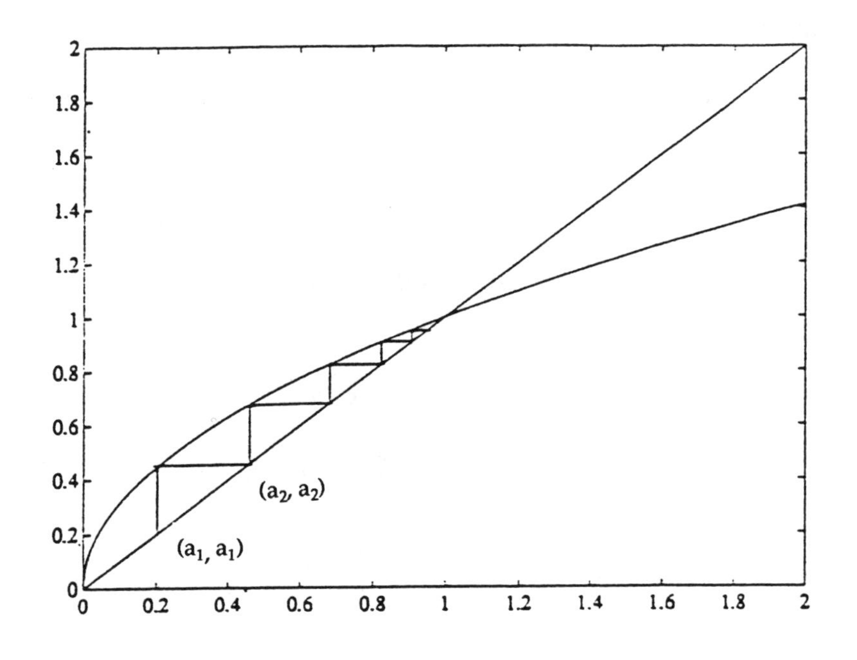

[MN]

16. a. If $a_n \to L$, then $L = 3L^2$ and so $L = 0$ or $L = \frac{1}{3}$.

b. If $0 < a_1 < \frac{1}{3}$, then $a_n \to 0$. If $a_1 = \frac{1}{3}$, then $a_n \to \frac{1}{3}$. Otherwise $a_n \to \infty$.

[MN]

17. a. If $a_n \to L$, then $L = B^L$ and so $B = L^{1/L}$.

b. The sequence converges for at least some values of a_1 when $B = 1.1$ or $B = 1.3$ but not when $B = 1.5$. So the cutoff for B is somewhere between 1.3 and 1.5. When B is less than this cutoff value, the graphs $y = B^x$ and $y = x$ intersect in two points. If a_1 is less than the x-coordinate of the right-hand intersection point, $\{a_n\}$ converges to the x-coordinate of the left-hand intersection point. For $B = 1.1$, $L \approx 1.11$; for $B = 1.3$, $L \approx 1.47$, as may be discovered by computing the first few terms of the sequence.

c. The slopes of $y = B^x$ and $y = x$ are equal when $\frac{d}{dx}B^x = \frac{d}{dx}x$. That is, when $B^x \ln B = 1$. Combining this with $B^x = x$, we see that $x = 1/\ln B$; so $1/\ln B = B^{1/\ln B} = e$, which implies $B = e^{1/e}$. For this value of B, $\{a_n\}$ converges if and only if $a_1 \le e$.

This question appeared in a somewhat different context in II.9.

[MP]

18. a. Divide $F_{n+1} = F_n + F_{n-1}$ by F_n and take the limit:

$$\frac{F_{n+1}}{F_n} = 1 + \frac{F_{n-1}}{F_n} = 1 + \frac{1}{F_n/F_{n-1}},$$
$$L = 1 + \frac{1}{L},$$
$$L^2 = L + 1.$$

L is the larger root.

b. Clearly true for $n = 1$. Assume true for some n. Then

$$\begin{aligned} x^{n+1} = xx^n &= x(xF_n + F_{n-1}) \\ &= (x+1)F_n + xF_{n-1} \\ &= x(F_n + F_{n-1}) + F_n \\ &= xF_{n+1} + F_n. \end{aligned}$$

c. $y^n - z^n = (yF_n + F_{n-1}) - (zF_n + F_{n-1}) = (y - z)F_n = \sqrt{5}\,F_n$ from which the conclusion follows.

d. $$\frac{F_{n+1}}{F_n} = \frac{\left(\frac{1+\sqrt{5}}{2}\right)^{n+1} - \left(\frac{1-\sqrt{5}}{2}\right)^{n+1}}{\left(\frac{1+\sqrt{5}}{2}\right)^{n} - \left(\frac{1-\sqrt{5}}{2}\right)^{n}} = \frac{\frac{1+\sqrt{5}}{2} - \left(\frac{1-\sqrt{5}}{1+\sqrt{5}}\right)^{n}\frac{(1-\sqrt{5})}{2}}{1 - \left(\frac{1-\sqrt{5}}{1+\sqrt{5}}\right)^{n}}$$
$$\longrightarrow \frac{1+\sqrt{5}}{2}.$$

[MN]

Section 2: Series of Numbers. Geometric Series

1. a. Yes. $\sum_{k=1}^{\infty} a_k = \lim_{n\to\infty} s_n = \lim_{n\to\infty}\left(5 - \frac{3}{n}\right) = 5$

   b. $\lim_{k\to\infty} a_k = 0$ for any convergent series.

   c. $s_{100} = 5 - \frac{3}{100} = 4.97$.

2. The series diverges since the partial sums are alternately 0 and 1.

3. The series converges since the partial sums are either 1 or $1 - \frac{1}{n}$.

4. $\sum_{k=1}^{n}\left(\frac{1}{\sqrt{k+1}} - \frac{1}{\sqrt{k+3}}\right) = \frac{1}{\sqrt{2}} + \frac{1}{\sqrt{3}} - \frac{1}{\sqrt{n+2}} - \frac{1}{\sqrt{n+3}}$ which converges to $\frac{1}{\sqrt{2}} + \frac{1}{\sqrt{3}}$.

5. $\sum_{k=1}^{n} \frac{1}{k^2+2k} = \sum_{k=1}^{n} \frac{1}{2}\left(\frac{1}{k} - \frac{1}{k+2}\right) = \frac{1}{2}\left(1 + \frac{1}{2} - \frac{1}{n+1} - \frac{1}{n+2}\right)$ which converges to $\frac{3}{4}$.

6. a. $\sum_{k=1}^{n}\left(\arctan(k+1) - \arctan k\right) = \arctan(n+1) - \arctan 1 \longrightarrow \frac{\pi}{2} - \frac{\pi}{4} = \frac{\pi}{4}$.

   b. $\sum_{k=1}^{n}\left(\ln(k+1) - \ln k\right) = \ln(n+1) - \ln 1 \longrightarrow \infty$.

7. a. $\frac{1}{1 - \sin x}$ when $x \neq \pi(n + 1/2)$, n any integer.

   b. $\frac{1}{1-x^2} - 1$ when $|x| < 1$.

   c. $\frac{1}{1 - \frac{1}{x}} - 1 - \frac{1}{x}$ when $|x| > 1$.

   d. $\frac{1}{1-(1-x)} = \frac{1}{x}$ when $|1-x| > 1$.

8. a. $\lim_{x\to 1^-} \frac{1}{1+x} = \frac{1}{2}$.

   b. No, its partial sums are alternately 0 and 1.

9. The ball falls 4 feet, then bounces up $\left(\frac{2}{3}\right)\cdot 4$ feet and falls the same distance, then bounces up $\left(\frac{2}{3}\right)^2\cdot 4$ feet and falls the same distance, and so on. The total distance is $4+2\cdot\left(\frac{2}{3}\right)\cdot 4+2\cdot\left(\frac{2}{3}\right)^2\cdot 4+2\cdot\left(\frac{2}{3}\right)^3\cdot 4+\cdots = 4+8\cdot\frac{2}{3}\left(1+\frac{2}{3}+\left(\frac{2}{3}\right)^2+\cdots\right) = 4+\frac{16}{3}\cdot\dfrac{1}{1-\frac{2}{3}}$ $= 20$ feet.

10. The fly and the further cyclist meet when the fly has flown 36 miles and the cyclist has cycled 24 miles, since $\frac{36}{15}=\frac{24}{10}$. That is, the fly has flown $\frac{3}{5}$ of the distance between them, and now the cyclists are $\frac{1}{5}\cdot 60$ miles apart since each has traveled $\frac{2}{5}$ of the distance. Again, the fly flies $\frac{3}{5}$ of this distance, which leaves the cyclists $\frac{1}{5^2}\cdot 60$ miles apart; and so on. In total, the fly flies

$$\frac{3}{5}\cdot 60+\frac{3}{5}\cdot\frac{1}{5}\cdot 60+\frac{3}{5}\cdot\frac{1}{5^2}\cdot 60+\frac{3}{5}\cdot\frac{1}{5^3}\cdot 60+\cdots = \frac{3}{5}\cdot 60\left(1+\frac{1}{5}+\frac{1}{5^2}+\frac{1}{5^3}+\cdots\right)$$
$$= \frac{3}{5}\cdot 60\cdot\frac{1}{1-\frac{1}{5}} = \frac{3}{5}\cdot 60\cdot\frac{5}{4} = 45 \text{ miles.}$$

There is an anecdote associated with this problem. According to one version of the story (which is quite possibly apocryphal), a prankster goes about asking colleagues to solve this problem. When he gets an answer, he asks how it has been obtained. Everyone replies that it takes three hours for the bikes to meet, during which time the fly has flown 45 miles. The prankster then says: "That shows that you're not a mathematician; a mathematician would have calculated the series." This goes on until the prankster comes to John von Neumann, a mathematician renowned for his ability to do complex calculations in his head, quickly and accurately. Von Neumann thinks about the question briefly, then gives the correct answer. Before the prankster has a chance to spring his joke, von Neumann leans back in his chair and says: "It was a very simple series."

11. This, of course, is one of Zeno's paradoxes, designed to support the view of the world held by Paramenides that the universe is an eternal and unchanging unity in which motion and change are illusions. Simple algebra shows that the hare catches up with the tortoise in 10/9 of a mile. The problem is worded so that this interval of distance is subdivided into infinitely many subintervals, as is the corresponding time interval. Specifically, the distance traveled by the hare to catch up with the tortoise is $1+\dfrac{1}{10}+\dfrac{1}{10^2}+\dfrac{1}{10^3}+\cdots = \dfrac{1}{1-\frac{1}{10}} = \dfrac{10}{9}$.

12. $\int_0^1 f(x)\,dx =$ sum of areas of triangles with height 1 and bases $\frac{1}{2}, \frac{1}{4}, \frac{1}{8}, \cdots$

$= \frac{1}{2}\left(\frac{1}{2} + \frac{1}{4} + \frac{1}{8} + \cdots\right) = \frac{1}{4} \cdot \dfrac{1}{1-\frac{1}{2}} = \frac{1}{2}.$

Alternatively, the answer can be found in a way analogous to the "non-series" solution of the fly-and-cyclist problem. Note that the area of each triangle is half of the area of the rectangle, bounded by $y = 0$ and $y = 1$, one of whose sides is the base of the triangle on the x-axis (and whose height is the height of the triangle). The union of all the triangles occupies half the square $[0,1] \times [0,1]$. Thus the value of the integral is $\dfrac{1}{2}$.

[MN]

13. a. This is an invitation to students to try an "experimental" approach to mathematics in which a conjecture is developed from examples before it is conclusively established.

b. $a_2 - a_1 = \frac{1}{2}a_1 + \frac{1}{2}a_0 - a_1 = -\frac{1}{2}(a_1 - a_0)$

$a_3 - a_2 = \frac{1}{2}a_2 + \frac{1}{2}a_1 - a_2 = -\frac{1}{2}(a_2 - a_1) = \dfrac{1}{2^2}(a_1 - a_0)$

$a_4 - a_3 = \frac{1}{2}a_3 + \frac{1}{2}a_2 - a_3 = -\frac{1}{2}(a_3 - a_2) = -\dfrac{1}{2^3}(a_1 - a_0)$

$\vdots \qquad \vdots$

$a_k - a_{k-1} = \dfrac{(-1)^{k-1}}{2^{k-1}}(a_1 - a_0)$

So $a_n = (a_n - a_{n-1}) + (a_{n-1} - a_{n-2}) + \cdots + (a_1 - a_0) + a_0 = a_0 + \sum_{k=1}^{n} \dfrac{(-1)^{k-1}}{2^{k-1}}(a_1 - a_0)$.

So $\lim_{n\to\infty} a_n = a_0 + (a_1 - a_0)\sum_{k=1}^{\infty} \dfrac{(-1)^{k-1}}{2^{k-1}} = a_0 + (a_1 - a_0) \cdot \dfrac{1}{1+\frac{1}{2}} = \frac{1}{3}a_0 + \frac{2}{3}a_1.$

[MN]

14. a. $S_n = 3 \cdot 4^n, \quad L_n = 1/3^n.$

b. $P_n = S_n \cdot L_n = 3 \cdot \left(\frac{4}{3}\right)^n \longrightarrow \infty.$

c. In going from T_n to T_{n+1}, we add S_n equilateral triangles with side length L_{n+1}. The area of an equilateral triangle with side length s is $\sqrt{3}s^2/4$. So the area added is $3 \cdot 4^n \cdot \dfrac{\sqrt{3}}{4}\left(\dfrac{1}{3^{n+1}}\right)^2 = \dfrac{\sqrt{3}}{12}\left(\dfrac{4}{9}\right)^n.$

d.
$$\begin{aligned}
A_n &= A_0 + \sum_{k=1}^{n}(A_k - A_{k-1}) \\
&= \frac{\sqrt{3}}{4} + \sum_{k=1}^{n} \frac{\sqrt{3}}{12}\left(\frac{4}{9}\right)^{k-1} \\
&= \frac{\sqrt{3}}{4} + \frac{\sqrt{3}}{12} \cdot \frac{1-\left(\frac{4}{9}\right)^n}{1-\frac{4}{9}} \\
&= \frac{\sqrt{3}}{4} + \frac{3\sqrt{3}}{20}\left(1-\left(\tfrac{4}{9}\right)^n\right) \longrightarrow \frac{\sqrt{3}}{4} + \frac{3\sqrt{3}}{20} = \tfrac{2}{5}\sqrt{3}.
\end{aligned}$$

So the Koch Snowflake has infinite length but encloses a region of finite area!

[MN]

15. a. $V_n = \displaystyle\int_{\pi(n-1)}^{\pi n} \pi e^{-2cx} \sin^2 x \, dx$ for $n \geq 1$.

b. $\dfrac{V_{n+1}}{V_n} = \dfrac{e^{-2c\pi(n+1)}}{e^{-2c\pi n}} = e^{-2c\pi}.$

c. Let $r = e^{-2c\pi}$, the ratio in b. Then:

$$\begin{aligned}
\text{Total Volume} &= V_1 + V_2 + V_3 + \cdots + V_n + \cdots \\
&= V_1 + rV_1 + r^2V_1 + \cdots + r^{n-1}V_1 + \cdots \\
&= \frac{V_1}{1-r} \\
&= \frac{\pi e^{-2c\pi}}{c(c^2+1)} \cdot \frac{e^{2c\pi}-1}{1-e^{-2c\pi}} \\
&= \frac{\pi}{c(c^2+1)}.
\end{aligned}$$

[MP]

16. a. True.

b. False, as when $a_n = \frac{1}{n}$.

c. False. The series diverges whenever $|x| \geq 1$.

d. False, as when $a_n = \frac{1}{n}$ and $b_n = -\frac{1}{n}$.

e. True. If $\sum(a_n + b_n)$ were to converge, then $\sum b_n = \sum[(a_n + b_n) - a_n]$ would converge.

f. False, as when $a_n = 1$ for all n. The *sequence* $\{a_n\}$ does converge.

g. False, as when $a_n = (-1)^n$ and the partial sums are all 0 and 1 (or -1 and 0, depending on the initial value of n).

h. True, since the partial sums are then monotonically increasing and bounded. Compare with part f.

i. False for almost all series. For example, try $a_n = b_n = 1/2^n$.

Section 3: Convergence Tests: Positive Series

1. Draw rectangles with bases along the intervals $[1,2]$, $[2,3]$, ..., $[n-1,n]$ and corresponding heights $f(2)$, $f(3)$, ..., $f(n)$. This yields the first of the following two inequalities

$$\sum_{k=2}^{n} a_k \le \int_1^n f(x)\,dx \le \sum_{k=1}^{n-1} a_k.$$

Then change the heights to $f(1)$, $f(2)$, ..., $f(n-1)$ for the second.

2. a. $\ln(n+1) = \int_1^{n+1} \frac{1}{x}\,dx \le \sum_{k=1}^{n} \frac{1}{k} = s_n.$ $\quad \ln n = \int_1^n \frac{1}{x}\,dx \ge \sum_{k=2}^{n} \frac{1}{k} = s_n - 1.$

   b. If $s_n > 10$, then $\ln n > 9$ and so $n > e^9 > 8000$.

   c. $\frac{\ln(n+1)}{\ln n} < \frac{s_n}{\ln n} < \frac{1}{\ln n} + 1$. Apply L'Hôpital's Rule and the squeeze principle.

   d. $(s_n - \ln n) - \big(s_{n+1} - \ln(n+1)\big) = \ln(n+1) - \ln n - \frac{1}{n+1} = \int_n^{n+1} \frac{1}{x}\,dx - \frac{1}{n+1} > 0.$

   e. By part *d* the sequence is decreasing, and by part a the sequence is bounded below by 0. So it has a limit.

3. Let the two ends of the top book have coordinates 0 and 1 (taking 1 to be the length of a book and taking the positive direction to be the direction in which the lower books are shifted relative to the upper books). Let c_n be the center of mass of the top n books; thus $c_1 = \frac{1}{2}$. Place the $(n+1)$st book beneath these with its ends at c_n and $c_n + 1$, so its center of mass is at $c_n + \frac{1}{2}$. Therefore

$$c_{n+1} = \frac{nc_n + 1 \cdot (c_n + \frac{1}{2})}{n+1} = c_n + \frac{\frac{1}{2}}{n+1}$$

and so $c_n = \frac{1}{2} \sum_{k=1}^{n} \frac{1}{k} \longrightarrow \infty$.

4. Since $\sin x \le x$ for positive x, $\sin(a_n) \le a_n$. Also, since $a_n \longrightarrow 0$ and $a_n \ge 0$, $\sin(a_n) \ge 0$ eventually. Therefore $\sum \sin(a_n)$ converges.

5. This result was actually known to Euler a century before the theorem on the number of primes was proved. The classical derivation involves infinite products and unique factorization. If one uses the approach suggested in the problem, $\sum \frac{1}{n \ln n}$ diverges by the integral test. So by the limit comparison test, $\sum \frac{1}{p_n}$ also diverges.

6. Most students are likely to believe that there are "more" integers that can be written without 0 than there are primes, but this exercise shows that this is false. There are 9 one-digit numbers that can be written without using 0. The sum of their reciprocals is less than 9. There are 81 two-digit numbers that can be written without using 0. Each of these is more than 10, so each reciprocal is less than $\frac{1}{10}$ and their sum is less than 81/10. Similarily the sum of the reciprocals of the three-digit numbers that can be written without 0 is less than $\frac{9^3}{100}$, and so forth. Adding all these up, we have a geometric series with ratio $\frac{9}{10}$, hence a convergent geometric series. In fact, the sum of this estimate is 90, and thus the sum of the reciprocals of the integers that can be written without 0 converges to a sum less than 90.

7. Such a series converges by comparison with a geometric series: $0 \le \frac{d_k}{10^k} \le \frac{9}{10^k}$.

8. a. Draw rectangles with bases along the intervals $[n, n+1]$, $[n+1, n+2]$, ... and corresponding heights $f(n+1)$, $f(n+2)$, ... This yields the first of the two inequalities. Then change the heights to $f(n)$, $f(n+1)$, ... for the second.

   b. This is just a rewriting of part a. It will be used three problems later.

9. a. For $n = 10^4$, $R_n \le \int_n^\infty \frac{1}{x^2}\,dx \le .0001$. $S_n \approx 1.6448$.

   b. For $n = 71$, $R_n \le \sum_{k=n+1}^{\infty} \frac{1}{k^3} \le \int_n^\infty \frac{1}{x^3}\,dx \le .0001$. $S_n \approx .5071$.

   c. If $n = 25$, $R_n \le \sum_{k=n+1}^{\infty} \left(\frac{2}{3}\right)^k = 2 \cdot \left(\frac{2}{3}\right)^n \le .0001$. $S_n \approx 1.7751$.

   d. For $n = 170$, $R_n \le \int_n^\infty \frac{\ln x}{x^3}\,dx = \frac{2\ln n + 1}{4n^2} \le .0001$. $S_n \approx .1980$.

   e. For $n = 25$, $R_n \le \sum_{k=n+1}^{\infty} \frac{\left(\frac{4}{3}\right)^k}{2^k} = \sum_{k=n+1}^{\infty} \left(\frac{2}{3}\right)^k = 2 \cdot \left(\frac{2}{3}\right)^n \le .0001$. $S_n \approx 6.0000$.

10. b. $e - \sum_{k=0}^{n} \frac{1}{k!} = \sum_{k=n+1}^{\infty} \frac{1}{k!} \le \sum_{k=n+1}^{\infty} \frac{1}{2^k} = \frac{1}{2^n} = 3/over24$ when $n = 3$.

So $1 + 1 + \frac{1}{2!} + \frac{1}{3!} < e < 1 + 1 + \frac{1}{2!} + \frac{1}{3!} + \frac{3}{24}$.

11. a. For $n = 100$, $|a_n| \le .0001$. $S_n + \int_n^\infty \frac{1}{x^2}\,dx \approx 1.6450$.

b. For $n = 33$, $|a_n| \le .0001$. $S_n + \int_n^\infty \frac{\ln x}{x^3}\,dx \approx .1982$.

c. For $n = 306$, $|a_n| \le .0001$. $S_n + \int_n^\infty \frac{1}{x(\ln x)^2}\,dx \approx 2.1098$.

12. a. $\sum_{k=1}^{\infty} b_k = \sum_{k=1}^{\infty} \left(\frac{1}{k} - \frac{1}{k+1}\right) = 1$. $a_k - b_k = \frac{1}{k^2} - \frac{1}{k(k+1)} = \frac{1}{k^2(k+1)} \le \frac{1}{k^3}$.

For $n = 71$, $R_n \le \sum_{k=n+1}^{\infty} \frac{1}{k^3} \le \int_n^\infty \frac{1}{x^3}\,dx \le .0001$.

So $\sum_{k=1}^{\infty} \frac{1}{k^2} \approx 1 + \sum_{k=1}^{n} (a_k - b_k) \approx 1.6448$.

b. $\sum_{k=1}^{\infty} b_k = 1$. $a_k - b_k = \frac{k}{k^3+1} - \frac{1}{k(k+1)} = \frac{k-1}{k(k+1)(k^2-k+1)} \le \frac{1}{(k-1)^3}$.

For $n = 72$, $R_n \le \sum_{k=n+1}^{\infty} \frac{1}{(k-1)^3} \le \int_n^\infty \frac{1}{(x-1)^3}\,dx \le .0001$.

So $\sum_{k=1}^{\infty} \frac{k}{k^3+1} \approx 1 + \sum_{k=1}^{n} (a_k - b_k) \approx 1.1115$.

c. $\sum_{k=1}^{\infty} b_k = \sum_{k=1}^{\infty} \left(\frac{\frac{1}{2}}{k} - \frac{1}{k+1} + \frac{\frac{1}{2}}{k+2}\right) = \frac{1}{4}$.

$a_k - b_k = \frac{1}{k^3} - \frac{1}{k(k+1)(k+2)} = \frac{3k+2}{k^3(k+1)(k+2)} \le \frac{3}{k^4}$.

For $n = 22$, $R_n \le \sum_{k=n+1}^{\infty} \frac{3}{k^4} \le \int_n^\infty \frac{3}{x^4}\,dx \le .0001$.

So $\sum_{k=1}^{\infty} \frac{1}{k^3} \approx \frac{1}{4} + \sum_{k=1}^{n} (a_k - b_k) \approx 1.2020$.

Section 4: Convergence Tests: All Series

1. $a_{n+1}/a_n = 1$ or $\frac{1}{2}$, alternately. So $\{a_{n+1}/a_n\}$ has no limit and is not bounded above by a number less than 1; thus, the ratio test is inconclusive. But the series is just $2\sum_{k=1}^{\infty} 1/2^k$, which is a geometric series converging to 2.

2. No, the terms go to 1, not 0. Consider asking your students whether $\sum(1 - 2^{-1/n})$ converges or diverges. (It diverges because $\lim_{n\to\infty} \frac{1-2^{-1/n}}{\frac{1}{n}} = \ln 2$ and the harmonic series $\sum \frac{1}{n}$ diverges.)

3. a. $a_n = \dfrac{1/2^n + 1}{1/2^n + 3} \longrightarrow \frac{1}{3}$.

   b. No. The terms do not go to 0.

   c. No. The terms do not go to 0.

4. a. $a_n = \dfrac{1/n + 1/n^2 2^n}{1 + 1/n^2 2^n} \longrightarrow 0$.

   b. No. Compare with the divergent series $\sum \frac{1}{n}$.

   c. Yes. The terms decrease in absolute value and have limit zero.

5. No. In absolute value, the terms are $\dfrac{n}{2n+1}$ which have limit $\frac{1}{2}$ not 0.

6. The terms do not decrease in absolute value but the partial sums are either 0 or equal the nth term (which tends to zero). The series therefore converges to 0.

7. a. False, as when $a_n = \frac{1}{2^n}$ for n odd and $\frac{1}{3^n}$ for n even.

   b. True.

   c. False, as when $a_n = \frac{(-1)^n}{n}$.

   d. False, as when $a_n = \frac{1}{n}$ for n odd and $\frac{1}{n^2}$ for n even.

   e. True.

   f. False, as when $a_n = \frac{(-1)^n}{\ln n}$.

   g. False, as when $a_n = \frac{(-1)^n}{\sqrt{n}}$.

   h. True. Since $a_n \to 0$, $a_n \leq 1$ for n sufficiently large; so $0 \leq a_n^2 \leq a_n$ for n large.

   i. False, as when $a_n = -n$ and $b_n = 0$.

   j. True.

   k. True by the limit comparison test.

   l. False, as when $a_n = \frac{1}{n^2}$ and $b_n = \frac{1}{n}$. (Thus remind students to beware of the limit 0 for a_n/b_n.)

   m. False, as when $a_n = (-1)^n 2^n$. (Thus remind students not to forget absolute value signs in the ratio test.)

8. A convergent alternating series $\sum (-1)^k a_k$, where $a_k \geq a_{k+1}$, satisfies

$$\left| \sum_{k=n+1}^{\infty} (-1)^k a_k \right| \leq a_{n+1}.$$

Therefore

$$\left| \ln(1+x) - \sum_{k=1}^{n} \frac{(-1)^{k+1} x^k}{k} \right| = \left| \sum_{k=n+1}^{\infty} \frac{(-1)^{k+1} x^k}{k} \right| \leq \frac{|x|^{n+1}}{n+1} \leq \frac{1}{n+1}$$

and so by letting $x \longrightarrow 1^-$, $\left| \ln 2 - \sum_{k=1}^{n} \frac{(-1)^{k+1}}{k} \right| \leq \frac{1}{n+1}$. Letting $n \longrightarrow \infty$ yields the desired result.

9. $$1 - \frac{1}{2} + \frac{1}{3} - \frac{1}{4} + \frac{1}{5} - \frac{1}{6} + \frac{1}{7} - \frac{1}{8} + \cdots$$
$$+\,0 + \frac{1}{2} + 0 - \frac{1}{4} + 0 + \frac{1}{6} + 0 - \frac{1}{8} + \cdots$$
$$= 1 + 0 + \frac{1}{3} - \frac{1}{2} + \frac{1}{5} + 0 + \frac{1}{7} - \frac{1}{4} + \cdots$$

It does not follow that $\ln 2 = \frac{3}{2}\ln 2$. A conditionally convergent series can have a different sum when its terms are added in a different order. Notice that in the series for $\frac{3}{2}\ln 2$, two positive terms precede each negative term.

10. Use the fact that $\lim\limits_{x\to 0} \dfrac{\sin x}{x} = 1$.

a. $\sum \sin \frac{1}{n}$ diverges by comparison with $\sum \frac{1}{n}$.

b. $\sum \frac{1}{n} \sin \frac{1}{n}$ converges by comparison with $\sum \dfrac{1}{n^2}$.

11. Since $\sum a_n$ converges, $a_n \longrightarrow 0$. So $|a_n| < 1$ for n sufficiently large, and $|a_n x^n| \le |x|^n$ for n large. Since $\sum |x|^n$ is a convergent geometric series, $\sum a_n x^n$ converges absolutely.

12. a. For $n = 4999$, $a_{n+1} \le .0001$. $S_n \approx .7853$.

b. For $n = 1382$, $a_{n+1} \le .0001$. $S_n \approx .5265$.

13. $$\left|\sum_{k=1}^{\infty}(-1)^k a_k - \left(\sum_{k=1}^{n}(-1)^k a_k + \frac{(-1)^{n+1}a_{n+1}}{2}\right)\right| = \left|\sum_{k=n+1}^{\infty}(-1)^k a_k - \frac{(-1)^{n+1}a_{n+1}}{2}\right|$$
$$= \frac{a_{n+1}}{2} - a_{n+2} + a_{n+3} - a_{n+4} + a_{n+5} - \cdots$$
$$= \tfrac{1}{2}(a_{n+1} - a_{n+2}) - \tfrac{1}{2}(a_{n+2} - a_{n+3}) + \tfrac{1}{2}(a_{n+3} - a_{n+4}) - \cdots$$
$$\le \tfrac{1}{2}(a_{n+1} - a_{n+2}),$$

where the inequality is exactly the one used in the preceding problem but now for the series $\sum(-1)^k(a_k - a_{k+1})$.

14. a. For $n = 49$, $\frac{1}{2}|a_{n+1} - a_{n+2}| \le .0001$. $S_n + \dfrac{(-1)^{n+1}a_{n+1}}{2} \approx .7854$.

b. For $n = 183$, $\frac{1}{2}|a_{n+1} - a_{n+2}| \le .0001$. $S_n + \dfrac{(-1)^{n+1}a_{n+1}}{2} \approx .9242$.

Section 5: Newton's Method

1. We use the tangent line to the graph of f at $x_0 = 2$ to find x_1: $2x_1 - 3 = 0$ implies $x_1 = \frac{3}{2}$.

2. a. (i) $x_1 \approx -1.8, \quad x_2 \approx -1.6.$

   (ii) $x_1 \approx 2.3, \quad x_2 \approx -1.$

   (iii) x_1 does not exist.

   b. For $x_0 = 0$ and $x_0 = 2$, $f'(x_0) = 0$ and so the tangent line never intersects the x-axis to produce x_1.

   c. $-2.5 \le x_0 \le \text{root} \approx -1.6$. The x_n move to the right but stay to the left of the root. (There are many other intervals besides this one, such as a small interval to the right of the minimum, but these are hard to identify.)

3. a. $x_{n+1} = x_n - \dfrac{x_n^{1/3}}{\frac{1}{3}x_n^{-2/3}} = x_n - 3x_n = -2x_n$. So the x_n keep doubling in magnitude (thus moving rapidly away from the root at $x = 0$) and changing signs.

   b. A graph shows the same behavior quite dramatically, although the precise doubling will not be obvious.

[MN]

4. $x_0 > 0$ leads to $\sqrt{2}$, $x_0 < 0$ leads to $-\sqrt{2}$, and $x_0 = 0$ leads nowhere since $f'(0) = 0$.

5. Small changes in x_0 can cause large changes in outcomes! In fact, observing the very different paths the various values of x_0 take is more interesting than the specific roots the paths converge to. This is easiest to appreciate with a software package that displays the paths on top of the graph of the function.

   a. 1.05 leads to the rightmost root; at the critical point 1, the method fails; 0.95 leads to the leftmost root; .911 and .91 both lead to $x_1 \approx -1$, the other critical point, and then go on to yield dramatically large (but dramatically different) values of x_2; .8893406567 is almost a period-2 point, making the method nearly circular for a while; 0.85 leads to the middle root. The roots are approximately -1.87939, 0.34730, 1.53209.

   b. Avoid critical points; choose x_0 close to the desired root by looking at the graph of f; choose x_0 so that $\{x_n\}$ forms a monotonic sequence, again judging from the graph.

[MN]

6. For $f(x) = x^3 - 3x + 1$, compute $F(x) = x - \dfrac{f(x)}{f'(x)}$. Then solve $F\big(F(x)\big) = x$ but reject those x for which $F(x) = x$.

7. a. (i) $x_1 = 0$ and x_2 is undefined.

   (ii) $x_2 = -\frac{1}{\sqrt{3}}$, $x_3 = 1/\sqrt{3}$, and so on. That is, $\frac{1}{\sqrt{3}}$ is a period-2 point.

   (iii) The x_n's get closer to 0 for a while, then get large suddenly, then again get closer to 0, and so on.

   b. $x^2 + 1 = 0$ has no real roots and no x_0 leads to a convergent sequence. Beware of applying Newton's method where it cannot succeed!

8. $x_{n+1} = x_n - \dfrac{x_n^2 - A}{2x_n} = x_n - \frac{1}{2}x_n + \dfrac{1}{2}\dfrac{A}{x_n} = \dfrac{1}{2}\left(x_n + \dfrac{A}{x_n}\right).$

9. $x_{n+1} = x_n - \dfrac{1/x_n - A}{-x_n^{-2}} = x_n + x_n - Ax_n^2 = x_n(2 - Ax_n).$

10. $y - f(x_1) = \dfrac{f(x_1) - f(x_0)}{x_1 - x_0}(x - x_1)$. When $y = 0$, $x = x_2 = x_1 - \dfrac{f(x_1)(x_1 - x_0)}{f(x_1) - f(x_0)}$.

11. Solve $f(r) = 0$, where $f(r) = \left(6000 - \dfrac{150}{r}\right)(1+r)^{48} + \dfrac{150}{r}$, by Newton's method. The answer is $r \approx 0.0077 = 0.77\%$.

12. The rightmost point on the graph of $y = a\cosh\left(\frac{x}{a}\right)$ has the coordinates $(50, a + 12)$, since the lowest point is $(0, a)$. So $a + 12 = a\cosh\left(\frac{50}{a}\right)$. Apply Newton's method to $f(a) = a + 12 - a\cosh\left(\frac{50}{a}\right)$. The answer is $a \approx 106.1$.

13. If r is the radius and h the height, then the total cost C is
$$C = 60(\pi r^2 + 2\pi r h) + 10(h + 2\pi r).$$
Since $\pi r^2 h = 1$, $h = \dfrac{1}{\pi r^2}$ and $C = 60\left(\pi r^2 + \dfrac{2}{r}\right) + 10\left(\dfrac{1}{\pi r^2} + 2\pi r\right)$.

$0 = \dfrac{dC}{dr} = 60\left(2\pi r - \dfrac{2}{r^2}\right) + 10\left(2\pi - \dfrac{2}{\pi r^3}\right)$ or $0 = 6\pi(\pi r^4 - r) + (\pi^2 r^3 - 1)$. Solve by Newton's method. Answer: $r \approx 0.65\,\text{ft}$, $h \approx 0.75\,\text{ft}$.

Section 6: Improper Integrals

1. a. Not all textbooks discuss these analogues of series tests. They are useful reinforcements. Suppose $0 \le f(x) \le g(x)$ whenever $x \ge a$. If $\int_a^\infty g(x)\,dx$ converges, so does $\int_a^\infty f(x)\,dx$, and if $\int_a^\infty f(x)\,dx$ diverges, so does $\int_a^\infty g(x)\,dx$. Another test that might not come as quickly to mind is the following: If $\int_a^\infty |f(x)|\,dx$ converges, so does $\int_a^\infty f(x)\,dx$. There is also a limit comparison test.

   b. Since $f \le g$, the area under the graph of f is less than or equal to the area under the graph of g. Thus, if the latter area is finite, so is the former.

2. a. Converges. Compare with $\dfrac{1}{x^4}$.

   b. Diverges. Compare with $\dfrac{1}{x^{1/2}}$.

   c. Converges. Compare with e^{-x}.

   d. Converges. Use $|\sin x| \le 1$.

3. a. $\displaystyle\int_0^\infty e^{-t}\,dt = 1.$

b. Let $u = t^{x-1}$, $dv = e^{-t}\,dt$. So

$$\begin{aligned}\Gamma(x) &= \int_0^\infty e^{-t}t^{x-1}\,dt \\ &= -e^{-t}t^{x-1}\Big|_0^\infty - \int_0^\infty (x-1)t^{x-2}(-e^{-t})\,dt \\ &= 0 + (x-1)\int_0^\infty e^{-t}t^{x-2}\,dt \\ &= (x-1)\Gamma(x-1).\end{aligned}$$

The limit $\lim_{t\to\infty} e^{-t}t^{x-1} = 0$ can be confirmed by L'Hôpital's Rule.

c. $\Gamma(n) = (n-1)!$ for all positive integers n.

d. For $t \geq 1$, $e^{-t}t^{-1/2} \leq e^{-t}$; for $0 \leq t \leq 1$, $e^{-t}t^{-1/2} \leq e^{-1}t^{-1/2}$. Since $\displaystyle\int_1^\infty e^{-t}\,dt$ and $\displaystyle\int_0^1 t^{-1/2}\,dt$ converge, so does $\displaystyle\int_0^\infty e^{-t}t^{-1/2}\,dt$.

e. Let $t = x^2$.

f. Some symbolic calculators will yield $\sqrt{\pi}$ exactly. Less sophisticated software should yield a good approximation by estimating an integral of the form $\displaystyle\int_\varepsilon^N e^{-t}t^{-1/2}\,dt$.

4. a. $\lim_{x\to 0} \dfrac{\sin x}{x} = 1.$

b. Some symbolic calculators will yield $\frac{\pi}{2}$ exactly. Less sophisticated software should yield a good approximation by estimating an integral of the form $\displaystyle\int_{\varepsilon}^{N} \frac{\sin x}{x}\,dx$. However, the convergence is slow! See the alternative solution for part c.

c. $\displaystyle\int_0^{\infty} \frac{\sin x}{x}\,dx = \sum_{k=0}^{\infty} \int_{k\pi}^{(k+1)\pi} \frac{\sin x}{x}\,dx = \sum_{k=0}^{\infty} \int_0^{\pi} \frac{(-1)^k \sin u}{u + k\pi}\,du$ where $u = x - k\pi$.

Although the convergent alternating series $\displaystyle\sum_{k=0}^{\infty} \frac{(-1)^k}{u + k\pi}$ does not converge absolutely, its terms can be added two at a time to produce an absolutely convergent series:

$$\left(\frac{1}{u} - \frac{1}{u+\pi}\right) + \left(\frac{1}{u+2\pi} - \frac{1}{u+3\pi}\right) + \left(\frac{1}{u+4\pi} - \frac{1}{u+5\pi}\right) + \cdots$$
$$= \frac{\pi}{u(u+\pi)} + \frac{\pi}{(u+2\pi)(u+3\pi)} + \frac{\pi}{(u+4\pi)(u+5\pi)} + \cdots.$$

Alternatively we can integrate by parts, letting $u = \dfrac{1}{x}$ so that $du = -\dfrac{dx}{x^2}$. The trick comes in chosing $v = 1 - \cos x$ for $dv = \sin x\,dx$. Then

$$\int_0^{\infty} \frac{\sin x}{x}\,dx = \frac{1-\cos x}{x}\bigg|_0^{\infty} + \int_0^{\infty} \frac{1-\cos x}{x^2}\,dx$$
$$= \int_0^{\infty} \frac{1-\cos x}{x^2}\,dx.$$

Since $\displaystyle\lim_{x\to 0} \frac{1-\cos x}{x^2} = \frac{1}{2}$, the integral on the right is not improper at $x = 0$. Furthermore it is the integral of a non-negative function, and $\dfrac{1-\cos x}{x^2} \le \dfrac{2}{x^2}$. It follows that this second integral converges (absolutely) and so the given integral converges. Also, numerical integration of $\displaystyle\int_0^{\infty} \frac{1-\cos x}{x^2}\,dx$ is a good deal more promising than numerical integration of $\displaystyle\int_0^{\infty} \frac{\sin x}{x}\,dx$.

Chapter VIII: Sequences and Series of Functions

Section 1: Sequences of Functions. Taylor Polynomials

1. a. The Taylor polynomial is the same as $f(x)$.

   b. If $m \geq n$, the Taylor polynomial is always the same as $f(x)$. If $m < n$, the Taylor polynomial is $f(x)$ minus its terms of degree greater that m.

2. If we approximate the position $f(t)$ after t seconds by the second-degree Taylor polynomial at $t = 0$, we have:

$$f(t) \approx f(0) + f'(0)t + \frac{f''(0)}{2}t^2.$$

So $f(t) - f(0) \approx 50t + \frac{6}{2}t^2$. Therefore $f(1) - f(0) \approx 50 + 3 = 53$ feet and $f(3) - f(0) \approx 150 + 27 = 177$ feet. But it would not be reasonable to approximate $f(3600) - f(0)$ in the same way, since the time interval is so large. Surely the car does not travel 39,060,000 feet = 7398 miles in one hour!

3. a. Mathematics is not always formulas. By drawing pictures, one can develop insights.

   b. $T_2(x) = 5 - \frac{1}{10}x^2$

   The vertex, of course, is the point $(0,5)$. The x-intercepts are $(-5\sqrt{2},0)$ and $(5\sqrt{2},0)$ which lies well outside the semicircle.

   c. In this case, the Taylor polynomial is:

$$3 - \frac{4}{3}(x-4) - \frac{25}{54}(x-4)^2.$$

   The graphs of f and both polynomials follow. Students may be surprised to see where the vertex and x-intercepts lie.

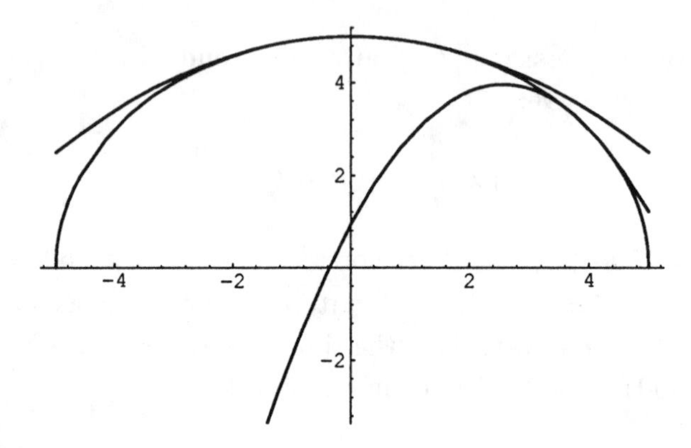

4. $\sin x \approx x - \frac{1}{6}x^3$ and $\cos x \approx 1 - \frac{1}{2}x^2$. So $\dfrac{(\sin x - x)^3}{x(1-\cos x)^4} \approx \dfrac{\left(-\frac{1}{6}x^3\right)^3}{x\left(\frac{1}{2}x^2\right)^4} = \dfrac{-x^9/6^3}{x^9/2^4} =$ $-2/over27$.

5. The second-degree Taylor polynomial of $\sin x$ at $x=0$ is $x + 0x^2/2 = x$.

6. The error is positive when f is concave up at $x=a$ $(f''(a) > 0)$. The tangent line is therefore below the graph of f near $x=a$.

7. a. Apply Taylor's remainder formula to the first-degree Taylor polynomial of $\ln(1+x)$ at $x=0$:

$$\ln(1+x) = x - \frac{x^2}{2(1+c)^2} \qquad \text{for some } c, \quad 0 < c < x.$$

So

$$\left|\ln(1+x) - x\right| = \left|\frac{-x^2}{2(1+c)^2}\right| \le \tfrac{1}{2}x^2.$$

b. Apply Taylor's remainder formula to the second-degree Taylor polynomial of $\sin x$ at $x=0$:

$$\sin x = x - \frac{0x^2}{2} - \frac{(\cos c)\, x^3}{6} \qquad \text{for some } c.$$

So

$$\left|\sin x - x\right| = \frac{|\cos c|\, |x|^3}{6} \le \tfrac{1}{6}|x|^3.$$

8. a. By graphing the difference, we see that the maximum error appears to occur at the right endpoint: $e - 2.5 \approx 0.218$.

b. For some c in $[-1,1]$, $\left|f(x) - T_2(x)\right| = \left|\dfrac{e^c x^3}{3!}\right| \le \tfrac{1}{6}e \approx 0.453$.

c. Since we do not know the value of c in part b, we use the inequality $e^c \le e$, which causes us to lose information. We thus get an upper bound for the error, but the actual error may be much smaller. In this example, it was less than half the error bound.

9. a. $\left|f(x) - T_6(x)\right| = \dfrac{|(\cos c)\, x^7|}{7!} \le \dfrac{\left(\frac{\pi}{4}\right)^7}{7!} \le 0.00004.$

b. $\sin\theta \approx \dfrac{\pi\theta}{180} - \tfrac{1}{6}\left(\dfrac{\pi\theta}{180}\right)^3 + \tfrac{1}{120}\left(\dfrac{\pi\theta}{180}\right)^5.$

10. $\left|\sin x - T_{2n}(x)\right| \le \dfrac{|x|^{2n+1}}{(2n+1)!}$. So $\left|\dfrac{\sin x}{x} - \dfrac{T_{2n}(x)}{x}\right| \le \dfrac{|x|^{2n}}{(2n+1)!} \le \dfrac{1}{(2n+1)!}$
since $0 \le x \le 1$. So

$$\left|\int_0^1 \frac{\sin x}{x}\,dx - \int_0^1 \frac{T_{2n}(x)}{x}\,dx\right| = \left|\int_0^1 \left(\frac{\sin x}{x} - \frac{T_{2n}(x)}{x}\right) dx\right| \le \int_0^1 \frac{dx}{(2n+1)!} = \frac{1}{(2n+1)!}.$$

For $2n+1 = 7$, $\dfrac{1}{(2n+1)!} \le 0.0002$. So $2n = 6$. So

$$\int_0^1 \frac{\sin x}{x}\,dx \approx \int_0^1 \frac{T_6(x)}{x}\,dx = \int_0^1 \left(1 - \frac{x^2}{3!} + \frac{x^4}{5!}\right) dx \approx 0.9461.$$

11. Both graphs look roughly like cubic polynomials on $[-1,1]$, and their cubic Taylor polynomials at $x=0$ are identical: $x - x^3/6$. Both polynomials have zero coefficients for x^4 (since both functions are odd), but thereafter their coefficients differ.

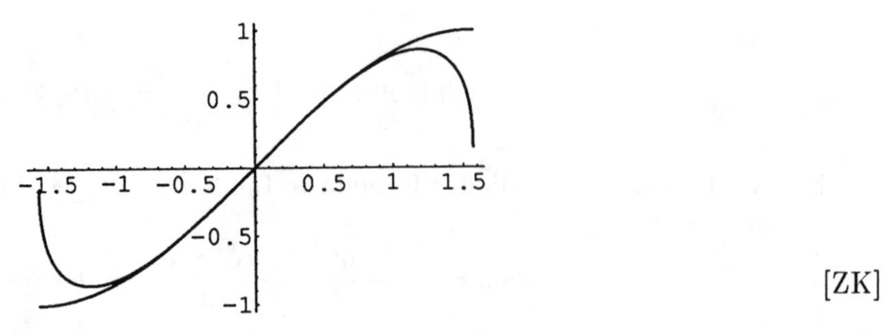

[ZK]

12. $\left|e^x - T_n(x)\right| = \dfrac{|e^c x^{n+1}|}{(n+1)!} \le \dfrac{e}{(n+1)!} \le \dfrac{3}{(n+1)!}$ if $0 \le x \le 1$.

If $n=7$, then $\dfrac{3}{(n+1)!} \le 0.0001$. So

$$e \approx T_7(1) = 1 + 1 + \tfrac{1}{2} + \tfrac{1}{6} + \tfrac{1}{24} + \tfrac{1}{120} + \tfrac{1}{720} + \tfrac{1}{5040} \approx 2.7183.$$

13. a. $\tan\left(\arctan\frac{1}{2} + \arctan\frac{1}{3}\right) = \dfrac{\frac{1}{2}+\frac{1}{3}}{1-\frac{1}{2}\cdot 3} = 1 = \tan\left(\frac{\pi}{4}\right)$.

c. $\pi = 4\left(\frac{1}{2}+\frac{1}{3}\right) - \frac{4}{3}\left(\left(\frac{1}{2}\right)^3 + \left(\frac{1}{3}\right)^3\right) + \frac{4}{5}\left(\left(\frac{1}{2}\right)^5 + \left(\frac{1}{3}\right)^5\right) - \frac{4}{7}\left(\left(\frac{1}{2}\right)^7 + \left(\frac{1}{3}\right)^7\right) + \cdots$.

You will need to continue to the 11th power for 5-place accuracy.

Section 2: Series of Functions. Taylor Series

1. Its derivative doesn't exist at $x = 0$. So $f'(0)$, the coefficient of x, doesn't exist.

2. a.
$$\begin{aligned} f(x) &= (1+x)^p \\ f'(x) &= p(1+x)^{p-1} \\ f^{(2)}(x) &= p(p-1)(1+x)^{p-2} \\ &\vdots \\ f^{(k)}(x) &= p(p-1)\cdots(p-k+1)(1+x)^{p-k} \end{aligned}$$

So $f(0) = 1$ and $f^{(k)}(0) = p(p-1)\cdots(p-k+1)$ for $k > 0$.

b. $f^{(k)}(x) = 0$ whenever $k > p$. So Taylor's remainder formula tells us that $(1+x)^p$ exactly equals its pth degree Taylor polynomial. Furthermore, for $k > 0$,

$$\frac{p!}{k!\,(p-k)!} = \frac{p(p-1)\cdots(p-k+1)(p-k)!}{k!\,(p-k)!} = \frac{p(p-1)\cdots(p-k+1)}{k!}.$$

3. Use $p = -\frac{1}{2}$ and $x = -\dfrac{v^2}{c^2}$ in the binomial series:

$$\left(1 - \frac{v^2}{c^2}\right)^{-1/2} - 1 = -\tfrac{1}{2}\left(-\frac{v^2}{c^2}\right) - \frac{\frac{1}{2}\left(-\frac{3}{2}\right)}{2}\left(-\frac{v^2}{c^2}\right)^2 + \cdots.$$

So $K = m_0c^2\left(\dfrac{v^2}{2c^2}\right) + \dfrac{m_0c^2 3v^4}{8c^4} + \cdots = \frac{1}{2}m_0v^2 + \dfrac{3m_0v^4}{8c^2} + \cdots$. Note that the first term is the Newtonian kinetic energy, and that when v is much less than c all the remaining terms are negligible since they contain a power of v^2/c^2, which is small.

4. Use $p = \frac{1}{2}$ and $x = \dfrac{R^2}{y^2}$ in the binomial series:

$$\sqrt{y^2 + R^2} - y = y\left[\left(1 + \frac{R^2}{y^2}\right)^{1/2} - 1\right] = y\left[\frac{1}{2}\frac{R^2}{y^2} + \frac{\frac{1}{2}\left(-\frac{1}{2}\right)}{2}\left(\frac{R^2}{y^2}\right)^2 + \cdots\right].$$

So

$$F(y) = \frac{2Qy}{R^2}\left[\frac{R^2}{2y^2} - \frac{R^4}{8y^4} + \cdots\right] = \frac{Q}{y} - \frac{QR^2}{4y^3} + \cdots.$$

When y is much larger than R, all terms after the first one are negligible since they contain a power of $\dfrac{R^2}{y^2}$ which is small. So $F(y) \approx \dfrac{Q}{y}$.

5. Since the series is alternating and decreasing in absolute value, the remainder is bounded by the next term:

$$\left|e^x - \sum_{k=0}^{n} \frac{x^k}{k!}\right| \le \frac{|x|^{n+1}}{(n+1)!} = \frac{1}{(n+1)!} \qquad \text{when} \quad x = -1.$$

When $n = 9$, $\dfrac{1}{(n+1)!} \le 5 \cdot 10^{-7}$. So $e^{-1} \approx \displaystyle\sum_{k=0}^{9} \frac{(-1)^k}{k!} \approx 0.3678792$. Then $e \approx 2.7182837$, an error of about $2 \cdot 10^{-6}$. Note that merely taking a reciprocal added to the error.

6. $$\int_0^1 \frac{\sin x}{x}\,dx = \int_0^1 \sum_{k=0}^{\infty} \frac{(-1)^k x^{2k+1}}{x(2k+1)!}\,dx$$
$$= \sum_{k=0}^{\infty} \frac{(-1)^k}{(2k+1)!} \int_0^1 x^{2k}\,dx$$
$$= \sum_{k=0}^{\infty} \frac{(-1)^k}{(2k+1)!(2k+1)}.$$

The remainder is bounded by the next term: $\dfrac{1}{(2n+3)!(2n+3)} \le 5 \cdot 10^{-7}$ when $2n+1 = 7$. So $\displaystyle\int_0^1 \frac{\sin x}{x}\,dx \approx 1 - \frac{1}{3! \cdot 3} + \frac{1}{5! \cdot 5} - \frac{1}{7! \cdot 7} \approx 0.9460828$.

7. N is an integer since all the denominators inside the brackets divide $m!$ and so can be canceled. e^{-1} is the sum of the alternating series $\displaystyle\sum_{k=0}^{\infty} \frac{(-1)^k}{k!}$, and so the remainder $\left|e^{-1} - \displaystyle\sum_{k=0}^{m} \frac{(-1)^k}{k!}\right|$ is bounded by $\dfrac{1}{(m+1)!}$. Therefore $|N| \le \dfrac{m!}{(m+1)!} = \dfrac{1}{m+1} < 1$. The only integer N satisfying $|N| < 1$ is $N = 0$. But if $N = 0$, e^{-1} would equal a partial sum of the alternating series, which cannot be true. This contradiction proves that e cannot have the form $\dfrac{m}{n}$.

8. After the integrand is approximated, the equation for the period becomes

$$5 = 4 \cdot \tfrac{3}{4} \int_0^{\pi/2} \left[1 + \tfrac{1}{2}a^2x^2 + \left(\tfrac{3}{8}a^4 - \tfrac{1}{6}a^2\right)x^4\right] dx = \tfrac{3}{2}\pi + \tfrac{1}{16}a^2\pi^3 + \tfrac{9}{1280}a^4\pi^5 - \tfrac{1}{320}a^2\pi^5.$$

This has the solution $a \approx 0.4503641203$. So $\theta = \left(\frac{180}{\pi}\right) 2 \arcsin a \approx 53^\circ$.

[MP]

Section 3: Power Series

1. The first series converges when $|x| < 1$, the second when $|x| > 1$. There is no x for which the entire computation makes sense.

2. a. Since $e^x = \sum_{k=0}^{\infty} \frac{x^k}{k!}$, $e^{2x} = \sum_{k=0}^{\infty} \frac{(2x)^k}{k!}$. So the coefficient of x^{100} is $\frac{2^{100}}{100!}$.

   b. $\frac{1 - \cos x}{x^2} = \sum_{j=0}^{\infty} (-1)^j \frac{x^{2j}}{(2j+2)!}$

   Calculating the derivative $f^{(100)}(x)$ eliminates the first 49 terms and evaluating the terms after the 50th term for $x = 0$ yields 0. Thus the value sought is the 100th derivative of $(-1)^{50} \frac{x^{(2)(50)}}{(102)!}$ which is a constant: $(100)!/(102)! = \frac{1}{(101)(102)}$.

3. Some hints to your students about this problem may be in order.

   a. Differentiate $\frac{1}{1-x} = \sum_{k=0}^{\infty} x^k$.

   b. Replace x by $-x^2$ in $e^x = \sum_{k=0}^{\infty} \frac{x^k}{k!}$.

   c. Replace x by $-x^2$ in $\frac{1}{1-x} = \sum_{k=0}^{\infty} x^k$. Then integrate.

   d. Replace x by $-x$ in $\frac{1}{1-x} = \sum_{k=0}^{\infty} x^k$. Then integrate and divide by x.

   e. Replace x by $\sqrt{x}$ in $\cos x = \sum_{k=0}^{\infty} \frac{(-1)^k x^{2k}}{(2k)!}$.

   f. Replace x by $-x^2$ and p by $-\frac{1}{2}$ in the binomial series for $(1+x)^p$. Then integrate.

4. a. $\tan x = x + \frac{1}{3}x^3 + \frac{2}{15}x^5 + \frac{17}{315}x^7 + \cdots$.

   c. The radius of convergence is $\frac{\pi}{2}$, the distance from $x = 0$ to the first singularity of $\tan x$. A graph makes this pretty obvious, as does the realization that where $\tan x$ approaches infinity (near $\pm\frac{\pi}{2}$) it cannot behave like a polynomial.

5. This is another problem where hints to your students may be approrpriate.

   a. xe^{x^2}. In the series for e^x, replace x by x^2 and multiply by x.

   b. $\dfrac{2}{(1+x)^3}$. Differentiate the series for $\dfrac{1}{1-x}$ twice and replace x by $-x$.

   c. $-\frac{1}{2}\ln(1-x^2)$. In the series for $\dfrac{1}{1-x}$, replace x by x^2, multiply by x, and integrate.

   d. $\dfrac{d}{dx}(x\cos x) = \cos x - x\sin x$. Multiply the series for $\cos x$ by x and differentiate.

6. a. $\displaystyle\sum_{k=1}^{\infty} kx^k = \frac{x}{(1-x)^2} = 2$ when $x = \frac{1}{2}$. Differentiate the series for $\dfrac{1}{1-x}$ and multiply by x.

   b. $\displaystyle\sum_{k=1}^{\infty} \frac{x^k}{k} = -\ln(1-x) = \ln 2$ when $x = \frac{1}{2}$. Integrate the series for $\dfrac{1}{1-x}$.

   c. $\displaystyle\sum_{k=1}^{\infty} \frac{x^k}{(k+1)!} = \frac{e^x - 1}{x} = \frac{e^2 - 1}{2}$ when $x = 2$.

7. a. $$\int_0^1 e^{-x^2}\,dx = \int_0^1 \sum_{k=0}^{\infty} \frac{(-x^2)^k}{k!}\,dx = \sum_{k=0}^{\infty} \int_0^1 \frac{(-1)^k x^{2k}}{k!}\,dx = \sum_{k=0}^{\infty} \frac{(-1)^k}{k!(2k+1)}.$$

So $$\left|\int_0^1 e^{-x^2}\,dx - \sum_{k=0}^{n} \frac{(-1)^k}{k!\,(2k+1)}\right| \le \frac{1}{(n+1)!\,(2n+3)} \le 5\cdot 10^{-7} \text{ when } n = 8.$$

So $$\int_0^1 e^{-x^2}\,dx \approx \sum_{k=0}^{8} \frac{(-1)^k}{k!\,(2k+1)} \approx 0.7468243.$$

b. The normal probability density function was introduced in VI.3. The calculations here parallel those in part a. If we make the change of variable $u = x/\sqrt{2}$, the integral becomes:

$$\frac{1}{\sqrt{\pi}}\int_0^{1/\sqrt{2}} e^{-u^2}\,du = \frac{1}{\sqrt{\pi}}\int_0^{1/\sqrt{2}} \sum_{k=0}^{\infty} \frac{(-u^2)^k}{k!}\,du$$

$$= \frac{1}{\sqrt{2\pi}} \sum_{k=0}^{\infty} \frac{(-1)^k}{k!(2k+1)2^k}.$$

$$\left|\int_0^1 \frac{1}{\sqrt{2\pi}} e^{-x^2/2}\,dx - \sum_{k=0}^{n} \frac{(-1)^k}{k!(2k+1)2^k}\right| \le \frac{1}{(n+1)!(2n+3)2^{n+1}\sqrt{2\pi}}$$

$$\le 5 \times 10^{-8} \text{ for } n = 6.$$

You may want to have your students generalize the calculation required here by defining $f(w) = \int_0^w \frac{1}{\sqrt{2\pi}} e^{-x^2/2}\,dx$, $w > 0$. For small values of w, i.e. $w < 3$, power series methods work well since convergence is rapid. For large values of w, integration by parts is useful: Let $u = \frac{1}{x}$, $dv = xe^{-x^2/2}\,dx$ and

$$f(w) = \frac{1}{\sqrt{2\pi}}\frac{1}{w}e^{-w^2/2} - \frac{1}{\sqrt{2\pi}}\int_w^{\infty} \frac{1}{x^2} e^{-x^2/2}\,dx.$$

The integral in this expression is positive and for large w, it is much smaller than the integral we began with. Note that $f(w) < \frac{1}{\sqrt{2\pi}}\frac{1}{w}e^{-w^2/2}$. If you wish, a second application of integration by parts yields an even smaller integral. Let $dv = xe^{-x^2/2}\,dx$ (so that $u = \frac{1}{x^3}$), and we obtain

$$f(w) = \frac{1}{\sqrt{2\pi}}\frac{1}{w}e^{-w^2/2} - \frac{1}{\sqrt{2\pi}}\frac{1}{w^3}e^{-w^2/2} + \frac{3}{\sqrt{2\pi}}\int_w^{\infty} \frac{1}{x^4} e^{-x^2/2}\,dx$$

with the following improvement on the value of f:

$$\frac{1}{\sqrt{2\pi}}\left\{\frac{1}{w} - \frac{1}{w^3}\right\}e^{-w^2/2} < f(w) < \frac{1}{\sqrt{2\pi}}\frac{1}{w}e^{-w^2/2}.$$

8. a. $\sum_{k=0}^{\infty} \frac{f^{(k)}(0)x^k}{k!} = 0 + 0x + 0x^2 + \cdots$ which has radius of convergence ∞ and sums to 0 for all x. But $f(x)$ equals this sum only for $x = 0$.

b. $$\begin{aligned} f'(0) &= \lim_{x\to 0} \frac{f(x) - f(0)}{x} \\ &= \lim_{x\to 0} \frac{e^{-1/x^2}}{x} \\ &= \lim_{y\to\pm\infty} \frac{y}{e^{y^2}} \quad (y = \tfrac{1}{x}) \\ &= \lim_{y\to\pm\infty} \frac{1}{2ye^{y^2}} \quad \text{(by L'Hôpital's Rule)} \\ &= 0. \end{aligned}$$

9. a. Let $f(x) = \frac{xe^{-x}}{1-e^{-x}} = \frac{x}{e^x - 1}$. $\lim_{x\to 0} f(x) = 1$, so the integral is not improper at $x = 0$. The series $\sum \frac{ne^{-n}}{1-e^{-n}}$ converges by limit comparison with $\sum \frac{n}{e^n}$. So $\int_0^\infty f(x)\,dx$ converges.

b. If $u = 1 - e^{-x}$, then $x = -\ln(1-u)$ and $dx = \frac{du}{1-u}$. Note that since $x \geq 0$, it follows that $0 \leq u < 1$. The integral becomes $-\int_0^1 \frac{\ln(1-u)}{u}\,du$.

c. $\ln(1-u) = -(u + \frac{u^2}{2} + \frac{u^3}{3} + \ldots)$, and the integral becomes: $\int_0^1 \sum_{k=0}^{\infty} \frac{u^k}{k+1}\,du =$ $\sum_{k=0}^{\infty} \frac{u^{k+1}}{(k+1)^2}\Big|_0^1 = \sum_{k=0}^{\infty} \frac{1}{k^2} = \frac{\pi^2}{6}$.

d. Approximately 10,000, as one might guess since $\int_n^\infty \frac{dx}{x^2} = \frac{1}{n}$.

Chapter IX: The Integral in $\mathbb{R}^2$ and $\mathbb{R}^3$

Section 1: Real-valued Functions of Two and Three Variables

1. a. Level curves 5

   b. Level curves 4

   c. Level curves 3

   d. Level curves 2

   e. Level curves 1

2. The corresponding level curves follow:

   a.

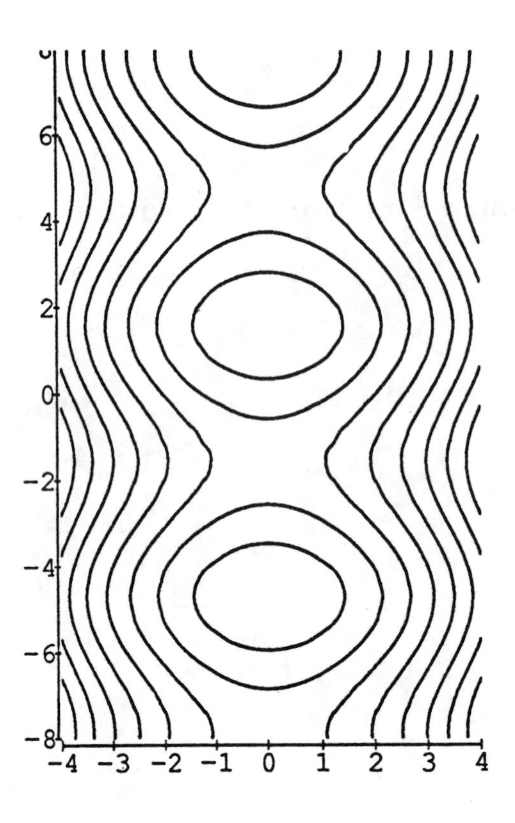

   b.

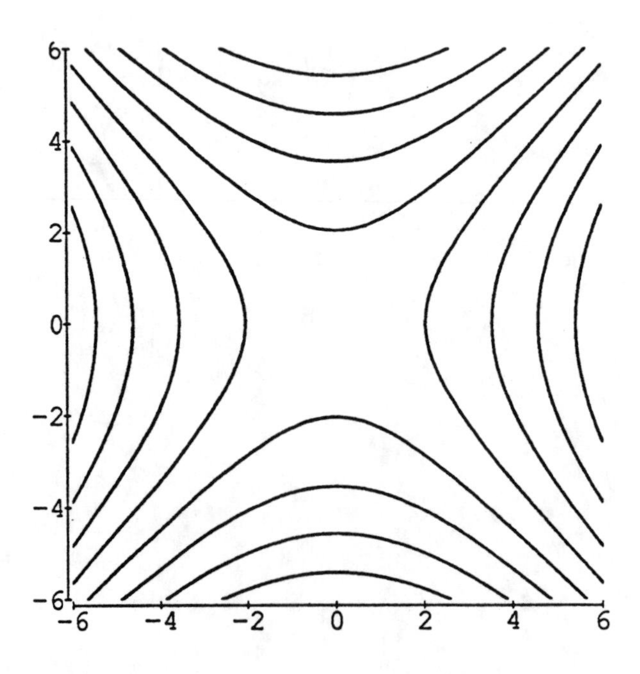

3. a.

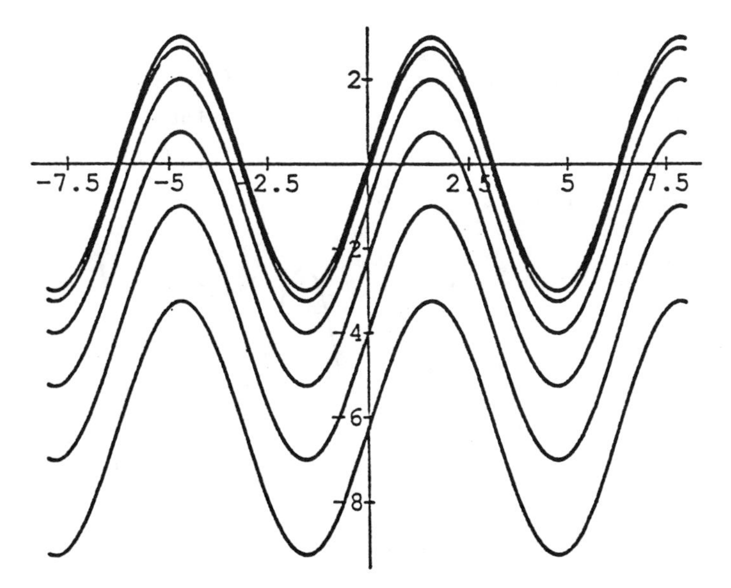

b.

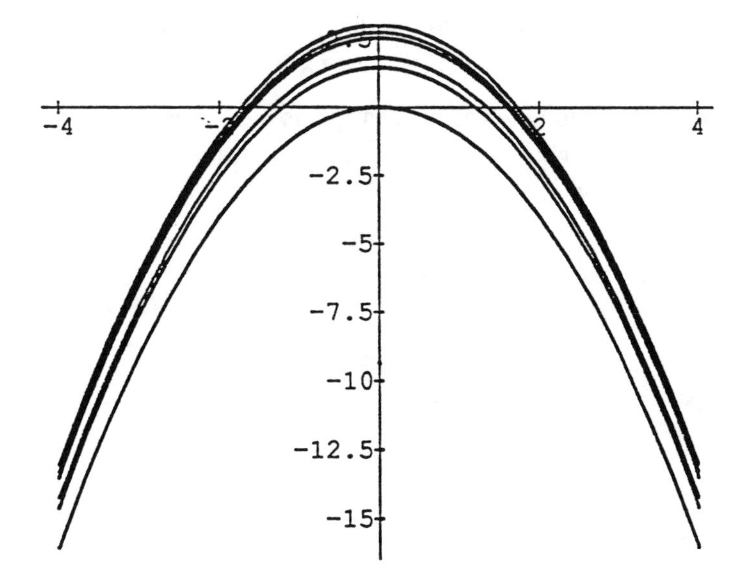

c. A level curve is obtained as the intersection of a surface and a plane z = constant, parallel to the xy-plane. This question is designed to reinforce students' geometric understanding of level curves as "slices" of a surface.

4. a. The line in the xy-plane containing the points $(0,0)$ and $(1,1)$ has the equation $x = y$. This is also the equation, in 3-space, of the plane PL_1. The curve of intersection, $PL_1 \cap S$, is given by the equation $y = 60x^2e^{-2x^2}$ and its graph follows:

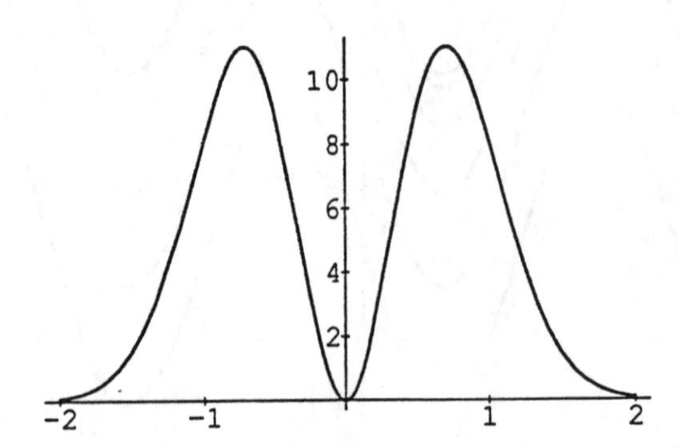

b. $PL_2 \cap S$ is the line in the xy-plane with the equation $y = 0$ (the x-axis).

c. The graph of $PL_3 \cap S$ is the mirror image, in the horizontal axis, of the graph of $PL_1 \cap S$ given in part a.

This question stimulates students to think in terms of the rate of change of a function of two variables in certain specified directions, and this leads naturally to a consideration of the directional derivative.

5. The function $w = u(x,t)$ describes a simple case of a "traveling wave." The points on the x-axis are vertically displaced as time elapses; $u(x,t)$ is the vertical displacement of the point x at time t. At time $t = 0$, the points are displaced as shown by $g(x,0) = f(x)$. See the sketch of $f(u)$. The sketches of part a show the "disturbance" at various times t, and the sketch of part b shows how the point $(6,t)$ is displaced as time elapses.

a.

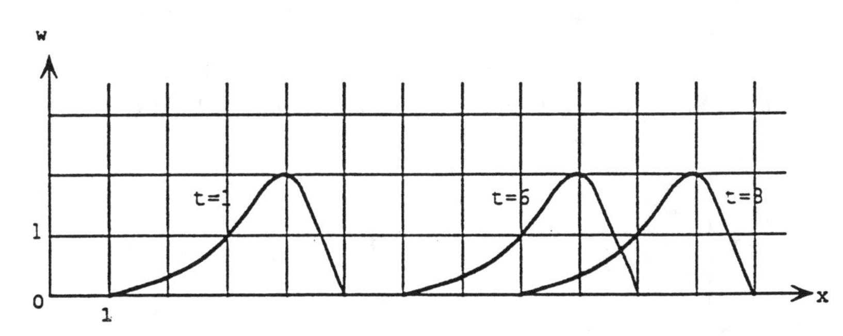

b.

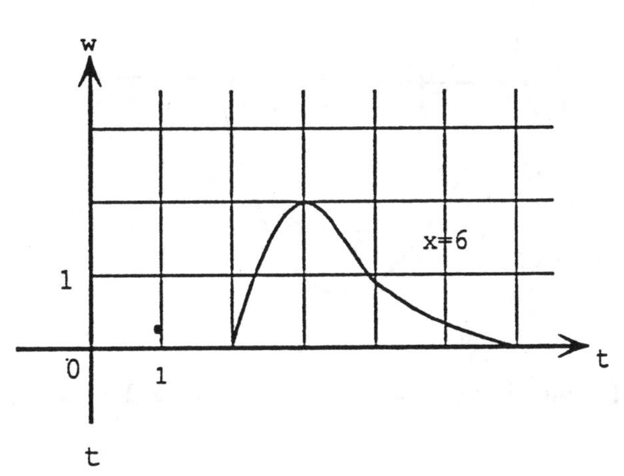

c.

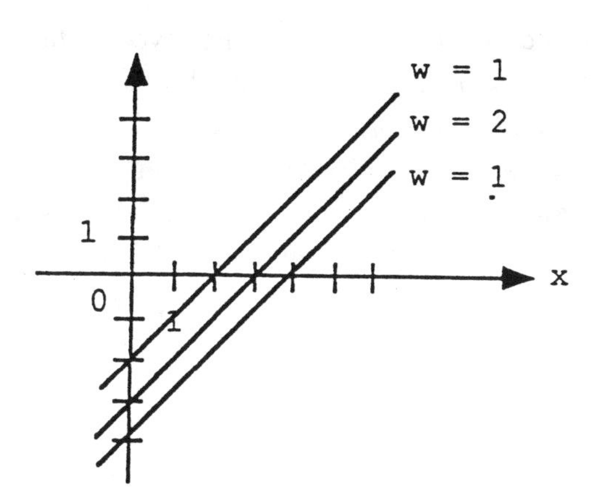

d. $t = 71$.

e. $x \notin [213.5, 215.8]$.

[CC]

6. a. $g(u) = 3$ yields $u = 0$ and $u \approx 5$. Corresponding to $u = 0$, we get a circle C: $x^2 + y^2 = 4$, and corresponding to $u = 5$, we get yet another circle C': $x^2 + y^2 = 9$. Note that since the domain of y is $[0, \infty)$, the domain of f is the complement of the open disk given by $x^2 + y^2 < 4$. The circle C is thus the "inner rim" of the domain of f.

[CC]

7. The graph of $f(2, y) = 8y^3 - 36y + 16y = 4(2y^3 - 5y)$ is given below.

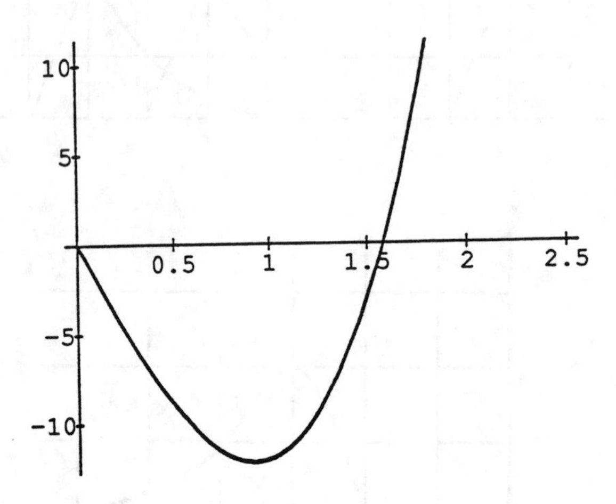

There is a lot of choice here; $P = (2, .5)$ and $Q = (2, 1.5)$ is one of the possibilities.

[CC]

8. The answers for parts a and b are the same. The level surface consists of the union of the coordinate planes: $x = 0$, $y = 0$, $z = 0$. In part c, $z = g(x, y) = 1/xy$. If $g(x, y) = 1$, then $y = 1/x$, and if $g(x, y) = 433$, then $y = 1/(433x)$. Both these curves are hyperbolas. In part d, S, a level *surface* of F, is given by a function of two variables. Specifying the value of z allows you to calculate one of these variables, say y, in terms of the other, x, and this gives the level *curves* requested in part c.

[CC]

Section 2: Definition of Double and Triple Integrals

1. a. $M_0 = 708$.

   b. $m_0 = 564$.

   c. $m_0 < V < M_0$.

   To justify this relationship, students may cite the fact that level curves of f are circles centered at the origin. The z-values of the circles decrease as the distance from the origin increases.

2. a. $M_1 = 677$; $m_1 = 605$.

   b. $m_0 < m_1 < V < M_1 < M_0$.

   c. $M_2 = 660.25$; $m_2 = 624.25$.

   d. $m_0 < m_1 < m_2 < V < M_2 < M_1 < M_0$.

   Let's hope that students do begin to appreciate how V is defined.

3. Regardless where D lies, it has an area of approximately 5.5 square units.

   a. Minimum $p = g\left(\frac{3}{2.25}\right) \approx 2.5$; maximum $p = g\left(\frac{3}{4}\right) \approx 3$.
   $m = (2.5)(5.5) = 13.75$; $M = (3)(5.5) = 16.5$.

   b. Minimum $p \approx g\left(\frac{3}{(.6)^2}\right) = g\left(\frac{3}{.36}\right) \approx .3$; maximum $p \approx g\left(\frac{3}{(2.8)^2}\right) = g\left(\frac{3}{7.84}\right) \approx 3.5$.
   $m = (.3)(5.5) = 1.65$; $M = (3.5)(5.5) = 19.25$.

[CC]

Section 3: Evaluation of Double Integrals

1. The definite integrals with the blanks filled in follow:

a. $\int_0^1 dy \int_y^{4y} f(x,y)\,dx + \int_1^2 dy \int_{4y-3}^{y+3} f(x,y)\,dx$

b. $\int_0^3 dx \int_{2x/5}^{5x/3} f(x,y)\,dy + \int_3^5 dx \int_{2x/5}^{(-3x+19)/2} f(x,y)\,dy$

c. $\int_0^4 dy \int_0^{\sqrt{4y-y^2}} f(x,y)\,dx$

d. $\int_{-1}^0 dx \int_{-x-1}^{\sqrt{1-x^2}} f(x,y)\,dy + \int_0^1 dx \int_{-\sqrt{1-x^2}}^{1-x} f(x,y)\,dy$

e. $\int_{-2}^{-1} dx \int_{-\sqrt{-x^2-2x}}^{\sqrt{-x^2-2x}} f(x,y)\,dy + \int_{-1}^1 dx \int_{-\sqrt{1-x^2}-1}^{\sqrt{1-x^2}+1} f(x,y)\,dy$
$+ \int_1^2 dx \int_{-\sqrt{2x-x^2}}^{\sqrt{2x-x^2}} f(x,y)\,dy$

2. This problem is structured like a popular TV quiz show, but it has a serious purpose: To get students to conceptualize about the physical and geometric significance of double and triple integrals. The examples given provide a template for other such problems.

a. What is the volume of the region in the region in the first octant below the plane $2x + 3y + z = 6$?

b. What is the mass of the ball of radius 1 where the density function is given by ρ?

c. What is the volume of the "ice cream-filled cone" cut out of the ball of radius 1 by the cone $z = r$?

3. $\int_0^1 \frac{dz}{1-xyz} = -\frac{\ln(1-xy)}{xy}$. $\ln(1-xy) = -\sum_{n=1}^{\infty} \frac{(xy)^n}{n}$ for $xy \in (-1,1)$ and, in this problem, $xy \in (0,1)$. The given triple integral is then equal to

$$\int_0^1 \int_0^1 \sum_{n=1}^{\infty} \frac{(xy)^{n-1}}{n}\,dy\,dx = \sum_{n=1}^{\infty} \frac{1}{n} \int_0^1 \frac{x^{n-1}}{n}\,dx = \sum_{n=1}^{\infty} \frac{1}{n^3}.$$

You may want to point out to your students that this characterization of $\sum \frac{1}{n^3}$ is particularly nice insofar as there is no closed-form expression for this summation.

4. a. $\int_0^4 \int_0^4 f(x,y)\,dy\,dx = \frac{1}{4}(1-e^{-4})^2 \approx .2409.$

   b. Using the symmetry of the function f, we get $4\int_0^c \int_0^c f(x,y)\,dy\,dx = (1-e^{-c})^2 = .5$. This implies that $c \approx 1.2279$.

   c. The probability is given by $4\int_0^\infty \int_0^\varepsilon f(x,y)\,dy\,dx = (1-e^{-\varepsilon})$ for a sufficiently small positive ε. If the average depth for a lot on Main Street is $\frac{1}{16}$ km, the probability is approximately .0606.

5. The dotted-line grid superimposed on the map suggests the subregions for which the temperature readings at the towns in the table can be used as local approximations. Since a division by the total area of the state is done in this problem, we can normalize distance measurements and take the area of each square subregion to be 1 unit. Regions which include parts of neighboring states and Lake Michigan have prorated areas. Subdivision has been used for regions with more than one town, with greater weight given towns more "representative" of the subregion in which they lie. Subregions for which there are no data, e.g. the triangular wedge between Hudson and La Crosse, have been ignored as has been the region north of Green Bay and east of Wausau (although it can be argued that the temperature in Green Bay can be used here).

   One teacher's reaction to this problem was:"Those of us in the rest of the country don't want to hear about *nice* days in Wisconsin – how about using that day back in January '79 when it was $-28°$ in Madison and something not quite believable in Rice Lake?" Feel free to change the temperatures given. You may want to change the state as well as the season. Clearly the most important part of students' response to this question is the justification they give for their calculations, rather than the calculations themselves, and the motivation this provides for the definition of average value of a function of two variables.

   Average temperature is approximately 79°.

6. a. $\dfrac{\displaystyle\int_0^{300}\int_{-60}^{150} T(x,y)\,dy\,dx}{(210)(300)} = \dfrac{492.135}{6.3} \approx 78°.$

b. The calculation involved in this problem makes it too tedious to assign if students do not have some electronic assistance at hand. If they do have a CAS, however, it illustrates the pedagogical advantages of a good software package which takes care of the calculational drudgery while leaving the structuring of the problem to the students.

Average temperature $= \dfrac{\text{Numerator}}{\text{Denominator}}$ where

$$\begin{aligned}\text{Numerator} &= \iint_{R_1} T(x,y)\,dA + \iint_{R_2} T(x,y)\,dA + \iint_{R_3} T(x,y)\,dA \\ &= (503.607)(10^4)\end{aligned}$$

and where the non-vertical, non-horizontal boundaries of the three subregions are given by the following equations:

$$\begin{aligned}\text{For } R_1 \quad & y = -\tfrac{14}{9}(x+10). \\ \text{For } R_2, \quad & y = -\tfrac{7}{15}x + 285.33. \\ \text{For } R_3, \quad & y = \tfrac{7}{6}x - 18.33.\end{aligned}$$

Denominator = Sum of areas of three trapezoids $\approx (6.3825)(10^4)$.
Average temperature $\approx 79°$.

7. a. This classic is sometimes omitted in the examples calculus students should see, and that is the reason for its inclusion in this collection.

$$\begin{aligned}\int_0^\infty\int_0^\infty e^{-(x^2+y^2)}\,dy\,dy &= \int_0^{\pi/2}\int_0^\infty e^{-r^2}\,r\,dr\,d\theta \\ &= \int_0^{\pi/2} d\theta\left(-\tfrac{1}{2}e^{-r^2}\right)\Big|_0^\infty \\ &= \tfrac{\pi}{4}\end{aligned}$$

b. $\displaystyle\int_0^\infty\int_0^\infty e^{-x^2-y^2}\,dy\,dx = \int_0^\infty e^{-x^2}\,dx \int_0^\infty e^{-y^2}\,dy = \left(\int_0^\infty e^{-x^2}\,dx\right)^2$

c. Since the given integral has the value $\frac{\pi}{4}$, $\displaystyle\int_0^\infty e^{-x^2}\,dx = \frac{\sqrt{\pi}}{2}$.

8. a. R is a parallelogram with corners $(0,0)$, $(1,2)$, $(1,3)$, $(0,1)$.

   b. The transformation T works as follows on the corners of R: $T(0,0) = (0,0)$; $T(1,2) = (1,0)$; $T(1,3) = (1,1)$; $T(0,1) = (0,1)$. The first of these equalities is satisfied for all choices of a, b, c, d. The second equality implies that $a = 1$, $c = 2$; the fourth implies that $b = 0$, $d = 1$. The third equality can be used to check the arithmetic. Note that since linear transformations map parallelograms onto parallelograms, T is determined by what it does to any three of the corners of R. $T^{-1} : x = u,\ y = 2u + v$.

   c. $|J| = \begin{vmatrix} 1 & 0 \\ 2 & 1 \end{vmatrix} = 1$. $\text{Area}(R) = \int_0^1 \int_0^1 dx\, dy = 1$.
   J is sometimes called the "stretching factor" in relating the areas of regions like R and S. In this case, no stretching occurs. You may want to ask your students to confirm this result by calculating the area of R by using the formula: area = (altitude)(length of base).

[GS]

9. a. The corners of R, a trapezoid, are: $(0,0)$, $(1,0)$, $(1,5)$, $(0,3)$.

   b. Let T^{-1} be the transformation defined by the two given equations. Then T is given by $u = x$, $v = \dfrac{y - ax}{b + cx}$. Note that $T(0,0) = (0,0)$ as long as $b \neq 0$; $T(1,0) = (1,0)$, which implies $a = 0$. $T(1,5) = (1,1)$, which implies that $b + c = 5$; $T(0,3) = (0,1)$, which implies that $b = 3$. The values of a, b, c which are required, therefore, are: $a = 0, b = 3, c = 2$. Note that the requirements of the transformation are consistent with the equation $x = u$.

   c. Parallelograms in the xy-plane will map onto S under linear transformations.

[GS]

10. This problem has been included because it gives students a geometric feel for the transformations encountered in changing from one set of coordinates to another in evaluating double integrals.

    a. Linear transformations, of which this and the one in part *b* below are examples, map parallelograms onto parallelograms. Moving in a counterclockwise fashion, we find that the corners of R are: $(0,0), (2,1), (3,3), (1,2)$. The equations of the sides of this parallelogram are (again moving in a counterclockwise fashion): $y = \frac{1}{2}x$, $y = 2x - 3$, $y = \frac{1}{2}x + \frac{3}{2}$, $y = 2x$.

    b. The corners of R are: $(0,0), (3,1), (5,2), (2,1)$. The equations of the sides are: $y = \frac{1}{3}x$, $y = \frac{1}{2}x - \frac{1}{2}$, $y = \frac{1}{3}x + \frac{1}{3}$, $y = \frac{1}{2}x$.

    c. The corners of R are: $(1,1), (e^2, e), (e^3, e^3), (e, e^2)$. The sides are parabolic arcs given by the following equations: $x = y^2$, $x^2 = e^3 y$, $y^2 = e^3 x$, $y = x^2$.

    d. The corners of R are: $(0,0), (0,-1), (2,0), (0,1)$. The sides are given by: $x = 0$ (negative y-axis), $\frac{x^2}{4} = 1 + y$, $\frac{x^2}{4} = 1 - y$, $x = 0$ (positive y-axis).

    e. The corners of R are: $(0,0), (1,0), (1,2), (0,1)$. The sides are: $y = 0$, $x = 1$, $x^2 = y - 1$, $x = 0$.

    f. The corners of R are: $(0,0), (1,0), (\cos 1, \sin 1)$. The sides are: $y = 0$, $x^2 + y^2 = 1$, $\frac{x}{\cos 1} = \frac{y}{\sin 1}$.

[GS]

11. a. The circle has a radius of a and its center at $(a, 0)$ in the xy-plane.

    b. The symmetry of the circle about the x-axis can be used to evaluate $\bar{r}$:

$$\bar{r} = \frac{\displaystyle\int_0^{\frac{\pi}{2}} d\theta \int_0^{2a\cos\theta} r(r\,dr)}{\displaystyle\int_0^{\frac{\pi}{2}} d\theta \int_0^{2a\cos\theta} r\,dr}$$

    c. The denominator in the equation for $\bar{r}$ in part *b* is half the area of the circle, and this is easily seen to be $\frac{1}{2}\pi a^2$. Some students may still choose to calculate the double integral although this requires a double angle substitution. The numerator is better behaved.

$$\bar{r} = \frac{16a^3/9}{a^2\pi/2} = \frac{32a}{9\pi}.$$

[GS]

12. a. Sometimes true. In the case of a rigid rotation of coordinates axes, for example; but it is certainly not always true. See part *b* below, which is related to this question.

b. Again, this is sometimes true, as in the case of rigid rotations; however, all that one can conclude from the equality of the two double integrals is that $f(\alpha(u,v),\beta(u,v))|J| = g(u,v)$ where $J = \dfrac{\partial(x,y)}{\partial(u,v)}$ is the Jacobian of the transformation. The point of this question is to get students to reflect on the role of that determinant in the change of variables. For many common transformations, $|J| \neq 1$, e.g. the change from rectangular to polar coordinates for which $|J| = r$.

c. 't ain't necessarily so. Consider the transformation T to polar coordinates and the regions R and $S = T(R)$ given by the sketch below:

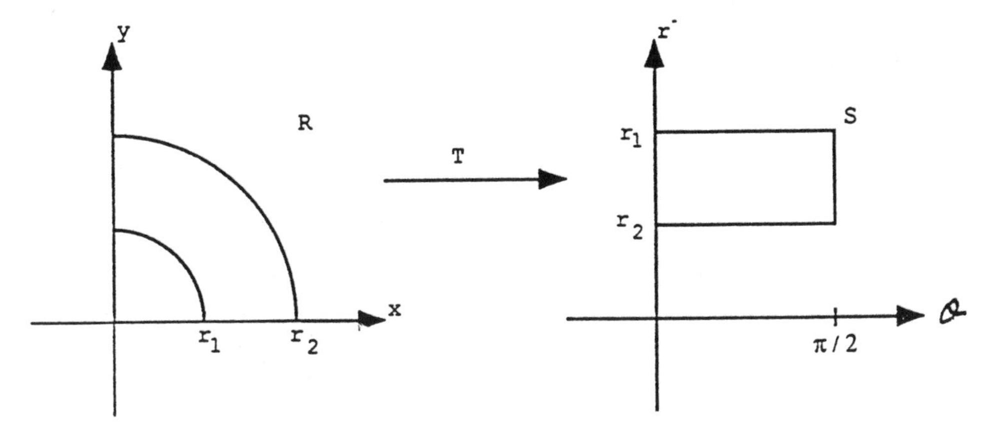

$\text{Area}(R) = \frac{\pi}{4}(r_2^2 - r_1^2)$; $\text{Area}(S) = \frac{\pi}{2}(r_2 - r_1)$.
$\text{Area}(R) = \text{Area}(S)$ if $r_1 + r_2 = 2$. Choose $r_1 = \frac{1}{2}$, $r_2 = \frac{3}{2}$.

13. a. This is sometimes true, but it depends on the location of R. The statement is false, for example, if R lies in the vertical strip for which $x \in (0,1)$.

b. The average value of f over R is given by $\dfrac{\iint_R f(x,y)\,dA}{\iint_R dA}$. Only if the area of R is 1 is the statement true.

c. The area of the polar "rectangle" is $\displaystyle\int_{\theta_1}^{\theta_2} d\theta \int_{r_1}^{r_2} r\,dr = \tfrac{1}{2}(r_2^2 - r_1^2)(\theta_2 - \theta_1)$.

The area of the quadrilateral (a trapezoid) is

$$\sqrt{2}\,\sqrt{1-\cos(\theta_2-\theta_1)}\,\frac{(r_1+r_2)}{2}(r_2-r_1) = \tfrac{\sqrt{2}}{2}\left(r_2^2 - r_1^2\right)\sqrt{1-\cos(\theta_2-\theta_1)}.$$

The area of the polar rectangle exceeds that of the trapezoid as long as $r_1 \le r_2$ and $\theta_1 \le \theta_2$. See the graphs of $x = \theta_2 - \theta_1$ and $y = \sqrt{2(1-\cos x)}$ below.

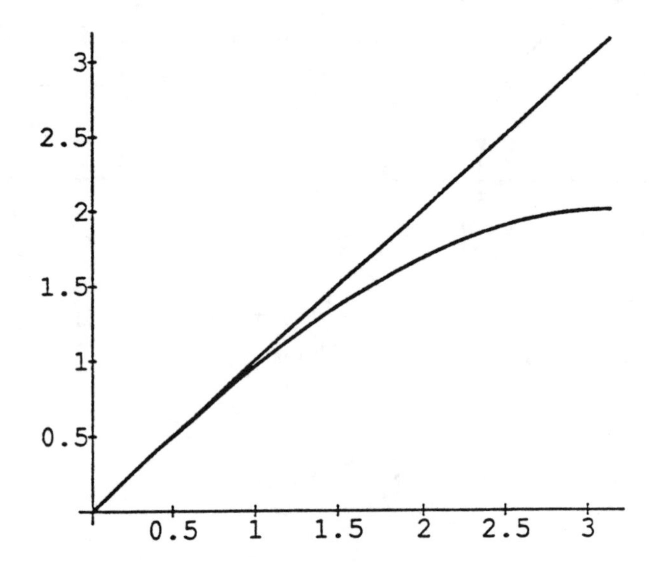

[GS]

Chapter X: Vectors and Vector Geometry

Section 1: Vectors

1. Suppose that the circle has radius R and is centered at the origin. Let $\mathbf{X}$ be a vector from the origin to one end of the diameter. Then $-\mathbf{X}$ is the vector to the other end of the diameter. Let $\mathbf{Y}$ be a vector from the origin to an arbitrary point on the circle. Then the inscribed angle θ is formed by the vectors $\mathbf{X}-\mathbf{Y}$ and $\mathbf{X}+\mathbf{Y}$. Furthermore

$$(\mathbf{X}-\mathbf{Y})\cdot(\mathbf{X}+\mathbf{Y}) = ||\mathbf{X}-\mathbf{Y}||||\mathbf{X}+\mathbf{Y}||\cos\theta$$

and

$$(\mathbf{X}-\mathbf{Y})\cdot(\mathbf{X}+\mathbf{Y}) = \mathbf{X}\cdot\mathbf{X}-\mathbf{Y}\cdot\mathbf{Y} = R^2 - R^2 = 0.$$

So $\cos\theta = 0$, and hence $\theta = \frac{\pi}{2}$.

2.

	Independent Variables	*Dependent Variables*
a.	scalar	scalar
b.	vector (3-dimensional)	scalar
c.	vector (2-dimensional)	scalar
d.	vector (3-dimensional)	scalar
e.	vector (2-dimensional)	vector (2-dimensional)
f.	vector (3-dimensional)	vector (3-dimensional)

3. The proof should look something like this: Let $\mathbf{X}$ and $\mathbf{Y}$ be two sides of a triangle, and let θ be the angle between them. The third side is $\mathbf{X}-\mathbf{Y}$. Then if $||\mathbf{X}|| = a$, $||\mathbf{X}|| = b$, and $||\mathbf{X}-\mathbf{Y}|| = c$, then

$$\begin{aligned} c^2 &= ||\mathbf{X}-\mathbf{Y}||^2 \\ &= (\mathbf{X}-\mathbf{Y})\cdot(\mathbf{X}-\mathbf{Y}) \\ &= ||\mathbf{X}||^2 + ||\mathbf{Y}||^2 - 2\mathbf{X}\cdot\mathbf{Y} \\ &= ||\mathbf{X}||^2 + ||\mathbf{Y}||^2 - 2||\mathbf{X}||||\mathbf{Y}||\cos\theta \\ &= a^2 + b^2 - 2ab\cos\theta. \end{aligned}$$

This again points out the value of vectors as a shorthand for geometry and trigonometry.

4. If we let A be the angle the vector makes with $\mathbf{i}$, B the angle it makes with $\mathbf{j}$, and C the angle it makes with $\mathbf{k}$, then

$$\cos A = \frac{a}{\sqrt{a^2+b^2}}, \quad \cos B = \frac{b}{\sqrt{a^2+b^2}}, \quad \text{and} \quad \cos C = 0.$$

Of course, $\cos^2 A + \cos^2 B + \cos^2 C = 1$. In this case, however, since $\cos C = 0$, we have $\cos^2 A + \cos^2 B = 1$. Since A and B are complementary angles, $\cos A = \sin B$ and $\sin A = \cos B$, so $\cos^2 A + \sin^2 A = 1 = \sin^2 B + \cos^2 B$. This is the trigonometric statement of the Pythagorean Theorem.

5. It is useful to distinguish between *geometric* methods and problems and *purely linear* methods and problems. The latter include anything which involves the equations of lines and planes, notions of collinearity and coplanarity (linear dependence and independence) and intersections of lines and planes. This is an example of a purely linear problem. Such problems may be solved by geometric means, but they may also be solved by purely linear methods. In this case, what is involved is solving a set of linear equations. The equation of a plane has the form $ax + by + cz = d$ where a, b, c, and d are constants. The coordinates of the three points A, B, and C can be substituted for x, y, and z in the equation of a plane. This gives three equations in the four unknowns a, b, c, and d, a system which can be solved by Gaussian elimination. Alternatively a geometric approach may be adopted. Let $\mathbf{V}$ be the vector from A to B and $\mathbf{W}$ be the vector from A to C. $\mathbf{V}$ and $\mathbf{W}$ lie in the plane to be determined and their cross product $\mathbf{N} = \mathbf{V} \times \mathbf{W}$ is normal to the plane. Let the point (x, y, z) lie in the plane, and let $\mathbf{X} = x\mathbf{i} + y\mathbf{j} + z\mathbf{k}$. Then $\mathbf{X} - \mathbf{V}$ also lies in the plane, and so $(\mathbf{X} - \mathbf{V}) \cdot \mathbf{N} = 0$ is the equation of the plane.

6. Like the problem above, both parts of this one are purely linear and hence have purely linear solutions. Possible geometric solutions follow.

   a. Let the three points be A, B, and C. Compute the vectors $\mathbf{V}$ and $\mathbf{W}$ as $B - A$ and $C - A$, respectively. If $\mathbf{V}$ is a scalar multiple of $\mathbf{W}$, the three points are collinear. Alternatively one could take the cross product $\mathbf{V} \times \mathbf{W}$. If $\mathbf{V} \times \mathbf{W} = \mathbf{0}$, then again the three points are collinear.

   b. Fix one of the four points, and let $\mathbf{X}$, $\mathbf{Y}$, and $\mathbf{Z}$ denote the vectors to each of the other three points. Then form the triple product $\mathbf{X} \cdot (\mathbf{Y} \times \mathbf{Z})$. If this triple product is 0, then the points in question are coplanar.

7. Let $\mathbf{A}_n = a_{1n}\,\mathbf{i} + a_{2n}\,\mathbf{j} + a_{3n}\,\mathbf{k}$ for $n = 1, 2, 3$, and let $\mathbf{C} = c_1\,\mathbf{i} + c_2\,\mathbf{j} + c_3\,\mathbf{k}$. If these vectors are regarded as column vectors, then the system can be written $x\mathbf{A}_1 + y\mathbf{A}_2 + z\mathbf{A}_3 = \mathbf{C}$. Take the scalar product of each side with $\mathbf{A}_2 \times \mathbf{A}_3$, yielding

$$x\mathbf{A}_1 \cdot (\mathbf{A}_2 \times \mathbf{A}_3) + y\mathbf{A}_2 \cdot (\mathbf{A}_2 \times \mathbf{A}_3) + z\mathbf{A}_3 \cdot (\mathbf{A}_2 \times \mathbf{A}_3) = \mathbf{C} \cdot (\mathbf{A}_2 \times \mathbf{A}_3).$$

But $\mathbf{A}_2 \cdot (\mathbf{A}_2 \times \mathbf{A}_3) = \mathbf{A}_3 \cdot (\mathbf{A}_2 \times \mathbf{A}_3) = 0$. So

$$x = \frac{\mathbf{C} \cdot (\mathbf{A}_2 \times \mathbf{A}_3)}{\mathbf{A}_1 \cdot (\mathbf{A}_2 \times \mathbf{A}_3)} = \frac{\begin{vmatrix} c_1 & a_{12} & a_{13} \\ c_2 & a_{22} & a_{23} \\ c_3 & a_{32} & a_{33} \end{vmatrix}}{\begin{vmatrix} a_{11} & a_{21} & a_{31} \\ a_{21} & a_{22} & a_{23} \\ a_{31} & a_{32} & a_{33} \end{vmatrix}}, \quad \text{etc.}$$

8. The Saudis, since the prevailing currents would most likely carry the oil onto their shores.

9. Midway between the western ends of Iran and Saudi Arabia.

10. The United Arab Emirates.

11. Lest we forget, the Beach Boys were into surfing, an activity which requires a strong onshore current. The onshore current is strongest along Saudi Arabia, with the southern portion of the United Arab Emirates and the western portion of Iraq also good surfing territory.

12. Any spill close off the southwestern coast of the United Arab Emirates figures to be washed ashore there. Iraq and Kuwait are relatively safe in this regard.

13. Off the border between Saudi Arabia and Qatar. There is virtually no current to speak of, so the rescuers can just come and get you.

14. It follows the contour of the shoreline. Note that it does not do this near the eastern portion of the coast of Iran.

Section 2: Velocity and Acceleration

1. Notice that $\mathbf{v}(t) = \frac{1}{t}\mathbf{i} - \frac{2}{t^2}\mathbf{j} + 3\mathbf{k}$. Consequently $\mathbf{x}(1) = -2\mathbf{i} + 4\mathbf{j} - 6\mathbf{k}$ and $\mathbf{v}(1) = \mathbf{i} - 2\mathbf{j} + 3\mathbf{k}$. Since the rocket will continue along the vector $\mathbf{v}(1)$ starting from the point $(-2, 4, -6)$, we see that $\mathbf{x}(1) + 2\mathbf{v}(1) = \mathbf{0}$ implies that the rocket will coast into the space station.

2. Equating the various components, we see that the only time at which they could collide is $t = 2$, and the position vector of each planet at that time is $4\mathbf{i}+8\mathbf{j}+4\mathbf{k}$. The velocity vectors are $\mathbf{x}'(t) = 2t\mathbf{i}+5\mathbf{k}$ and $\mathbf{y}'(t) = -3\mathbf{i}+4t\mathbf{j}+2t\mathbf{k}$. Consequently, $\mathbf{x}'(2) = 4\mathbf{i}+5\mathbf{k}$ (and so the velocity of Earth is $\sqrt{41}$), and $\mathbf{y}'(2) = -3\mathbf{i} + 8\mathbf{j} + 4\mathbf{k}$ (so the velocity of Xyra is $\sqrt{89}$). Clearly Xyra is moving faster. If A is the angle at which they collide, then $\mathbf{x}'(2) \cdot \mathbf{y}'(2) = \|\mathbf{x}'(2)\|\|\mathbf{y}'(2)\| \cos A$. Therefore, $\cos A = \frac{8}{\sqrt{3649}} \approx .1324$ and the angle of collision is about $83^\circ\, 23'$.

3. Since the scalar product $\mathbf{X}(t) \cdot \mathbf{X}'(t) = 0$, we must have

$$x\frac{dx}{dt} + y\frac{dy}{dt} = 0.$$

Therefore $x\,dx = -y\,dy$. This is a well-known separable equation, which can be integrated to yield $\frac{1}{2}x^2 = -\frac{1}{2}y^2 + C$ where C is a constant of integration. It is easily seen that the point $(x(t), y(t))$ must lie on a circle. Note that this result holds even if the motion is not uniform!

4. a. The expression for $\mathbf{x}$ involves the parametric representation of an ellipse with verticies at $(a, 0)$ and at $(0, b)$ in the plane of motion. The constant r is the number of radians traversed in one year. Also $r = d\theta/dt$ where θ is the angle between the position vector and the vector $\mathbf{i}$. Differentiating the expression for $\mathbf{x}$, we obtain

$$\mathbf{v}(t) = -ra(\sin rt)\,\mathbf{i} + rb(\cos rt)\,\mathbf{j}.$$

Note that $|\mathbf{v}(t)| = r\sqrt{a^2 + b^2}$, a constant.

b. $\mathbf{a}(t) = -r^2 a(\cos rt)\,\mathbf{i} - r^2 b(\sin rt)\,\mathbf{j} = -r^2\mathbf{x}(t)$. So the acceleration continually points in the direction opposite to that of the position vector; i.e., towards the origin.

The relevance of this problem to the Theory of Gravitation is worth pointing out. Kepler's First Law (which is a consequence of Newton's Theory of Gravitation) states that the planets move in elliptical orbits with the sun at one of the foci. The basic Theory of Gravitation states that the gravitational force between two masses is attractive, and is exerted on the line joining the two masses. This problem therefore shows that a planet does *not* move at constant angular velocity relative to the center of its elliptical orbit.

5. Differentiating $\mathbf{v}(t)\cdot\mathbf{v}(t) = c^2$, we obtain $\mathbf{v}(t)\cdot\mathbf{v}'(t) + \mathbf{v}'(t)\cdot\mathbf{v}(t) = 2\mathbf{v}(t)\cdot\mathbf{a}(t) = 0$. This problem is straightforward, but students may attempt to make things more difficult for themselves by writing the radius, velocity, and acceleration vectors in coordinate form. This is useful for some problems, but others will be more easily solvable with a coordinate-free approach.

6. a. Straightforward.

b. Since the wheel has radius 1, the point on the end of the spoke has radius vector $\mathbf{x}(t) = (\cos bt)\,\mathbf{i} + (\sin bt)\,\mathbf{j}$. Differentiating twice, we see that $\mathbf{v}_w = -b(\sin bt)\,\mathbf{i} + b(\cos bt)\,\mathbf{j}$ and that $\mathbf{a}_w = -b^2(\cos bt)\,\mathbf{i} - b^2(\sin bt)\,\mathbf{j}$.

c. Differentiating $\mathbf{r}(t)$, we obtain $\mathbf{v}(t) = (s\cos bt - bst\sin bt)\,\mathbf{i} + (s\sin bt + bst\cos bt)\,\mathbf{j} = s(\cos bt)\,\mathbf{i} + s(\sin bt)\,\mathbf{j} + st\mathbf{v}_w$.

d. Differentiating $\mathbf{v}(t)$, we obtain

$$\mathbf{a}(t) = (-2bs\sin bt - b^2 st\cos bt)\,\mathbf{i} + (2bs\cos bt - b^2 st\sin bt)\,\mathbf{j} = 2s\mathbf{v}_w + st\mathbf{a}_w.$$

This is not a difficult problem, but the term $2s\mathbf{v}_w$ which appears, and is known as the Coriolis acceleration, has profound physical significance. The Coriolis acceleration is the result of two interacting motions, the movement of the bug and the rotation of the wheel. On the rotating Earth, an object moving in the northern hemisphere is deflected clockwise by Coriolis acceleration, and counterclockwise in the southern hemisphere. Coriolis acceleration significantly affects such important phenomena as wind movement in the atmosphere and the oceanic rotary movements known as gyres.

Section 3: Arc Length

1. Suppose that we parametrize the helix by using a parameter t such that $t = 1$ corresponds to a complete turn. Then, if the helix has radius R and height H and makes N turns, the obvious parametrization is $x = R\cos 2\pi t$, $y = R\sin 2\pi t$, $z = Ht/N$, for $0 \le t \le N$. The arc length is given by $s = \displaystyle\int_0^N \sqrt{4\pi^2R^2 + (H/N)^2}\,dt = \sqrt{4\pi^2R^2N^2 + H^2}$. When $R = 5$, $H = 4$, and $N = 3$, the length of the spring is $\sqrt{900\pi^2 + 15}$, and when $R = 4$, $H = 3$, and $N = 5$, the length of the spring is $\sqrt{1600\pi^2 + 9}$. The second spring is longer. Note this is a physically intuitive result; the more turns, the longer the spring. When partial derivatives are introduced, it is worth revisiting this problem and taking the partial derivatives of the spring length with respect to the variables R, N, and H, and interpreting them.

2. It is easy to see geometrically that the area under a curve has *both* an upper bound (consisting of the sum of the upper rectangles) and a lower bound (consisting of the sum of the lower rectangles). Therefore, if the sums of the upper rectangles and the sums of the lower rectangles converge to the same quantity, this quantity must be the area under the curve. However, the arc length of a curve has only a *lower* bound in terms of the sum of polygonal line segments. Consequently the integral used to compute the arc length of a curve provides a sensible definition of length. Concepts students have already encountered, such as volume and work, involve both lower and upper bounds. The area of a surface of revolution, however, is analogous to arc length in that the integral used to compute it again provides a sensible definition of area for such a surface.

3. a. The curve is a circular helix. Elementary calculus and the definition of arc length shows that $s(t) = 5t$, so $t = \frac{1}{5}s$. Therefore $\mathbf{x}(s) = \cos\left(\frac{3}{5}s\right)\mathbf{i} + \sin\left(\frac{3}{5}s\right)\mathbf{j} + \left(\frac{4}{5}s\right)\mathbf{k}$.

   b. Differentiating shows that the unit tangent vector $\mathbf{T}(s)$ is given by $\mathbf{T}(s) = -\frac{3}{5}\sin\left(\frac{3}{5}s\right)\mathbf{i} + \frac{3}{5}\cos\left(\frac{3}{5}s\right)\mathbf{j} + \frac{4}{5}\mathbf{k}$. At this stage, students might verify that $\mathbf{T}(s)$ really is a unit vector. In theory, of course, it should be a unit vector, but theory is sometimes a little overwhelming, and it is nice to see that it works for a specific example. Differentiating again yields $\dfrac{d\mathbf{T}}{ds} = -\frac{9}{25}\cos\left(\frac{3}{5}s\right)\mathbf{i} - \frac{9}{25}\sin\left(\frac{3}{5}s\right)\mathbf{j}$. The length of this vector is $\frac{9}{25}$, so the unit normal is $\mathbf{N}(s) = -\cos\left(\frac{3}{5}s\right)\mathbf{i} - \sin\left(\frac{3}{5}s\right)\mathbf{j}$. Therefore $\mathbf{x}(s) + \mathbf{N}(s) = \frac{4}{5}s\,\mathbf{k}$, which lies on the z-axis.

   c. The answer is no. From the curve, the center of this circle lies in the direction of N, which points towards the z-axis, but is at a distance of the radius of curvature away. The radius of curvature is $\dfrac{25}{9}$, which is much larger than 1.

Chapter XI: The Derivative in Two and Three Variables

Section 1: Partial Derivatives

1. a. The partial derivative with respect to x gives, on some day t, the rate of change of temperature as the depth increases.

   b. The partial derivative with respect to t gives, at some fixed depth x, the rate of change of temperature as the number of days increases.

   The additional questions will help students understand this model better, but they are not essential to this problem and may be omitted.

   c. The period is clearly 365 days.

   d. This term causes a phase shift because the temperature below ground will lag behind the temperature at the surface.

   e. This factor reduces the amplitude of fluctuation of the temperature as depth increases.

2. This problem is similar to the preceding problem. It could be assigned after explaining that problem to see if students now understand partial derivatives.

   a. The partial derivative with respect to x gives, at a fixed time t, the rate of change of intensity as the depth increases.

   b. The partial derivative with respect to t gives, at a fixed depth x, the rate of change of intensity as the time changes.

   c. I_0 is the maximum intensity of the light on the surface of the water at midday.

   d. This factor reduces the maximum intensity as the depth increases.

3. Since the plane is perpendicular to the y-axis, the y-coordinate is held fixed in each case. Therefore students need to evaluate f_x at the point in question.

   a. $f_x(1, 2) = 12$

   b. $f_x\left(\frac{\pi}{4}, \frac{\pi}{2}\right) = \frac{\sqrt{2}}{2}$

   c. $f_x(2, 3) = 9e^6$

   This question can be asked, of course, in terms of the intersection of a given surface with a plane perpendicular to the x-axis through a given point.

4. a. The correct choice is D. Students might need an additional hint that they just follow the curve that is the "front" of the surface.

   b. The correct choice is also D. This time they are to follow the curve which bounds the right of the surface.

   c. The correct answer could be written using the wording of choice B in part a. This time it is a little more difficult to determine which curve to follow.

5. a. The cross-section for $x = 3$ is the seventh curve from the right of the figure, and the cross-section for $y = 1$ is the seventh curve from the back of the figure.

   b. Students should have little difficulty in drawing a line that is tangent to these curves. Check to see if they are labeled correctly.

   c. In parametric form L_1 is: $x = 3$, $y = t + 1$, $z = 2t + 1$, and L_2 is: $x = 3 + t$, $y = 1$, $z = -\pi t + 1$.

   d. The tangent plane is the plane containing the two tangent lines that intersect at the point $(3, 1, 1)$.

6. The function used to create these level curves is $z = y^2 - x^2$.

   a. $f_x(1,1) = -2$ and $f_y(1,1) = 2$.

   b. $f_x(0,0) = f_y(0,0) = 0$. This might be a place to begin talking about saddle points.

7. a. $f(.6, .8) \approx 5.59$ hundreds of feet $= 559$ feet.

   b. $f_x(.6, .8) = 2.54$. Students can illustrate this by drawing the tangent line to the third curve from the "back of the box" at $(.6, .8)$, which is the curve when $y = .8$.

   c. $f_y(.6, .8) = 2(2.54) = 5.08$. Students can illustrate this by drawing the tangent line to the sixth curve from the "right of the box" at $(.6, .8)$, which is the curve when $x = .6$.

   d. It is clearly steeper to go north because of the factor 2 times y in the equation for the elevation.

   e. The incline going west is given by $-f_x(.6, .8) = -2.54$, and going south is given by $-f_y(.6, .8) = -5.08$.

   f. Students should give answers like "southeast" or "northwest". Better answers might be "east-southeast" or "west-northwest".

8. a. The level curves will be ellipses with center $(1,0)$ which is approximately Iowa City. These level curves are the isobars for that day.

   b. Approximate answers are:

$$\text{From Des Moines to Iowa City,} \quad \frac{P(1,0)-P(0,0)}{1-0} = .1;$$

$$\text{from Des Moines to Ames,} \quad \frac{P(0,.25)-P(0,0)}{.25-0} = -.05;$$

$$\text{from Des Moines to Ottumwa,} \quad \frac{p(.5,.5)-p(0,0)}{.71} = .035.$$

   c. The maximum occurs at the point $(1,0)$ which is approximately the location of Iowa City. The location of the minimum is less obvious but appears to occur at the northwest corner of the state. We are constrained to stay inside Iowa in this problem, and that offers you an opportunity to discuss constrained optimization.

9. a. $\dfrac{\partial P}{\partial x} = 2xe^{3/y}$ and $\dfrac{\partial P}{\partial y} = -\dfrac{3x^2}{y^2}e^{3/y} + \dfrac{2}{\sqrt{y}}$.

   b. $\dfrac{\partial P}{\partial y}$ gives the rate of change of profit (marginal profit) for an increase in the selling price of whiteboards.

   c. Yes, $\dfrac{\partial P}{\partial y}$ can be negative and that means profits will decrease as the price of whiteboards is increased.

   d. No, $\dfrac{\partial P}{\partial x}$ cannot be negative. This means that the more we charge for blackboards, the more we make in profit. That further implies that this function could not really model the profit of any manufacturing process.

10. a. $\dfrac{\partial f}{\partial x} = \dfrac{x^3y}{e^y}$, which describes the rate of change of crimes committed as the number of unemployed people x increases. $\dfrac{\partial f}{\partial y} = \frac{1}{4}x^4e^{-y}(1-y)$, which describes the rate of change of crimes committed as the size of the police force y increases.

    b. $\dfrac{\partial f}{\partial x}$ is always positive, which implies that the number of crimes committed will increase if the number of unemployed people increases.

    c. $\dfrac{\partial f}{\partial y}$ is negative when $y > 1$. This implies that the number of crimes committed will decrease if the size of the police force increases.

11. It is easy to check that the number of bus riders is given by $f(x,y) = \dfrac{100y^2}{x}$ where $x > 0$ and $y > 0$.

    a. $f\left(\frac{1}{2}, 5\right) = 5000$, which is the number of riders if the cost of a ride is 50 cents and the cost of parking is \$5.

    b. $\dfrac{\partial f}{\partial x} = -\dfrac{100y^2}{x^2}$, and $\dfrac{\partial f}{\partial y} = \dfrac{200y}{x}$.

    c. $\dfrac{\partial f}{\partial x}$ measures the change in the number of riders when the fare increases. It is negative because an increase in fare will decrease the number of riders.

    d. $\dfrac{\partial f}{\partial y}$ measures the change in the number of riders when the price of downtown parking increases. It is positive because an increase in the parking fee would increase the number of riders on the bus.

Section 2: Gradient and Directional Derivatives

1. This problem is intended to give students a feel for level curves and the fact that the maximum rate of change of the elevation of the surface is perpendicular to the level curve.

   a. The path is a level curve on the surface.

   b. They will need to turn 90° either to the right or the left.

   c. They will need to turn 90° to the left if they turned right to answer part *b*, or turn right if the turned left to answer part *b*. Alternatively they could answer that they turned 180° from the direction of steepest incline up the hill.

   d. The direction of maximum rate of change of the elevation of the surface is perpendicular to the level curve. The theorem may be stated in the form that the gradient of a function at a point is normal to the level curve through that point.

2. a. The vectors are to be drawn on the figure.

   b. Since $\nabla f(-1,0) = (3,-3)$, $(\frac{1}{\sqrt{2}}, -\frac{1}{\sqrt{2}})$ is the direction in which f increases most rapidly.

   c. The other vector is tangent to the level curve at $(-1,0)$. So the rate of change of f in this direction is 0.

3. The first step is to find $\nabla f(.6,.8) = (2.54, 5.08) = 2.54(1,2)$, which is the direction of maximum increase of f at $(.6,.8)$. Therefore, the directions specified by the vectors $(2,-1)$ and $(-2,1)$, which are perpendicular to $\nabla f(.6,.8)$, will keep you at the same elevation.

4. a. Answers will vary depending upon the terminology of the textbook students are using. One possible answer: The directional derivative is the magnitude of the gradient vector times $\cos\theta$, where θ is the angle between the gradient vector and the direction of the directional derivative.

   b. The directional derivative is maximum in the direction of the gradient, minimum in the opposite direction, 0 if the direction is perpendicular to the direction of the gradient, and $\frac{1}{2}$ its maximum value for $\theta = 60°$, since $\cos 60° = \frac{1}{2}$, either left or right of the direction of the gradient.

   The second part of this question is closely related to Problem 1. If you didn't assign that problem earlier and if the weather is nice, this would be a good time for students to take a short walk.

5. This is an easy arithmetic problem.
$f_x(8,5) \approx \frac{.2}{.01} = 20$, $f_y(8,5) \approx -\frac{.1}{.02} = -5$, $\nabla f(8,5) \approx (20,-5)$, and
$D_{\mathbf{u}} f(8,5) = \nabla f(8,5) \cdot \nabla \mathbf{u} \approx 8$.

6. a.

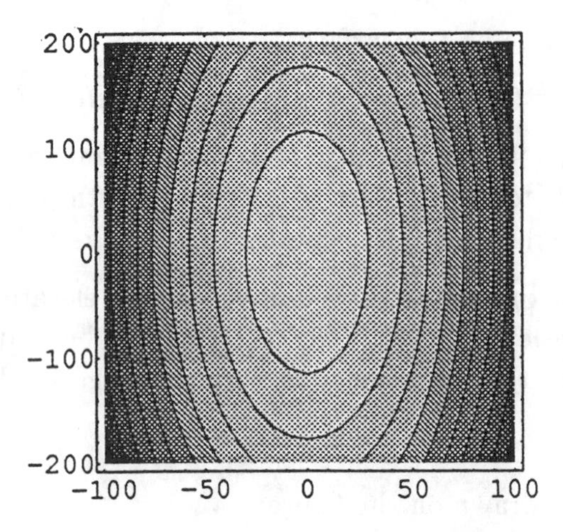

b. Jimmy is in over his head! $f(25,90) \approx 82.5$ft.

c. $f_x(25,90) = -.75$

d. The directional derivative is $(-.75,-.18) \cdot \left(\frac{1}{\sqrt{2}}, -\frac{1}{\sqrt{2}}\right) \approx -.4$.

e. In the direction of the gradient, which is $(-.75,-.18)$.

f. To reach shallow water as quickly as possible, Jimmy should swim in the direction opposite to that of the gradient.

7. a. $\nabla F = (f_x, f_y, -1)$, and therefore the first two components of these vectors are the same. They are different in that ∇f gives the direction of maximum increase in f, while ∇F gives the direction of a (downward) normal vector to the surface.

b. The normal vector is $2\,\mathbf{i} - 3\,\mathbf{j} - \mathbf{k}$.

8. Yes. If $\mathbf{u} = a\,\mathbf{i} + b\,\mathbf{j}$, then the value of one of the directional derivatives is $af_x + bf_y$. Similarly, if $\mathbf{v} = c\,\mathbf{i} + d\,\mathbf{j}$, then the value of the other directional derivative is $cf_x + df_y$. Therefore we have two linear equations in two unknowns to solve to find ∇f at this point.

Section 3: Equation of the Tangent Plane

1. Answers to this question will vary depending on how the author of your text approachs the problem of tangent planes. This question is based upon the more fundamental question "What information is needed to determine a plane?" You might want to start this section with this general question and then ask the more specific question at a later time.

   a. One answer is to know a vector normal to the plane and a point on the plane. If the surface is given by $F(x, y, z) = 0$, then the gradient ∇F evaluated at the point of tangency is the key to finding the equation of the tangent plane.

   b. A plane is determined by three non-collinear points, two intersecting lines, a line and a point not on the line, or two parallel lines. Students should account for all of these possibilities to answer this part of the question.

2. This is a continuation of Problem 5 in Section 2, but it could be assigned even if that problem were not.

   a. On the basis of the information given, the answer is: $z - 33.1 = 20(x-8) - 5(y-5)$. Students who pass a plane through the three given points will get the same answer. Be on the look-out for what strategy your students employ! At this point, you may want to return to Taylor series and introduce your students to the approximation of functions of several variables by Taylor polynomials in several variables and how this generalizes the way the tangent plane $\nabla F(P) \cdot (\mathbf{X} - \mathbf{P}) = 0$ approximates a surface $F(\mathbf{X}) = 0$ at a point P.

   b. Students should use the equation of the tangent plane from part a here.

$$f(8.01, 5.02) \approx 33.1 + .2 - .1 = 33.2$$

3. This problem arises in different engineering contexts: Removing overburden to uncover coal for strip mining, or leveling the side of a hill to build a house.

   a. The equation of the tangent plane is $z = f(x, y) = 720 + .1x + .08y$.

   b. This part is not essential to answering the real question of the problem, but it can be used to get a better "feel" for level curves and the volume involved. The graph is below.

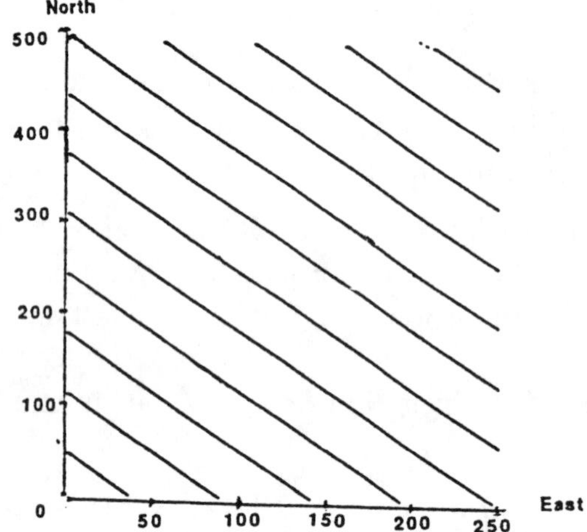

   c. The maximum, which is 785, occurs at $(250, 500)$. The average value is $\frac{1}{2}(720 + 785) = 752.5$ feet. Students may learn more by attempting to do a volume calculation by a double integral:

$$\int_0^{500} \int_0^{250} f(x, y)\, dx\, dy$$

   The average elevation is then volume(overburden)/area(region)= 752.5 feet. If this argument is introduced, then you should consider asking what it has to do with the argument involving averages in an effort to motivate the definition of the average of a function.

   d. The overburden is $(752.5 - 720) \cdot 250 \cdot 500 = 4{,}062{,}500$ cubic feet.

   e. We need to solve the equation:

$$\mathrm{L} : 720 + 0.1x + .08y = 752.5$$

   for one variable in terms of the other in order to obtain a lower limit of integration for the double integral. If we solve for x in terms of y, we obtain $x = 325 - 0.8y$, and the amount of overburden to be moved is given by:

$$\int_{75/0.8}^{500} \int_{325-0.8y}^{250} (f(x, y) - 752.5)\, dx\, dy = 715{,}169\,\mathrm{ft}^3$$

The lower limit of integration for y is obtained by noting that, in the equation for L, when $x = 250$, $y = 75/0.8$.

4. The answer is $z = 0$. If students have trouble with this problem, encourage them to sketch the surface S.

Section 4: Optimization

Calculus texts tend to have a good selection of problems for this section, so only a few problems are included here.

1. If students have access to a CAS, then you could ask them to graph these functions. There are, of course, many other functions that can be added to this list.

   a. Minimum of 0 at $(0,0)$.

   b. Minimum of 1 at $(0,0)$.

   c. Maximum of 1 at $(0,0)$.

   d. Minimum of -1 at $\left(0,\frac{3\pi}{2}\right)$ and an infinite number of other points, and a maximum of 1 at $\left(0,\frac{\pi}{2}\right)$ and other points.

   e. There is no maximum or minimum since the range of f is the real line.

   f. Clearly there is a maximum of 1 at $(0,0)$. Less clearly, there is a minimum value of approximately -0.067 located at the four points $(2.36,0)$, $(-2.36,0)$, $(0,2.36)$, and $(0,-2.36)$. It takes a CAS to find these points. A cross section of f along the x-axis is given below. (A cross section along the y-axis would look the same.)

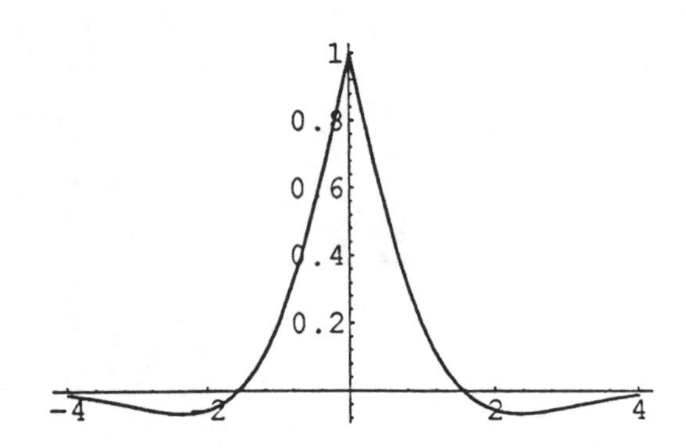

2. a. The only critical point is at $(0,0)$: there are none in the interior of the square.

   b. On the various parts of the boundary we have the following:

$$\begin{aligned}
x = 0&: \text{ maximum of } 0, \text{ and minimum of } -3\\
x = 1&: \text{ maximum of } 4\tfrac{1}{3}, \text{ and minimum of } 3\\
y = 0&: \text{ maximum of } 4, \text{ and minimum of } 0\\
y = 1&: \text{ maximum of } 3, \text{ and minimum of } -3
\end{aligned}$$

   c. The absolute maximum is clearly $4\frac{1}{3}$, and the absolute minimum is -3.

3. a. $-\nabla T\left(\frac{1}{2},\frac{1}{3}\right) = \left(-\frac{1}{3},1\right)$ is the direction of quickest cooling.

   b. If the bug goes either in the direction $(3,1)$ or in the opposite direction $(-3,-1)$, both of which are perpendicular to $\nabla T\left(\frac{1}{2},\frac{1}{3}\right)$, the temperature will remain constant.

   c. The only critical point is $(0,0)$ where T takes the value 0. On the boundary $y = x$, 0 is again the maximum at $(0,0)$. However, on the boundary $y = x^2$, the maximum is $T\left(\frac{1}{3},\frac{1}{9}\right) = \frac{1}{27}$. This is, therefore, the maximum value of T for the region.

   d. $\dfrac{dT}{dt} = \left(\frac{1}{3},-1\right)\cdot\left(\frac{2}{\sqrt{5}},-\frac{1}{\sqrt{5}}\right) = \frac{\sqrt{5}}{3}.$

4. a. Your students will probably use a straight-forward geometric approach to this question. The vector $(1,2,-1)$ is perpendicular to S and so the line $L:\ (x,y,z) = t(1,2,-1)$ passes through the origin and intersects S at the point $(1,2,-1)$. The distance between the origin and S is, therefore, $\sqrt{1+4+1} = \sqrt{6}$. This approach is actually an instance of using Lagrange multipliers, a technique exploited below.

   b. S is a sphere of radius 3, centered at the origin. The (unique) distance between the origin and any point of S is 3. Note that the position vector of any point of S is perpendicular to S.

(i)

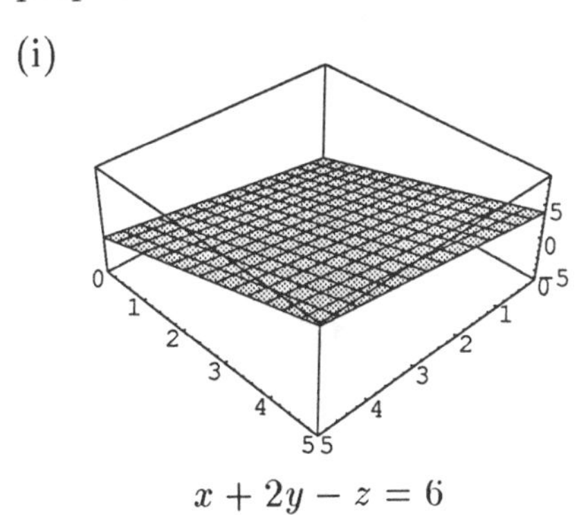

$x + 2y - z = 6$

(ii)

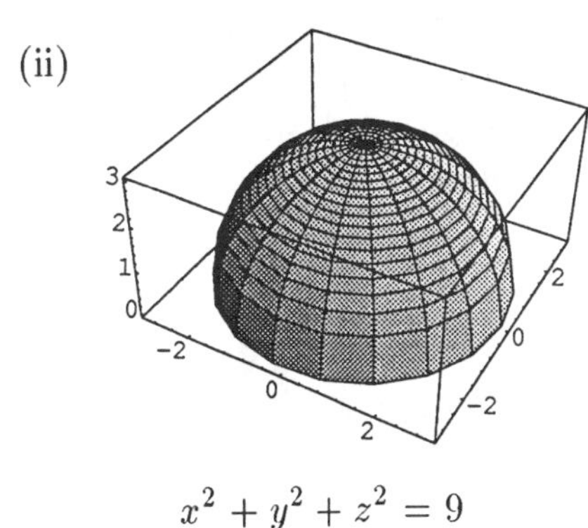

$x^2 + y^2 + z^2 = 9$

5. a. Let us first try a definition-inspired approach to this question. The square of the distance between the origin and a point P in S is given by $x^2 + y^2 + z^2$ where $z = 4/xy$. Let $f(x,y) = x^2 + y^2 + 16/x^2y^2$. Setting f_x and f_y equal to 0 and simplifying yields the following equations: $x^4y^2 = 16$ and $x^2y^4 = 16$ which are satisfied at four points of S: (s,s,s), $(s,-s,-s)$, $(-s,-s,s)$, and $(-s,s,s)$ where $s = \sqrt[3]{4}$. It is geometrically clear that at these points, the distance between the origin and S is shortest. This distance is $\sqrt{(3)(2^{4/3})} \approx 2.7495$. If we use the technique of Lagrange multipliers, then we need to solve a system of equations:

$$\nabla f = \lambda \nabla g$$
$$g(x,y,z) = 4$$

where $f(x,y,z) = x^2 + y^2 + z^2$ and $g(x,y,z) = xyz$. This system becomes: $(2x, 2y, 2z) = \lambda(yz, xz, xy)$, $xyz = 4$. It yields, of course, the same answers as before. The algebra here is probably best left to a CAS.

b. A CAS is necessary here since both the techniques in part c lead to equations which are difficult to solve. Let $f(x,y) = x^2 + y^2 + (2 - \cos x - \sin y)^2$. Then the system $f_x(x,y) = 0$, $f_y(x,y) = 0$ becomes:

$$x + \sin x(2 - \cos x - \sin y) = 0$$
$$y - \cos y(2 - \cos x - \sin y) = 0$$

An approximate solution of this system of equations is: $x = 0$, $y = 0.47872$. The corresponding third coordinate is $z = 2 - \cos 0 - \sin 0.47872 \approx 0.539355$ and the distance between the point with coordinates $(0, 0, 47872, 0.539355)$ and the origin is approximately 0.721165. Geometric considerations make it clear that this is the shortest distance between S and the origin. Students should appreciate that the principles they used previously to obtain the shortest distance are the same principles used here. A CAS is required only to cope with the algebraic difficulties of solving two simultaneous equations. If students do have access to a software package which solves simultaneous equations, the scope of this sort of problem can be broadened considerably.

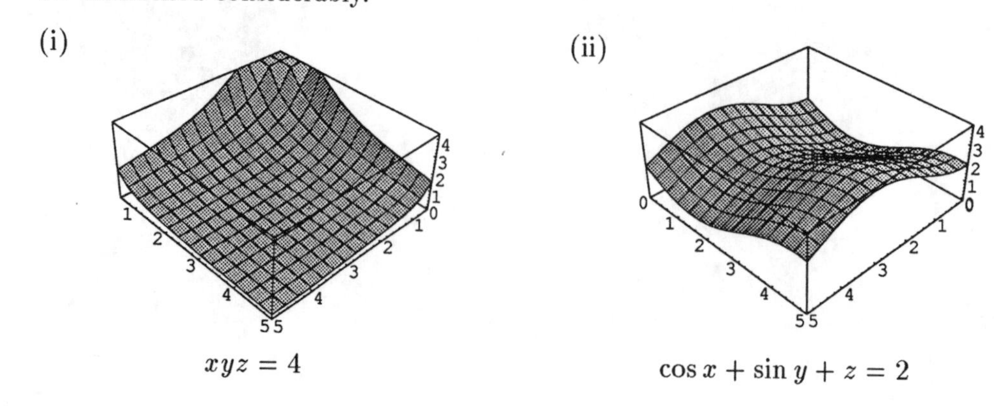

$xyz = 4$ $\qquad$ $\cos x + \sin y + z = 2$

Chapter XII: Line Integrals

Section 1: Line Integrals

1. a. Approximating the integral with a Riemann sum,

$$\int_{C_1} f(x,y)\,ds \approx \tfrac{1}{2}(4+2+0+(-1)) = \tfrac{5}{2}.$$

   b. $\displaystyle\int_{C_2} f(x,y)\,ds \approx \tfrac{1}{2}(2+1-1-2) = 0.$

   c. This one requires a bit more thought since the length of the base of each line segment used to form the Riemann sum is now $\frac{1}{\sqrt{2}}$.

$$\int_{C_3} f(x,y)\,ds \approx \tfrac{1}{\sqrt{2}}(4+1+1+0) = 3\sqrt{2}.$$

2. Since $f(x,y) = k$ for all (x,y) on the curve C, $\displaystyle\int_C f(x,y)\,ds = \int_C k\,ds = kA(C)$, where $A(C)$ is the arc length of C.

3. $\displaystyle\int_a^b \sqrt{1+\left(f'(t)\right)^2}\,dt$

4. The sketch below suggests the formula required:

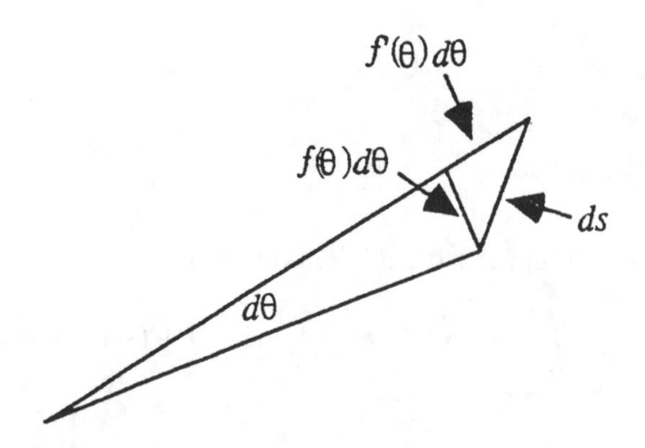

In a more algebraic vein, parametrization of the curve in rectangular coordinates is

$$\gamma(\theta) = \big(f(\theta)\cos\theta, f(\theta)\sin\theta\big) \quad (\theta \in [a,b]).$$

Hence,

$$\begin{aligned}
\|\gamma'(\theta)\|^2 &= \big\|\big(f'(\theta)\cos\theta - f(\theta)\sin\theta, f'(\theta)\sin\theta + f(\theta)\cos\theta\big)\big\|^2 \\
&= \big(f'(\theta)\big)^2\cos^2\theta + \big(f(\theta)\big)^2\sin^2\theta - 2f'(\theta)f(\theta)\cos\theta\sin\theta \\
&\quad + \big(f'(\theta)\big)^2\sin^2\theta + \big(f(\theta)\big)^2\cos^2\theta + 2f'(\theta)f(\theta)\cos\theta\sin\theta \\
&= \big(f'(\theta)\big)^2 + \big(f(\theta)\big)^2.
\end{aligned}$$

Therefore, the length of the curve is $\displaystyle\int_a^b \sqrt{\big(f'(\theta)\big)^2 + \big(f(\theta)\big)^2}\,d\theta.$

5. a. This is really an easy problem as long as students don't get intimidated by the fact that f is defined piecewise. $\displaystyle\int_{C_1} f(x,y)\,ds = \int_{-2}^{3} f(t,0)\,dt = 3\cdot 2 + 2\cdot 1 = 8.$

b. Let $\gamma(t) = (1+\cos t, \sin t)$, $t \in [0,\pi]$, be the parametrization of C_2. C_2 intersects the line $x+y=1$ at $\big(1-\frac{1}{\sqrt{2}}, \frac{1}{\sqrt{2}}\big) = \gamma\big(\frac{3\pi}{4}\big)$. Therefore

$$\int_{C_2} f(x,y)\,ds = \int_0^{3\pi/4} 1\,dt + \int_{3\pi/4}^{\pi} 2\,dt = \frac{3\pi}{4} + \frac{\pi}{2} = \frac{5\pi}{4}.$$

6. a. A parametrization of the path of a bead is $\gamma(z) = (z, 1-z)$ ($z \in [0,1]$), where the first coordinate represents the horizontal distance from the pole and the second the height above the ground. If the bead is a distance y above the ground, in a short time dt the bead will have traveled $ds = \sqrt{2g(1-y)}\,dt$ meters. Since $ds = \sqrt{2}\,dz$, the time it takes the bead to travel down the wire is

$$\int_0^1 \frac{1}{\sqrt{2gz}} \sqrt{2}\,dz = \frac{2}{\sqrt{g}} \approx .64\,\text{sec.}$$

An alternative method is to note the magnitude of the component of the gravitational acceleration vector in the direction of the wire is $g/\sqrt{2}$. Since the wire is $\sqrt{2}$ meters long, the time T it takes to travel the wire satisfies the equation $\frac{1}{2}(g/\sqrt{2})T^2 = \sqrt{2}$, so $T = 2/\sqrt{g}$ as above.

b. A parametrization for the bead's path is now $\gamma(z) = (1-\cos z, 1-\sin z)$, $z \in \left[0, \frac{\pi}{2}\right]$, so the time it takes to travel down the wire is $\displaystyle\int_0^{\pi/2} \frac{1}{\sqrt{2g\sin z}}\,dz \approx .59\,\text{sec.}$

c. This might make a good project for an interested student. Other paths that can be tried include a parabolic path and, of course, a cycloid.

7. a. By symmetry, the time to travel this path is just twice the time to travel from the center to the final destination. If you are at a distance r from the center of the swamp, in a small time dt you will travel approximately $ds = v(r)\,dt$ miles. Thus, $dt = ds/v(r)$, so the time it takes to travel this path is $2\int_0^2 \frac{1}{5+10s}\,ds = \frac{\ln 5}{5} \approx$ .3219 hours. On the semicircular path, you would travel 2π miles at a constant speed of 25 MPH, so it would take you $\frac{2\pi}{25} \approx .2513$ hours to travel this way. Hence the semicircular path is faster.

   b. By symmetry, the time to travel the whole path is just double the time it takes to travel the first half of the path. A parametrization for the first half of the path is $\gamma(z) = (2z-2, -z)$, $z \in [0,1]$, so $\|\gamma'(z)\| = \sqrt{5}$. At the point $\gamma(z)$, you are $\sqrt{(2z-2)^2 + (-z)^2} = \sqrt{5z^2 - 8z + 4}$ miles from the center of the swamp, so your speed is $1/\left(5 + 10\sqrt{5z^2 - 8z + 4}\right)$ MPH. Hence, the time it takes to travel the first half of the path is

$$\int_0^1 \frac{1}{5 + 10\sqrt{5z^2 - 8z + 4}}\sqrt{5}\,dz \approx .1328\,\text{hours}.$$

   So the time it takes to travel the whole path is about .2656 hours, just a bit more than the semicircular path.

This problem leads to a natural question (which in turn leads to the calculus of variations): what is the fastest possible path? This might be a good project for a bright student.

8. a. True. This is the most basic property of line integrals with respect to arc length.

b. False. In fact, for any continuous parametrization $\gamma_1 : [a,b] \to \mathbb{R}^n$, define $\gamma_2 : [a,b] \to \mathbb{R}^n$ by $\gamma_2(t) = \gamma_1(a+b-t)$, so γ_2 is an orientation-reversing reparametrization of γ_1. Then for any continuous vector field $\mathbf{f}$, $\int_{\gamma_1} \mathbf{f}(\mathbf{r}) \cdot d\mathbf{r} = -\int_{\gamma_2} \mathbf{f}(\mathbf{r}) \cdot d\mathbf{r}$.

c. True. If γ is a parametrization of C, the assumption is that $\mathbf{f}(\gamma(t)) \cdot \gamma'(t) = 0$, which immediately implies that the given line integral is zero.

d. True. Let γ be a parametrization of C. Since $\mathbf{f}$ is parallel to the tangent vectors of C, $\mathbf{f}(\gamma(t)) \cdot \gamma'(t) = \|f(\gamma(t))\| \|\gamma'(t)\|$. The statement now follows from the definitions of the two integrals.

e. True. γ goes around the unit circle once while α goes around it twice in the same direction, so the integral on the right must be twice the one on the left. Alternatively one can make the substitution $u = 2t$ in the second integral.

f. True. This is essentially a question from Chapter V, only asked here for a function of 2 variables. The argument parallels what was given by way of commentary for the problem in Chapter V. Suppose $f(x_0, y_0) \neq 0$ for some (x_0, y_0); without loss of generality we can assume $f(x_0, y_0) > 0$. Then by continuity there must be an open region R for which $f(x,y) > 0$ for all $(x,y) \in R$. Hence for any curve C contained in R, $\int_C f(s)\, ds$ is positive, a contradiction.

9. a. $\int_C \mathbf{f}(\mathbf{r}) \cdot d\mathbf{r} = 0$ since the tangent vectors to C are orthogonal to $f(x, y)$.

   b. The integral is positive since the x components of the tangent vectors of C_1 have the same sign as the x component of the horizontal vector field f.

   c. These two integrals are equal. Line integrals of f depend only on the x-coordinates of the endpoints of the curve, which are the same for these two curves. To show directly that these integrals are equal, note that the x component of $\mathbf{f}$ appears to be a function of x alone, say $\mathbf{f}(x, y) = (g(x), 0)$. Hence

$$\int_{C_1} \mathbf{f}\, d\mathbf{r} = \int_0^1 (g(2t), 0) \cdot (2, 1)\, dt = \int_0^1 2g(2t)\, dt$$
$$= \int_0^2 g(u)\, du = \int_0^2 (g(u), 0) \cdot (1, 0)\, du = \int_{C_2} \mathbf{f}\, d\mathbf{r}.$$

   d. It appears that this field is conservative. Intuitively, any line integral of $\mathbf{f}$ depends only on the x-coordinates of the endpoints of the curve. More rigorously, let $f(x, y) = (g(x), 0)$ as in part c, and let G be an antiderivative of g. Let $\phi(x, y) = G(x)$. Then $\nabla\phi = (g(x), 0) = f(x, y)$.

10. a. The line integral is positive since the tangent vectors of C_1 point in the positive x direction and the x component of the vector field is positive on C_1.

   b. This line integral is negative since the tangent vectors to C_2 point more or less in the direction opposite to that of the vector field.

   c. No. The result of part *b* shows immediately that this field is not conservative. Intuitively the field is not conservative since it has a clear "circulation."

11. a. Positive.

   b. Zero, by the symmetry of the field.

   c. Zero. The tangent vectors to C_3 are always perpendicular to the radial vector field $\mathbf{h}$ (also, this field is conservative; see part *d*).

   d. This vector field is conservative. More generally, any radial vector field (or central force field, as physicists would say) is conservative. In fact, it is easy to see that the value of $\int_C \mathbf{h}(\mathbf{r})\, d\mathbf{r}$ depends only on distance of the endpoints of C to the origin. Another example of a radial vector field is given in Problem 2 of the next section.

Section 2: Conservative Vector Fields and Green's Theorem

1. a. Since

$$\frac{\partial}{\partial y}(2x+y)\cos(x^2+xy) = \cos(x^2+xy) - x(2x+y)\sin(x^2+xy)$$
$$= \frac{\partial}{\partial x}[x\cos(x^2+xy)+1],$$

f is a conservative vector field.

b. This is an integral that wouldn't be fun to do by brute force. Let C' be the straight line segment running between the endpoints of C, namely $(0,0)$ and $(0,2)$. A parametrization of C' is $\alpha(t) = (0,t)$ $(0 \le t \le 2)$. Since **f** is conservative,

$$\int_C \mathbf{f}(\mathbf{r}) \cdot d\mathbf{r} = \int_{C'} \mathbf{f}(\mathbf{r}) \cdot d\mathbf{r} = \int_0^2 (t\,\mathbf{i} + 1\,\mathbf{j}) \cdot (0\,\mathbf{i} + 1\,\mathbf{j})\,dt = 2.$$

Alternatively one can find a potential function for **f**. One such function is $\phi(x,y) = \sin(x^2+xy)+y$. Hence $\int_C \mathbf{f}(\mathbf{r}) \cdot d\mathbf{r} = \phi(0,2) - \phi(0,0) = 2 - 0 = 2$.

c. Even though **g** is *not* conservative, it can be split up into a conservative part plus a part whose integral is easy to do:

$$\int_C \mathbf{g}(\mathbf{r}) \cdot d\mathbf{r} = \int_C \mathbf{f}(\mathbf{r}) \cdot d\mathbf{r} + \int_C x\,\mathbf{j} \cdot d\mathbf{r} = 2 + \int_0^\pi \sin t \cos t\,dt = 2 + \tfrac{\pi^2}{2}.$$

2. a. $\dfrac{\partial P}{\partial y} = \dfrac{3xy}{(x^2+y^2)^{5/2}} = \dfrac{\partial Q}{\partial x}$.

b. The potential functions are of the form $\phi(x,y) = \dfrac{1}{\sqrt{x^2+y^2}}$ plus an arbitrary constant.

c. Let ϕ be the potential function found in part *b*. Then $\int_C \mathbf{f}(\mathbf{r}) \cdot d\mathbf{r} = \phi(3,4) - \phi(1,0) = \frac{1}{5} - 1 = -\frac{4}{5}$.

3. a. This is routine.

b. The potential functions are of the form $\phi(x,y,z) = \dfrac{1}{\sqrt{x^2+y^2+z^2}}$ plus an arbitrary constant.

c. This field could be the gravitational field generated by a point mass at the origin or the electric field generated by a negative point charge at the origin.

4. It seems obligatory to put some version of this problem in every collection of problems on line integrals.

a. $\dfrac{\partial P}{\partial y} = \dfrac{y^2 - x^2}{(x^2+y^2)^2} = \dfrac{\partial Q}{\partial x}$.

b. A parametrization of such a circle is $\gamma(t) = (k\cos t, k\sin t)$, $0 \le t \le 2\pi$. Hence,

$$\int_C P\,dx + Q\,dy = \int_0^{2\pi} \frac{1}{k^2}(-k\sin t, k\cos t)\cdot(-k\sin t, k\cos t)\,dt = \int_0^{2\pi} dt = 2\pi.$$

c. The condition $\dfrac{\partial P}{\partial y} = \dfrac{\partial Q}{\partial x}$ must hold for *all* points inside the curve C. Here there is only one point (the origin) where this does not hold, but that is enough.

d. Since $\dfrac{\partial P}{\partial y} = \dfrac{\partial Q}{\partial x}$ for all points inside C_1, the integral is zero.

e. Let C be a circle centered at the origin oriented clockwise such that C_2 is contained in the interior of C. Then by part *b*, $\int_C \mathbf{f}(\mathbf{r})\cdot d\mathbf{r} = -2\pi$. Let C_3 be the closed curve consisting of the upper halves of C and C_2 connected along the x-axis and let C_4 be the lower halves of C and C_2 connected along the x-axis (see the diagram below). Then by part *d*, $\int_{C_3} \mathbf{f}(\mathbf{r})\cdot d\mathbf{r} = 0 = \int_{C_4} \mathbf{f}(\mathbf{r})\cdot d\mathbf{r}$. But

$$\int_C \mathbf{f}(\mathbf{r})\cdot d\mathbf{r} + \int_{C_2} \mathbf{f}(\mathbf{r})\cdot d\mathbf{r} = \int_{C_3} \mathbf{f}(\mathbf{r})\cdot d\mathbf{r} + \int_{C4} \mathbf{f}(\mathbf{r})\cdot d\mathbf{r} = 0.$$

Hence $\int_{C_2} \mathbf{f}(\mathbf{r})\cdot d\mathbf{r} = 2\pi$.

5. Let P and Q be the x and y components of $\mathbf{F}$ respectively. Then $\dfrac{\partial P}{\partial y} = 2xyf(x^2+y^2) = \dfrac{\partial Q}{\partial x}$. Hence $\mathbf{F}$ is conservative. A potential function for $\mathbf{F}$ is $\frac{1}{2}G(x^2+y^2)$, where G is an antiderivative of f.

6. The result follows from the equalities $\nabla(fg) = f\nabla g + g\nabla f$ and $\int_C \nabla(fg) \cdot d\mathbf{r} = 0$. This problem illustrates once again the importance of integration by parts which students effectively use in proving the equality given in the problem. The boundary terms usually present in an integration by parts cancel since C is a closed curve and the boundary terms have the same value at the beginning and at the end of C.

7. a. This is not necessarily true. Problem 2 gives an example where this is true while Problem 4 gives an example where this is false.

   b. This is not necessarily true. For example, let $\mathbf{f}(x, y) = (y^2, 0)$ and let C be the unit circle. Then it can be easily verified that $\int_C \mathbf{f}(\mathbf{r}) \cdot d\mathbf{r} = 0$ and yet $\mathbf{f}$ is not a conservative field anywhere.

   c. True.

   d. False. For example, let $\mathbf{f}(x, y, z) = x\,\mathbf{j}$. Then $\nabla \times \mathbf{f} = \mathbf{k} \neq \mathbf{0}$ for all $(x, y, z) \in \mathbb{R}^3$, but $\int_C \mathbf{f}(\mathbf{r}) \cdot d\mathbf{r} = 0$ for any closed curve that lies in the plane $\{(x, y, z) \mid x = 0\}$.

8. a. By Green's Theorem, $\int_C x\,dy = \iint_D 1\,dx\,dy$ = area of D, where D is the region bounded by C.

   b. By part a, the area inside the ellipse is $\int_C x\,dy = \int_0^{2\pi} ab\cos^2 t\,dt = \pi ab$.

   c. Both these integrals are equal to $\iint_D 1\,dx\,dy$, where D is the region bounded by C.

9. Let D_r be the disk of radius r centered at the origin. By Green's Theorem,

$$\lim_{r\to 0} \frac{1}{\pi r^2} \int_{C_r} P\,dx + Q\,dy = \lim_{r\to 0} \frac{1}{\pi r^2} \iint_{D_r} \left(\frac{\partial Q}{\partial x} - \frac{\partial P}{\partial y}\right) dx\,dy.$$

When r is small, $\iint_{D_r} \left(\frac{\partial Q}{\partial x} - \frac{\partial P}{\partial y}\right) dx\,dy$ is approximately the value of the integrand at zero times the area of D_r (this could be made rigorous by using the Mean Value Theorem). Hence the limit is $\dfrac{\partial Q}{\partial x}(0,0) - \dfrac{\partial P}{\partial y}(0,0)$.